T0360858

POMERANCHUK 100

POMERANCHUK 100

A I Alikhanov Institute of Theoretical and Experimental Physics (ITEP)

Moscow 5 – 6 June 2013

Editors

Alexander Gorsky
Mikhail Vysotsky

A I Alikhanov Institute of Theoretical and
Experimental Physics (ITEP), Russia

World Scientific

NEW JERSEY · LONDON · SINGAPORE · BEIJING · SHANGHAI · HONG KONG · TAIPEI · CHENNAI

Published by

World Scientific Publishing Co. Pte. Ltd.

5 Toh Tuck Link, Singapore 596224

USA office: 27 Warren Street, Suite 401-402, Hackensack, NJ 07601

UK office: 57 Shelton Street, Covent Garden, London WC2H 9HE

British Library Cataloguing-in-Publication Data
A catalogue record for this book is available from the British Library.

POMERANCHUK 100

ISBN 978-981-4616-84-3

Printed in Singapore

PREFACE

This Volume is dedicated to the 100-th anniversary of Isaak Pomeranchuk, one of the most universal theorist of the last century, who founded the Theory Division of ITEP. Pomeron – the vacuum pole – enters the physics language forever. Isaak Pomeranchuk has contributed to the astrophysics and theory of nuclear reactors, to the theory of liquid helium and theory of high energy scattering.

The mini-Conference devoted to the anniversary took place in ITEP at June 5–6, 2013. The papers included in this Volume are contributed by the participants of the Conference as well as one of the winners of the Pomeranchuk Prize 2013, Mikhail Shifman. We have tried to follow the Pomeranchuk's unity of physics and the papers cover the diverse topics. Very roughly we could divide them into Phenomenology of the QED and the Standard Model, Quantum Field Theory, Astrophysics and Cosmology. The papers of I. Dremin, B. Kerbikov, V. Neznamov, V. Shevchenko, V. Zakharov in collaboration with V. Kirilin and A. Sadofyev, M. Vysotsky and S. Godunov cover the different aspects of the strong and electromagnetic interactions. The papers of A. Belavin and B. Mukhametzhanov, A. Gorsky and K. Bulycheva, N. Nekrasov, M. Shifman and A. Yung, A. Smilga, K. Zarembo and J. Russo discuss the different aspects of the Quantum Field Theory and in particular focus on the new methods developed last decade to describe the systems at strong coupling in universal manner. The contributions of S. Blinnikov, A. Dolgov, I. Kriplovich and A. Rudenko are devoted to the astrophysical and cosmological aspects of the modern physics.

We hope that the readers of this Volume will feel the beauty of the modern physics in its unity. We would like to thank K. Bulycheva and S. Godunov for the help in preparation of this Volume.

Alexander Gorsky
Mikhail Vysotsky

CONTENTS

DOUGLAS APPROACH
TO LIOUVILLE MINIMAL GRAVITY

A. BELAVIN

L.D. Landau Institute for Theoretical Physics, 142432 Chernogolovka, Russia
Moscow Institute of Physics and Technology, 141700 Dolgoprudny, Russia
Institute for Information Transmission Problems, 127994, Moscow, Russia

B. MUKHAMETZHANOV*

L.D. Landau Institute for Theoretical Physics, 142432 Chernogolovka, Russia
Department of Physics, Harvard University, 02138 Cambridge, USA
*E-mail: baur@itp.ac.ru

We give a brief review of Douglas approach to calculating the correlation numbers in Liouville Minimal Gravity. Most of the material is covered in.[1,2]

1. Introduction

Minimal Liouville gravity was invented and partly solved by V. Knizhnik, A. Polyakov and A. Zamolodchikov in 1987.[3] It is an example of Liouville gravity with matter sector being the (q, p) Minimal Model of CFT .

One of the main problems in Minimal Liouville Gravity is to find n-point correlation functions given by the integrals over the moduli space of Riemann surfaces with n punctures is still open.

$$Z_{m_1 n_1 \ldots m_N n_N} = \langle O_{m_1 n_1} \ldots O_{m_N n_N} \rangle, \qquad O_{m,n} = \int_M \mathcal{O}_{m,n} \qquad (1)$$

where $\mathcal{O}_{m,n}$ are the observables to be introduced below. Some progress in this direction was achieved in[4] using Liouville Higher equations of motion[5] of Alexei Zamolodchikov.

It is convenient to define the generating function of the correlation numbers

$$Z_L(\lambda) = \langle \exp \sum_{m,n} \lambda_{m,n} O_{m,n} \rangle, \qquad (2)$$

$$Z_{m_1 n_1 \ldots m_N n_N} = \frac{\partial}{\partial \lambda_{m_1 n_1}} \cdots \frac{\partial}{\partial \lambda_{m_N n_N}} \Bigg|_{\lambda=0} Z_L(\lambda) \qquad (3)$$

The generating function $Z_L(\lambda)$ can be considered as the partition function of the perturbed theory. The coupling constants $\lambda_{m,n}$ are the coordinates on the space of these perturbed Minimal Liouville Gravity theories.

On the other hand there is another, discrete, approach to two-dimensional gravity, which was invented and developed in.[6–12] This approach is realized through Matrix Models. The fluctuating two-dimensional surfaces are approximated by an ensemble of graphs. The continuous geometry is restored in the scaling limit, when large size graphs dominate.

These two approaches are based on the same idea of two-dimensional fluctuating geometry and had been expected to give the same results.

In 1989 Douglas discovered that the partition function in the discrete approach satisfies the KdV and so-called "string equation".

The KdV times $\tau_{m,n}$ of the generalized KdV hierarchy play a role of the perturbation parameters $\lambda_{m,n}$ in Minimal Liouville gravity. Their gravitational dimensions coincide. However a direct identification of the times $\tau_{m,n}$ and coupling constants $\lambda_{m,n}$ leads to inconsistencies and actually the relation between them is non-linear.

In Minimal Gravity one has conformal and fusion selection rules. For instance, the one-point correlation numbers of all operators (except unity operator) must be zero, the two-point correlation numbers must be diagonal.

There exist similar restrictions on higher point correlation numbers. They are not satisfied in the Douglas approach if we identify $\tau_{m,n}$ and $\lambda_{m,n}$. This problem was pointed out and partly solved by Moore, Seiberg and Staudacher in.[13] The idea is that due to possible contact terms in OPE the times $\tau_{m,n}$ in the Douglas approach and coupling constants $\lambda_{m,n}$ in Minimal Liouville Gravity are related in a non-linear fashion

$$\tau_{m,n} = \lambda_{m,n} + \sum_{m_1 n_1 m_2 n_2} C_{m,n}^{m_1 n_1 m_2 n_2} \lambda_{m_1 n_1} \lambda_{m_2 n_2} + \ldots \qquad (4)$$

and an appropriate choice of such substitution allowed them to reach the coincidence up to 2-point correlation numbers in $(2, 2s+1)$ Minimal Gravity.

After the 4-point correlation numbers in Minimal Gravity had been calculated,[4] it became possible to make new explicit checks against Douglas equation approach. It was performed for $(2, 2s + 1)$ Minimal Gravity and the full correspondence between Douglas equation approach and Minimal Liouville Gravity was reached for this case in.[1]

Our aim is to generalize the results for $(2, 2s + 1)$ case to the case of $(3, 3s + 1)$ and $(3, 3s + 2)$ Minimal Gravity. Our analysis is based on the following assumptions:

- There exist special coordinates $\tau_{m,n}$ in the space of perturbed Minimal Liouville Gravities such that the partition function of Minimal Liouville Gravity satisfies the Douglas string equation.
- String equation define the partition function as logarithm of Sato's tau-function of the dispersionless generalized KdV hierarchy with the initial conditions defined by the Douglas string equation.
- The times $\tau_{m,n}$ are related to the natural coordinates $\lambda_{m,n}$ on the space of the perturbed Minimal Liouville Gravities by a non-linear transformation.
- We establish the form of this transformation $\tau(\lambda)$ by the requirement that the correlation numbers, which are the coefficients of the expansion of the partition function in the coordinates $\lambda_{m,n}$, satisfy the conformal and fusion selection rules once we pick an appropriate solution of Douglas string equation.
- To perform this program we get a convenient expression for the partition function using the relation of String equation with a Frobenius manifold structure.

2. Minimal Models of CFT

The Minimal Liouville gravity consists of Liouville theory of the field ϕ and some matter sector which is taken to be a (q, p) Minimal Model of CFT.

The Minimal Model $\mathcal{M}_{q,p}$ has primary fields, which are enumerated by the Kac table: $\Phi_{m,n}$, where $m = 1, \ldots, q - 1$ and $n = 1, \ldots, p - 1$. Only a half of the fields $\Phi_{m,n}$ are independent.

$$\Phi_{m,n} = \Phi_{q-m,p-n} \tag{5}$$

We consider the Minimal Models $\mathcal{M}_{2,2s+1}$ and $\mathcal{M}_{3,3s+p_0}$, where $p_0 = 1, 2$. In these models it suffices to consider the fields from the first raw

$\Phi_k = \Phi_{1,k+1}$. In the model $(2, 2s+1)$ the independent fields are Φ_k, $0 \le k \le s-1$. In the model $(3, 3s+p_0)$ the independent fields are Φ_k, $0 \le k \le p-2$. The operator product expansion (OPE) for these fields is

$$[\Phi_{k_1}][\Phi_{k_2}] = \sum_{k=|k_1-k_2|:2}^{I(k_1,k_2)} [\Phi_k] \tag{6}$$

where $[\Phi_k]$ denotes the contribution of the irreducible Virasoro representation with the highest state Φ_k. Summation goes from $k = |k_1 - k_2|$ till $I(k_1, k_2)$ with the step 2 where

$$I(k_1, k_2) = \min(k_1 + k_2, 2p - k_1 - k_2 - 4). \tag{7}$$

The small conformal group and OPE put strong constraints on correlation functions. For instance constraints for one- and two-point correlation functions are

$$\langle \Phi_k(x) \rangle = 0, \qquad k \ne 0 \tag{8}$$

$$\langle \Phi_{k_1}(x_1)\Phi_{k_2}(x_2) \rangle = 0, \qquad k_1 \ne k_2. \tag{9}$$

For higher correlation numbers we use the OPE fusion to bring it to the two-point correlation numbers and then we use the conformal rules. For instance the three-point correlation functions satisfy

$$\langle \Phi_{k_1}\Phi_{k_2}\Phi_{k_3} \rangle = 0, \quad k_3 > I(k_1, k_2) = \begin{cases} k_1 + k_2, & k_1 + k_2 \le p - 2 \\ 2p - k_1 - k_2 - 4, & k_1 + k_2 > p - 2 \end{cases}$$

where we assume that $k_1 \le k_2 \le k_3$. For four-point correlation functions we have

$$\langle \Phi_{k_1}\Phi_{k_2}\Phi_{k_3}\Phi_{k_4} \rangle = 0, \qquad k_4 > I(I(k_1, k_2), k_3) \tag{10}$$

where we assume that $k_1 \le k_2 \le k_3 \le k_4$ and

$$I(I(k_1, k_2), k_3) = \begin{cases} \min\left(k + k_3, 2(p-2) - k - k_3\right), & k \le p - 2, \\ \min\left(k - k_3, 2(p-2) - k + k_3\right), & k > p - 2, \end{cases} \tag{11}$$

where $k = k_1 + k_2$. These and similar equations we call *selection rules*.

3. Liouville Field Theory

The Polyakov's continuous approach to two-dimensional quantum gravity is defined through the path integral over two-dimensional Riemannian metrics $g_{\mu\nu}$ interacting with some conformal matter. In conformal gauge $g_{\mu\nu} = e^{\phi}\hat{g}_{\mu\nu}$ it leads to Liouville theory.

The observables of the (q, p) are cohomologies of an appropriate BRST operator. They are enumerated as primary fields in corresponding Minimal Model and denoted as $O_{m,n}$. Explicitly

$$O_{m,n} = \int_{x \in M} \mathcal{O}_{m,n}(x), \qquad \mathcal{O}_{m,n}(x) = \Phi_{m,n}(x) e^{2b\delta_{m,n}\phi(x)} \sqrt{\hat{g}} d^2 x \quad (12)$$

where $\Phi_{m,n}$ – primary fields of the (q, p) CFT Minimal Model, considered in the previous section. The numbers $\delta_{m,n}$ are the so-called gravitational dimensions

$$\delta_{m,n} = \frac{p + q - |pm - qn|}{2q}. \quad (13)$$

The operators $O_{m,n}$ have the scaling property

$$O_{m,n} \sim \mu^{-\delta_{m,n}}. \quad (14)$$

and also satisfy the same selection rules as $\Phi_{m,n}$ do.

We define the correlation numbers in Minimal Liouville Gravity as

$$Z_{m_1 n_1 \ldots m_N n_N} = \langle O_{m_1,n_1} \ldots O_{m_N,n_N} \rangle \quad (15)$$

The generating function of these correlation numbers is the main object of our study

$$Z_L(\lambda) = \left\langle \exp \sum_{m,n} \lambda_{m,n} O_{m,n} \right\rangle. \quad (16)$$

4. Contact Terms

The N-point correlation numbers on genus zero Riemann surface involve integration over $N - 3$ points on the 2d surface M

$$Z_{m_1 n_1 \ldots m_N n_N} = \int \langle \mathcal{O}_{m_1,n_1}(x_1) \ldots \mathcal{O}_{m_N,n_N}(x_N) \rangle d^2 x_1 \ldots d^2 x_{N-3} \quad (17)$$

In this expression one can encounter contact delta-like terms when two or more points x_i are coincident. Such terms are not controlled by CFT and thus make the definition of correlation numbers ambiguous.

The ambiguity in contact terms leads to the fact that we can add to the n-point correlation numbers some k-point correlation numbers

$$\langle O_{m_1,n_1} O_{m_2,n_2} \rangle \to \langle O_{m_1,n_1} O_{m_2,n_2} \rangle + \sum_{m,n} A_{m,n}^{(m_1 n_1)(m_2 n_2)} \langle O_{m,n} \rangle. \quad (18)$$

Such a substitution is equivalent to a change of coupling constants in the generating function

$$\lambda_{m,n} \to \lambda_{m,n} + \sum_{m_1,n_1,m_2,n_2} A_{m,n}^{(m_1 n_1)(m_2 n_2)} \lambda_{m_1 n_1} \lambda_{m_2 n_2}. \quad (19)$$

Thus, any addition of contact terms is equivalent to some non-linear polynomial change of coupling constants

$$\lambda_{m,n} \to A\mu^{\delta_{m,n}} + \lambda_{m,n} + \sum_{m_1,n_1} \sum_{m_2,n_2} C_{m,n}^{(m_1 n_1)(m_2 n_2)} \lambda_{m_1 n_1} \lambda_{m_2 n_2} +$$

$$+ \sum_{m_1,n_1} \sum_{m_2,n_2} \sum_{m_3,n_3} C_{m,n}^{(m_1 n_1)(m_2 n_2)(m_3 n_3)} \lambda_{m_1 n_1} \lambda_{m_2 n_2} \lambda_{m_3 n_3} + \dots$$

In MLG we can impose a strong restriction on this change of coupling constants because they have certain mass dimension

$$\lambda_{m,n} \sim \mu^{\delta_{m,n}} \quad (20)$$

Therefore we can demand all the terms to have the same dimensions as that of $\lambda_{m,n}$

$$\delta_{m,n} = \delta_{m_1,n_1} + \delta_{m_2,n_2} + \delta_{m_3,n_3} + \dots \quad (21)$$

However in the Liouville Minimal Gravity such resonance conditions are very common and a freedom of the change still remains.

We use the coordinates which natural in the discrete approach to evaluate the free energy $\mathcal{F}(t)$ from the Douglas equation, then we make such a change of variables $t = t(\lambda)$ that the selection rules are satisfied.

5. 3- and 4-Point Functions of MLG

The explicit formulae for three-point[14] and four-point[4] correlation numbers in MLG are known. We will compare these results with those from the Douglas string equation.

To make sensible comparisons we write down the quantities which do not depend on the normalizations of operators and correlators

$$\frac{\langle\langle O_{m_1,n_1} O_{m_2,n_2} O_{m_3,n_3} \rangle\rangle^2}{\prod_{i=1}^3 \langle\langle O^2_{m_i,n_i} \rangle\rangle} = \frac{\prod_{i=1}^3 |m_i p - n_i q|}{p(p+q)(p-q)} \tag{22}$$

$$\frac{\langle\langle O_{m_1,n_1} O_{m_2,n_2} O_{m_3,n_3} O_{m_4,n_4} \rangle\rangle}{\left(\prod_{i=1}^4 \langle\langle O^2_{m_i,n_i} \rangle\rangle\right)^{\frac{1}{2}}} = \frac{\prod_{i=1}^4 |m_i p - n_i q|}{2p(p+q)(p-q)} \times$$

$$\left(\sum_{i=2}^4 \sum_{r=-(m_1-1)}^{m_1-1} \sum_{t=-(n_1-1)}^{n_1-1} |(m_i-r)p - (n_i-t)q| - m_1 n_1 (m_1 p + n_1 q) \right) \tag{23}$$

where $\langle\langle \ldots \rangle\rangle = \frac{\langle \ldots \rangle}{\langle 1 \rangle}$. Sums over r, t in the last formula are with the step 2.

The four-point correlation numbers were obtained under certain assumptions. In the particular case when $m_i = 1, i = 1, \ldots, 4$ it looks as follows (suppose $n_1 \leq \cdots \leq n_4$)

$$n_1 + n_4 \leq n_2 + n_3.$$

6. Douglas String Equation

Due to Douglas[15] the free energy of the matrix model, corresponding to the partition function of (q, p) Minimal Gravity on the sphere, is described in the following way. One introduces two polynomials $Q(y)$ and $P(y)$

$$Q(y) = y^q + \sum_{\alpha=1}^{q-1} u_\alpha(x) y^{q-\alpha-1}, \tag{24}$$

$$P(y) = \left(\frac{1}{q}\left(1 + \frac{p}{q}\right) Q^{\frac{p}{q}} + \sum_{\alpha=1}^{q-1} \sum_{k=1}^{} \frac{1}{q}\left(k + \frac{\alpha}{q}\right) t_{k,\alpha} Q^{k+\frac{\alpha}{q}-1} \right)_+ \tag{25}$$

where $Q^{\frac{a}{q}}$ is understood as series expansion in y and $(\ldots)_+$ means that only non-negative powers of p are taken in this expansion. Then one takes these polynomials to satisfy the so called *string equation*

$$\{P, Q\} = \frac{\partial P}{\partial x} \frac{\partial Q}{\partial y} - \frac{\partial P}{\partial y} \frac{\partial Q}{\partial x} = 1 \tag{26}$$

The free energy $\mathcal{F}(t)$ satisfies the equation

$$\frac{\partial^2 \mathcal{F}}{\partial x^2} = u_1^*$$ (27)

where u_α^* is an appropriate solution of the string equation.

The first term in (25) describes the critical point of the matrix model and corresponds to the Minimal Liouville Gravity. Other terms in (25) correspond to the Minimal Liouville Gravity perturbed by primary operators. In (25) we did not indicate the interval in which the index k is varied. We will do this later from the condition that the perturbations by the times $t_{k,\alpha}$ correspond to the coupling constants $\lambda_{m,n}$ of the perturbation of the Minimal Liouville Gravity by all primary operators from (q,p) minimal CFT.

It was shown in[16] that one can take the first integral of the string equation. After this integration the Douglas string equation (26) becomes[16]

$$\frac{\partial S}{\partial u_\alpha(x)} = 0$$ (28)

where

$$S[u_\alpha(x)] = S_{s+1,p_0} + \sum_{\alpha=1}^{q-1} \sum_{k=0} t_{k,\alpha} S_{k,\alpha}$$ (29)

where $(q,p) = (q, sq + p_0)$, $1 \leq p_0 \leq q - 1$, $t_{0,1} = x$ and we introduced a notation

$$S_{k,\alpha} = Res(Q^{k+\frac{\alpha}{q}})$$ (30)

Here Q is the polynomial (24) and the residue is taken at $y = \infty$. The Douglas equation (26) or equivalently the least action principle equation (29) might have many different solutions. Later we are going to pick a particular solution with special properties which allows us to satisfy conformal selection rules.

7. Gravitational Dimensions from Douglas Approach

Now we want to ask in what region we should vary the index k in order to make the correspondence with the perturbations by all the primary operators in MLG.

To identify the free energy of String equation and the partition function of the Minimal Liouville Gravity on the sphere, we first analyse the dimensions. The Liouville Gravity partition function has the gravitational dimension

$$Z_L \sim \mu^{\frac{p+q}{q}}$$

where μ is the cosmological constant.

On the other hand the equation $\frac{\partial^2 \mathcal{F}}{\partial x^2} = u_1^*$ gives that

$$Z \sim x^2 u_1 \sim y^{2(p+q)}$$

where the scaling dimension of x is determined from the equation $\{P, Q\} = 1$ and scaling dimension of u^α from the constraint that all terms in Q are of the same dimension

$$x \sim y^{p+q-1}, \qquad u_\alpha \sim y^{\alpha+1}$$

Thus if we want $\mathcal{F} \sim Z_L$ we get

$$y \sim \mu^{\frac{1}{2q}}$$

This determines

$$t_{k,\alpha} \sim \mu^{\frac{s+1-k}{2} + \frac{p_0 - \alpha}{2q}}.$$

In particular $t_{s-1,p_0} \sim \mu$. We want to identify these times and their dimensions with the dimensions of the coupling constants $\lambda_{m,n}$ in the MLG.

In the (q, p) MLG one has the following scaling dimensions of $\lambda_{m,n}$ $(1 \le m \le q-1; 1 \le n \le p-1)$

$$\delta_{m,n} = \frac{p + q - |pm - qn|}{2q}.$$

To find the relation between $t_{k,\alpha}$ and $\lambda_{m,n}$ in Minimal Gravity we can use the fact that the expression for the "action" determines the dimensions of $t_{k,\alpha}$. Let's write it in the form $S = Res\left(Q^{\frac{p+q}{q}} + \sum_{m,n} \tau_{m,n} Q^{a_{m,n}}\right)$. Then the dimension of time $\tau_{m,n}$ is (since $Q \sim y^q \sim \mu^{\frac{1}{2}}$)

$$\tau_{m,n} \sim \mu^{\frac{1}{2}\left(\frac{p+q}{q} - a_{m,n}\right)}. \tag{31}$$

Now we see that in order for the dimensions of the times $\tau_{m,n}$ to coincide with those of $\lambda_{m,n}$, we need to put $a_{m,n} = \frac{|pm - qn|}{q}$.

8. Transformation from KdV-Frame to CFT-Frame

So we get the expression for the action, in which the correspondence with the operators from MLG is evident

$$S = Res \left(Q^{\frac{p+q}{q}} + \sum_{m=1}^{q-1} \sum_{n=1}^{p-1} \tau_{m,n} Q^{\frac{|pm-qn|}{q}} \right). \tag{32}$$

The dimensions of times $\tau_{m,n}$ coincide with the dimensions of $\lambda_{m,n}$. However the variables themselves do not necessarily coincide but can have a non-linear relation like

$$\tau_{m,n} = C_{m,n} \mu^{\delta_{m,n}} + \lambda_{m,n} + \sum_{m_1,n_1} C_{m,n}^{m_1 n_1} \mu^{\delta_{m,n} - \delta_{m_1,n_1}} \lambda_{m_1,n_1} +$$

$$+ \sum_{m_1,n_1} \sum_{m_2,n_2} C_{m,n}^{(m_1 n_1)(m_2 n_2)} \mu^{\delta_{m,n} - \delta_{m_1,n_1} - \delta_{m_2,n_2}} \lambda_{m_1 n_1} \lambda_{m_2 n_2} + \dots \tag{33}$$

where the terms in the r.h.s are chosen in such a way that they have the same scaling dimension as $\lambda_{m,n}$. And the cosmological constant μ appears in the non-negative integer powers.

These relations for the series of $(2, 2s+1)$ Minimal Gravities were found a few years ago in[1] and for the case $(3, 3s + p_0)$ in.[2] They are defined from the conditions which will be mentioned below.

9. String Equation and Frobenius Manifold Structure

The free energy $\mathcal{F}$ in the Douglas approach is related with the suitable solution of String equation u_1^* as

$$\frac{\partial^2 \mathcal{F}}{\partial x^2} = u_1^* \tag{34}$$

To get us a more explicit expression for $\mathcal{F}$ we use the connection between String equation and Frobenius manifold structure.

Let $\mathcal{A}$ be an algebra of polynomials modulo Q', where prime is the derivative over y

$$\mathcal{A} = C[y]/Q'. \tag{35}$$

Also define a bilinear form on the space of such polynomials

$$(P_1(y), P_2(y)) := Res \left(\frac{P_1(y) P_2(y)}{Q'} \right), \tag{36}$$

where the residue is taken at $p = \infty$. Introducing some basis ϕ_α in $\mathcal{A}$

$$\phi_\alpha \phi_\beta = C^\gamma_{\alpha\beta} \phi_\gamma \quad \mod (Q'). \tag{37}$$

one have that

$$Res \frac{\phi_\alpha \phi_\beta \phi_\gamma}{Q'} = C^\delta_{\alpha\beta} \cdot Res \frac{\phi_\delta \phi_\gamma}{Q'} = C^\delta_{\alpha\beta} g_{\delta\gamma} = C_{\alpha\beta\gamma} \tag{38}$$

where the indices are raised and lowered by the metric $g_{\alpha\beta} = (\phi_\alpha, \phi_\beta)$.

There are two useful bases ϕ_α. The first one is the monomials y^α, and another one is

$$\phi_\alpha = \frac{\partial Q}{\partial v^\alpha} \tag{39}$$

where $v^\alpha = -\frac{q}{q-\alpha} Res Q^{\frac{q-\alpha}{q}}$ are flat coordinates on M. In this basis $g_{\alpha\beta} = \delta_{\alpha+\beta,q}$ and there exists such a function F that

$$C_{\alpha\beta\gamma} = \frac{\partial^3 F}{\partial v^\alpha \partial v^\beta \partial v^\gamma}. \tag{40}$$

and the structure constants $C_{\alpha\beta\gamma}$ are evaluated in the basis $\frac{\partial Q}{\partial v^\alpha}$. These formulae provide Frobenius manifold structure for $\mathcal{A}$ case. We claim that

$$\mathcal{F} = \frac{1}{2} \int_0^{u*} C^{\beta\gamma}_\alpha \frac{\partial S}{\partial u^\beta} \frac{\partial S}{\partial u^\gamma} du^\alpha \tag{41}$$

To check this we need to verify two points. The first is that one-form $\Omega = C^{\beta\gamma}_\alpha \frac{\partial S}{\partial v^\beta} \frac{\partial S}{\partial v^\gamma} dv^\alpha$ is closed. This is verified by taking de Rham differential of Ω and using the associativity of $\mathcal{A}$ together the recursion relation for $S_{k,\alpha}$

$$C^\gamma_{\alpha\beta} C^\phi_{\gamma\delta} = C^\phi_{\alpha\gamma} C^\gamma_{\beta\delta},$$

$$\frac{\partial^2 S_{n,\alpha}}{\partial v^\beta \partial v^\gamma} = C^\delta_{\beta\gamma} \frac{\partial S_{n-1,\alpha}}{\partial v^\delta}$$

The second point is that $\frac{\partial^2 \mathcal{F}}{\partial x^2} = u_1^*$. This is verified by direct differentiation of $\mathcal{F}$ and using that $t_{0,1} = x$ and $C^\beta_{\alpha 1} = \delta^\beta_\alpha$.

We assume that after inserting an appropriate resonance relations $t(\lambda)$ and picking up an appropriate solution of the Douglas string equation, the free energy coincides with the Liouville Minimal Gravity partition function $\mathcal{F}(t(\lambda))$

$$\mathcal{F}(t(\lambda)) = Z_L(\lambda) \qquad (42)$$

Having the explicit expression for the partition function we can calculate the correlation numbers in two steps. We make the resonance substitution $t(\lambda)$ in the partition function and then just take derivatives and take all $\lambda_{m,n} = 0$ except for the cosmological constant $\lambda_{1,1} = \mu$.

$$Z_{m_1 n_1 \ldots m_N n_N} = \left. \frac{\partial}{\partial \lambda_{m_1 n_1}} \cdots \frac{\partial}{\partial \lambda_{m_N n_N}} \right|_{\substack{\lambda_{m,n}=0 \\ \text{for } (m,n) \neq (1,1)}} \mathcal{F}[t(\lambda)]. \qquad (43)$$

The resonance relations $t(\lambda)$ and the solution of the Douglas string equation are defined from the requirement that the correlation numbers (43) satisfy conformal selection rules.

10. $(2, p)$ MLG

Firstly we are going to review, as a simple example, $(2, 2s + 1)$ Minimal Liouville Gravity which was studied in.[1]

In this case we have the polynomial Q and the free energy

$$Q = y^2 + u, \qquad \mathcal{F} = \frac{1}{2} \int_0^{u_*} S_u^2(u) du \qquad (44)$$

where u_* is the solution of the equation

$$S_u \equiv \frac{\partial S}{\partial u} = u^{s+1} + \sum_{n=1}^{s} \tau_{1,n} u^{s-n} = 0 \qquad (45)$$

and we have changed the normalization of the times $\tau_{m,n}$ for later convenience. To get the generating function for the correlation numbers one inserts the resonance relations

$$\tau_{1,k+1} = \lambda_k + \sum C_k^{k_1 \ldots k_n} \lambda_{k_1} \ldots \lambda_{k_n} \qquad (46)$$

into the partition function and polynomial S_u. The result is of the form

$$\mathcal{F} = Z_0 + \sum_{k=1}^{s-1} \lambda_k Z_k + \frac{1}{2} \sum_{k_1,k_2=1}^{s-1} \lambda_{k_1} \lambda_{k_2} Z_{k_1 k_2} + \ldots \qquad (47)$$

$$S_u = S_u^0 + \sum_{k=1}^{s-1} \lambda_k S_u^k + \frac{1}{2} \sum_{k_1,k_2=1}^{s-1} \lambda_{k_1} \lambda_{k_2} S_u^{k_1 k_2} + \dots \tag{48}$$

Also from the original form of the polynomial S_u one finds that

$$
\begin{aligned}
S_u^0(u) &= u^{s+1} + B\mu u^{s-1} + C\mu^2 u^{s-3} + \dots \\
S_u^k(u) &= A_k u^{s-k-1} + B_k \mu u^{s-k-3} + C_k \mu^2 u^{s-k-5} + \dots \\
S_u^{k_1 k_2}(u) &= A_{k_1 k_2} u^{s-k_1-k_2-3} + B_{k_1 k_2} \mu u^{s-k_1-k_2-5} + \\
&\quad C_{k_1 k_2} \mu^2 u^{s-k_1-k_2-k_3-7} + \dots
\end{aligned}
\tag{49}
$$

where all the polynomials have certain parity since $\mu \sim u^2$. As above we find that the dimensions in this case are

$$\lambda_k \sim \mu^{\frac{k+2}{2}} \tag{50}$$

$$\mathcal{F} \sim \mu^{\frac{2s+3}{2}} \tag{51}$$

$$Z_{k_1\dots k_n} \sim \mu^{\frac{2s+3-\sum(k_i+2)}{2}} \tag{52}$$

As usually in the spirit of the scaling theory of criticality, we are interested only in the singular part of the partition function and disregard the regular part as non-universal.

Notice that $Z_{k_1\dots k_n}$ is always singular if $\sum k_i$ is even. On the other hand when $\sum k_i$ is odd and additionally $\sum_{i=1}^{n} k_i \leq 2s + 3 - 2n$, then the correlation number $Z_{k_1\dots k_n}$ involves only non-negative integer powers of μ and thus is non-singular. This inequality always holds for one- and two-point correlation numbers. So we shall consider the sector of odd $\sum k_i$ only starting from the three point correlation numbers.

It is convenient to switch to dimensionless quantities

$$s_k = \frac{g_k}{g} u_0^{-(k+2)} \lambda_k \tag{53}$$

$$S_u(u) = g u_0^{s+1} Y_u(u/u_0) \tag{54}$$

$$\mathcal{F} = g^2 u_0^{2s+3} \mathcal{Z} \tag{55}$$

where $u_0 = u_*(\lambda = 0) \sim \mu^{\frac{1}{2}}$, $g_k = \frac{(p-k-1)!}{(2p-2k-3)!!}$ and $g = \frac{(p+1)!}{(2p+1)!!}$.
Then one has

$$\mathcal{Z} = \frac{1}{2} \int_0^{x_*} Y_u^2(x)\,dx \tag{56}$$

where $x = \frac{u}{u_0}$ and $x_* = x_*(s)$ is an appropriate zero of the polynomial $Y_u(x)$. Notice that $x_*(s = 0) = 1$. Similarly to dimensional quantities one has expansions

$$\mathcal{Z} = \mathcal{Z}_0 + \sum_{k=1}^{s-1} s_k \mathcal{Z}_k + \frac{1}{2} \sum_{k_1,k_2=1}^{s-1} s_{k_1} s_{k_2} \mathcal{Z}_{k_1 k_2} + \ldots \tag{57}$$

$$Y_u = Y_u^0 + \sum_{k=1}^{s-1} s_k Y_u^k + \frac{1}{2} \sum_{k_1,k_2=1}^{s-1} s_{k_1} s_{k_2} Y_u^{k_1 k_2} + \ldots \tag{58}$$

$$Y_u^0(x) = C_0 x^{s+1} + C_0' x^{s-1} + \ldots \tag{59}$$

$$Y_u^k(x) = C_k x^{s-k-1} + C_k' x^{s-k-3} + \ldots \tag{60}$$

$$Y_u^{k_1 k_2}(x) = C_{k_1 k_2} x^{s-k_1-k_2-3} + C_{k_1 k_2}' x^{s-k_1-k_2-5} + \ldots \tag{61}$$

Using the partition function we find one- and two-point correlation numbers

$$\mathcal{Z}_k = \int_0^1 dx Y_u^0(x) Y_u^k(x) \tag{62}$$

$$\mathcal{Z}_{k_1 k_2} = \int_0^1 dx (Y_u^{k_1}(x) Y_u^{k_2}(x) + Y_u^0(x) Y_u^{k_1 k_2}(x)) \tag{63}$$

where the one-point correlation numbers are singular only for even k and the two-point correlation numbers are singular only for even $k_1 + k_2$.

Fusion rules demand that one-point correlation numbers for $k \neq 0$ and two point numbers when $k_1 \neq k_2$ are zero. The second term in the two point numbers is actually absent because of (62). Also it is convenient here to introduce a new variable y instead of x

$$\frac{y+1}{2} = x^2, \qquad dx = \frac{dy}{2\sqrt{2}(1+y)^{\frac{1}{2}}} \tag{64}$$

In terms of the variable y the polynomials $Y_u^0, Y_u^k, \ldots$ will contain all powers instead of going with step 2. And the shift was made in order for the interval of integration to be $[-1, 1]$ instead of $[0, 1]$. Thus the fusion rules condition become an orthogonality condition on the polynomials Y_u^0, Y_u^k

$$\mathcal{Z}_k = \int_{-1}^{1} \frac{dy}{2\sqrt{2}(1+y)^{\frac{1}{2}}} Y_u^0(x) Y_u^k(x) = 0, \quad k \neq 0 \tag{65}$$

$$\mathcal{Z}_{k_1 k_2} = \int_{-1}^{1} \frac{dy}{2\sqrt{2}(1+y)^{\frac{1}{2}}} Y_u^{k_1}(x) Y_u^{k_2}(x) = 0, \quad k_1 \neq k_2 \tag{66}$$

Together with the condition $Y_u^0(1) = 0$, it determines the polynomials Y_u^0 and Y_u^k:

s odd	$Y_u^0(y) = P_{\frac{s+1}{2}}^{(0,-\frac{1}{2})}(y) - P_{\frac{s-1}{2}}^{(0,-\frac{1}{2})}(y)$
s even	$Y_u^0(y) = x\left(P_{\frac{s}{2}}^{(0,\frac{1}{2})}(y) - P_{\frac{s-2}{2}}^{(0,\frac{1}{2})}(y) \right)$
s+k odd	$Y_u^k(y) = P_{\frac{s-k-1}{2}}^{(0,-\frac{1}{2})}(y)$
s+k even	$Y_u^k(y) = x P_{\frac{s-k-2}{2}}^{(0,\frac{1}{2})}(y)$

where $P_n^{(a,b)}$ is Jacobi polynomial.

Due to the relation between Jacobi polynomials and Legendre polynomials P_n

$$P_n^{(0,-\frac{1}{2})}(2x^2 - 1) = P_{2n}(x) \tag{67}$$

$$x P_n^{(0,\frac{1}{2})}(2x^2 - 1) = P_{2n+1}(x) \tag{68}$$

it is of course in agreement with earlier results.

Taking the third derivatives of $\mathcal{Z}$ we get the three-point correlation numbers

$$\mathcal{Z}_{k_1 k_2 k_3} = -\left. \frac{Y_u^{k_1} Y_u^{k_2} Y_u^{k_3}}{\frac{dY_u^0}{dx}} \right|_{x=1} + \int_0^1 dx Y_u^{k_1 k_2} Y_u^{k_3} =$$

$$-\frac{1}{p} + \int_0^1 dx Y_u^{k_1 k_2} Y_u^{k_3} \tag{69}$$

where we used the properties of Jacobi polynomials to get: $Y_u^k(x = 1) = 1$, $\left. \frac{dY_u^0}{dx} \right|_{x=1} = \frac{1}{2s+1} = \frac{1}{p}$. Besides we assume that $k_1, k_2 \leq k_3$.

As we will see the first term reproduces the expression from Minimal Gravity and the role of the second term is to kill the first term when the fusion rules are violated.

Also we need not care about the case of odd $k_1+k_2+k_3$. When $k_1+k_2+k_3$ is odd and $< p$ the fusion rules are violated but $\mathcal{Z}_{k_1 k_2 k_3}$ is non-singular. And if $k_1 + k_2 + k_3$ is odd and $\geq p$ the integral term is automatically zero. Thus we focus on the case when $k_1 + k_2 + k_3$ is even. Fusion rules demand

$$\int_0^1 dx Y_u^{k_1 k_2} Y_u^{k_3} = \begin{cases} \frac{1}{p} & \text{if} \quad k_1 + k_2 < k_3 \\ 0 & \text{if} \quad k_1 + k_2 \geq k_3 \end{cases} \tag{70}$$

Again, switching to variables y we get

$(s + k_1 + k_2)$ odd	$Y_u^{k_1 k_2}(y) = \frac{1}{p} \sum_{n=0}^{\frac{s-k_1-k_2-3}{2}} (4n+1) P_n^{(0,-\frac{1}{2})}(y)$
$(s + k_1 + k_2)$ even	$Y_u^{k_1 k_2}(y) = \frac{x}{p} \sum_{n=0}^{\frac{s-k_1-k_2-4}{2}} (4n+3) P_n^{(0,\frac{1}{2})}(y)$

Now we can compare the quantities independent on the normalization in the Douglas equation approach with those in MLG.

$$\frac{(\mathcal{Z}_{k_1 k_2 k_3})^2 \mathcal{Z}_0}{\prod_{i=1}^3 \mathcal{Z}_{k_i k_i}} \tag{71}$$

When the fusion rules for three-point numbers are satisfied we have $\mathcal{Z}_{k_1 k_2 k_3} = -\frac{1}{2s+1}$ and this quantity gives

$$\frac{(\mathcal{Z}_{k_1 k_2 k_3})^2 \mathcal{Z}_0}{\prod_{i=1}^3 \mathcal{Z}_{k_i k_i}} = \frac{\prod_{i=1}^3 (2s - 2k_i - 1)}{(2s+3)(2s+1)(2s-1)} \tag{72}$$

which coincides with the value from Minimal Gravity (22).

Direct calculation of 4-point correlator gives

$$\mathcal{Z}_{k_1 k_2 k_3 k_4} = \left(-\frac{\frac{d^2 Y_u^0}{dx^2}}{(\frac{dY_u^0}{dx})^3} + \frac{\sum_{i=1}^4 \frac{dY_u^{k_i}}{dx}}{(\frac{dY_u^0}{dx})^2} - \frac{\sum_{i<j} Y_u^{k_i k_j}}{\frac{dY_u^0}{dx}} \right) \Bigg|_{x=1} +$$

$$+ \int_0^1 dx (Y_u^{k_1 k_2} Y_u^{k_3 k_4} + Y_u^{k_1 k_3} Y_u^{k_2 k_4} + Y_u^{k_1 k_4} Y_u^{k_2 k_3}) +$$

$$+ \int_0^1 dx (Y_u^{k_1 k_2 k_3} Y_u^{k_4} + Y_u^{k_1 k_2 k_4} Y_u^{k_3} + Y_u^{k_1 k_3 k_4} Y_u^{k_2} + Y_u^{k_2 k_3 k_4} Y_u^{k_1})$$

The only role of the terms in the third line again is to satisfy the fusion rules by cancelling the unwanted terms in the first two lines when the fusion rules are violated. So, taking only the part which contributes when the fusion rules are satisfied and switching to the variable y instead of x

$$\mathcal{Z}_{k_1 k_2 k_3 k_4} = \left(-\frac{16\frac{d^2 Y_u^0}{dy^2} + 4\frac{dY_u^0}{dy}}{(4\frac{dY_u^0}{dy})^3} + \frac{\sum_{i=1}^{4} 4\frac{dY_u^{k_i}}{dy}}{(4\frac{dY_u^0}{dy})^2} - \frac{\sum_{i<j} Y_u^{k_i k_j}}{4\frac{dY_u^0}{dy}} \right)\Bigg|_{y=1} +$$

$$+ \int_{-1}^{1} \frac{dy}{2\sqrt{2}(1+y)^{\frac{1}{2}}} (Y_u^{k_1 k_2} Y_u^{k_3 k_4} + Y_u^{k_1 k_3} Y_u^{k_2 k_4} + Y_u^{k_1 k_4} Y_u^{k_2 k_3})$$

Using the properties of Jacobi polynomials we obtain

$$\mathcal{Z}_{k_1 k_2 k_3 k_4} = \frac{1}{2(2s+1)^2} \left(-(s-1)(s+2) - 2 + \sum_{i=1}^{4} F(k_i+1) - \right.$$

$$\left. - F(k_{(12|34)}) - F(k_{(13|24)}) - F(k_{(14|23)}) \right)$$

where

$$F(k) = (s-k-1)(s-k-2), \qquad k_{(ij|lm)} = \min(k_i+k_j, k_l+k_m) \quad (73)$$

This is again in agreement with MLG.

The general conjecture obtained in[1] is

$$\mathcal{Z} = \frac{1}{2} \int_{0}^{x*} Y_u^2(x) dx$$

$$Y_u(x) = \sum_{N=0}^{\infty} \sum_{k_1 \dots k_N = 1}^{s-1} \frac{s_{k_1} \dots s_{k_N}}{N!} \left(\frac{d}{dx} \right)^{N-1} P_{s-\sum k - N}(x)$$

where $P_n(x)$ are the Legendre polynomials. The correlation numbers are given by the derivatives of the partition function

$$Z_{k_1 \dots k_n} = \frac{\partial}{\partial s_{k_1}} \dots \frac{\partial}{\partial s_{k_n}} \Bigg|_{\substack{s_k = 0 \\ k \neq 0}} \mathcal{Z}$$

11. $(3, p)$ MLG

In the case $(3, p)$ where $p = 3s + p_0$, $p_0 = 1, 2$, the polynomials Q and S are

$$Q = y^3 + uy + v,$$

$$S(u,v) = Res \left(Q^{\frac{p}{3}+1} + \sum_{n=1}^{s} \tau_{1,n} Q^{s-n+\frac{p_0}{3}} + \sum_{n=s+1}^{p-1} \tau_{1,n} Q^{n-s-\frac{p_0}{3}} \right)$$

$$S(u,v) = S_{s+1,p_0} + \sum_{k=0}^{s-1} t_k S_{s-k-1,p_0} + \sum_{k=s}^{p-2} t_k S_{k-s+1,-p_0}$$

where by definition $t_k = \tau_{1,k+1}$ and $S_{k,\alpha} = ResQ^{k+\frac{\alpha}{3}}$.

One has to choose an appropriate solution (u_*, v_*) of the string equations

$$S_u = 0$$

$$S_v = 0$$

Using explicit expressions for $C^\alpha_{\beta\gamma}$ we find the free energy for this model

$$\mathcal{F} = \frac{1}{2} \int_{\gamma(\lambda)} \left((S_u^2 - \frac{u}{3} S_v^2) du + 2 S_u S_v dv \right)$$

where the contour $\gamma(\lambda)$ goes from $(u,v) = (0,0)$ to $(u,v) = (u_*(\lambda), v_*(\lambda))$.
The scaling analysis gives

$$p \sim \mu^{\frac{1}{6}}, \quad u \sim \mu^{\frac{1}{3}}, \quad v \sim \mu^{\frac{1}{2}} \quad Z \sim \mu^{1+\frac{p}{3}}, \quad S \sim \mu^{\frac{p+4}{6}}$$

and the dimensions of the times t_k

$$t_k \sim \mu^{\frac{p+3-|p-3(k+1)|}{6}} \sim \begin{cases} \mu^{\frac{k+2}{2}}, & 0 \leq k \leq s-1 \\ \mu^{s-\frac{k}{2}+\frac{p_0}{3}}, & s \leq k \leq p-2 \end{cases}$$

The admissible resonances are

$$t_k = \lambda_k + c_k \mu^{\frac{k+2}{2}} + \sum_{\substack{l=1 \\ (k-l) \in 2Z}}^{s-1} \beta_{kl} \mu^{\frac{k-l}{2}} \lambda_l + \sum C_k^{k_1 k_2} \lambda_{k_1} \lambda_{k_2} \mu^{\frac{k-k_1-k_2-2}{2}} + \ldots,$$

$$0 \leq k \leq s-1$$

$$t_k = \lambda_k + \sum_{\substack{l=k+2 \\ (l-k) \in 2Z}}^{3s+\alpha-2} \beta_{kl} \mu^{\frac{l-k}{2}} \lambda_l + \sum C_k^{k_1 k_2} \lambda_{k_1} \lambda_{k_2} \mu^{\frac{-k+k_1-k_2-2}{2}} + \ldots,$$

$$s \leq k \leq p-2$$

where the terms on r.h.s. are non-zero only if the cosmological constant μ appears in them in an integer non-negative power. After substitution of the times $t_k(\lambda)$ in free energy one gets the n-point correlation number as

$$Z_{k_1\ldots k_n} = \frac{\partial}{\partial \lambda_{k_1}} \cdots \frac{\partial}{\partial \lambda_{k_n}} \bigg|_{\lambda=0} \mathcal{F}$$

It will be crucial to know the solution (u_*, v_*) of the string equations at $\lambda_{m,n} = 0$. We will show that one of the equations is always satisfied by such a v_* that $v_*(\lambda = 0) = 0$. From the dimensional analysis

$$S^0 \sim \mu^{\frac{p+4}{6}}$$

$$S^{k_1\ldots k_n} \sim \mu^{\frac{p+4}{6} - \sum_{i=1}^{n} \frac{p+3-|p-3(k_i+1)|}{6}}$$

Each of the functions $S^{k_1\ldots k_n}$ is a polynomial in the variables u, v, μ. Thus

$$S^{k_1\ldots k_n} = \sum_{M,N,K} u^M v^N \mu^K \sim \mu^{\frac{p+4}{6} - \sum_{i=1}^{n} \frac{p+3-|p-3(k_i+1)|}{6}}$$

$$\frac{M}{3} + \frac{N}{2} + K = \frac{p+4}{6} - \sum_{i=1}^{n} \frac{p+3-|p-3(k_i+1)|}{6}$$

It is equivalent to

$$p + \sum_{i=1}^{n} k_i - N = 2M + 2N + 6K + \sum_{i=1}^{n}(p+3-|p-3(k_i+1)|+k_i)$$

The right hand side is even. Thus $(p + \sum_{i=1}^{n} k_i - N)$ is also even. Consequently the functions $S^{k_1\ldots k_n}$ have definite parities with respect to v

$$S^0(u, -v) = (-1)^p S^0(u, v)$$

$$S^{k_1\ldots k_n}(u, -v) = (-1)^{p+\sum_i k_i} S^{k_1\ldots k_n}(u, v)$$

Thus, as promised, $v = 0$ is always a solution at $\lambda = 0$. Note also that when $\lambda_{m,n} = 0$ the contour of integration follows from $(0, 0)$ to $(u_*(\lambda = 0), v_*(\lambda = 0)) = (u_0, 0)$ and thus can be picked to lie on u-axis. This important fact effectively reduces the problem of finding the polynomials depending on two variables to the problem involving only the polynomials of one variable.

The functions $S^{k_1 \cdots k_n}(u, 0)$ are polynomials in u and μ. The degree of these polynomials can be defined by dimensional analysis. Also the powers of u in these polynomials go with the step 3 since $\mu \sim u^3$.

We disregard all the correlation numbers which are proportional to the integer powers of the cosmological constant μ as non-universal. Their dimensions can also be defined from the dimension of the partition function and of the times $\lambda_{m,n}$.

Now we are going to evaluate zero-, one-, two-, three-point correlation numbers. When evaluating one- and two-point numbers, we do not need to differentiate over the limits of integration since these terms give zero because of the string equation. When taking $\lambda = 0$ after differentiating over λ's , one finds that many of the functions $S_u^{k_1 \cdots k_n}$, $S_v^{k_1 \cdots k_n}$ are odd in v and thus vanish at $v = 0$. Thus the functions $Z_0, Z_k, Z_{k_1 k_2}$ are equal respectively to:

	p even	p odd
	$\frac{1}{2} \int_0^{u_0} (S_u^0)^2 du$	$-\frac{1}{6} \int_0^{u_0} (S_v^0)^2 u du$
	$\int_0^{u_0} S_u^0 S_u^k du$	$-\frac{1}{3} \int_0^{u_0} S_v^0 S_v^k u du$
k_1, k_2 even	$\int_0^{u_0} (S_u^{k_1} S_u^{k_2} + S_u^0 S_u^{k_1 k_2}) du$	$-\frac{1}{3} \int_0^{u_0} (S_v^{k_1} S_v^{k_2} + S_v^0 S_v^{k_1 k_2}) u du$
k_1, k_2 odd	$\int_0^{u_0} (S_u^0 S_u^{k_1 k_2} - \frac{u}{3} S_v^{k_1} S_v^{k_2}) du$	$\int_0^{u_0} (S_u^{k_1} S_u^{k_2} - \frac{u}{3} S_v^0 S_v^{k_1 k_2}) du$
$k_1 + k_2$ odd	$Z_{k_1 k_2} = 0$	$Z_{k_1 k_2} = 0$

where all the polynomials are taken at $v = 0$ since the contour of integration goes along the u-axis.

Now the analysis is similar to the case of Lee-Yang series $(2, 2s + 1)$. We switch from $S(u, 0)$ and u to dimensionless quantities $Y(x)$ and $x = \frac{u}{u_0}$ respectively and from λ_k to dimensionless s_k and use the variable y

$$\frac{y + 1}{2} = x^3, \qquad dx = \frac{dy}{3 \sqrt[3]{2}(1 + y)^{\frac{2}{3}}}$$

From the selection rules we get

p even	$Y_u^0 = x^{2(p_0-1)}(P_{\frac{s-p_0+2}{2}}^{(0,\frac{2}{3}(2p_0-3))}(y) - P_{\frac{s-p_0}{2}}^{(0,\frac{2}{3}(2p_0-3))}(y))$
p odd	$Y_v^0 = x^{2-p_0}(P_{\frac{s+p_0-1}{2}}^{(0,\frac{2}{3}(1-p_0))}(y) - P_{\frac{s+p_0-3}{2}}^{(0,\frac{2}{3}(1-p_0))}(y))$
$(p+k)$ even, $k < s$	$Y_u^k = x^{2(p_0-1)} P_{\frac{s-k-p_0}{2}}^{(0,\frac{2}{3}(2p_0-3))}(y)$
$(p+k)$ odd, $k < s$	$Y_v^k = x^{2-p_0} P_{\frac{s-k+p_0-3}{2}}^{(0,\frac{2}{3}(1-p_0))}(y)$
$(p+k)$ even, $k \geq s$	$Y_u^k = x^{2(2-p_0)} P_{\frac{k-s+p_0-2}{2}}^{(0,\frac{2}{3}(3-2p_0))}(y)$
$(p+k)$ odd, $k \geq s$	$Y_v^k = x^{p_0-1} P_{\frac{k-s-p_0+1}{2}}^{(0,\frac{2}{3}(p_0-2))}(y)$

where all the polynomials are evaluated at $v = 0$.
For the three-point correlation numbers a direct calculation gives

$$Z_{k_1k_2k_3} = (S_u^{k_1}S_u^{k_2} - \frac{u_0}{3}S_v^{k_1}S_v^{k_2})\partial_{k_3}u_* + S_u^{k_1}S_v^{k_2}\partial_{k_3}v_* +$$

$$\int_0^{u_0} du(S_u^{k_1}S_u^{k_2k_3} - \frac{u}{3}S_v^{k_1}S_v^{k_2k_3}) + \int_0^{u_0} du(S_u^0 S_u^{k_1k_2k_3} - \frac{u}{3}S_v^0 S_v^{k_1k_2k_3}) + \text{perm.}$$

$$(74)$$

Taking derivative of the string equations one gets at $s_k = 0$

$$S_u^k + S_{uu}^0 \partial_k u_* + S_{uv}^0 \partial_k v_* = 0$$

$$S_v^k + S_{vu}^0 \partial_k u_* + S_{vv}^0 \partial_k v_* = 0$$

So using the parities of polynomials $S^{k_1\ldots k_n}$ one has at $v = 0$

	p even	p odd
k even	$\partial_k u_* = -\dfrac{S_u^k}{S_{uu}^0}$	$\partial_k u_* = -\dfrac{S_v^k}{S_{uv}^0}$
k odd	$\partial_k v_* = -\dfrac{S_v^k}{S_{vv}^0}$	$\partial_k v_* = -\dfrac{S_u^k}{S_{uv}^0}$

where for each case only non-zero derivative is represented. For instance for even p and k we have $\partial_k v_* = 0$.

The simplest three-point correlation numbers to analyse here is when p is even and all k are even. In this case

$$Z_{k_1 k_2 k_3} = -\frac{S_u^{k_1} S_u^{k_2} S_u^{k_3}}{S_{uu}^0}\Bigg|_{u=u_0} +$$
$$\int_0^{u_0} S_u^0 S_u^{k_1 k_2 k_3} du + \int_0^{u_0} (S_u^{k_1} S_u^{k_2 k_3} + S_u^{k_2} S_u^{k_1 k_3} + S_u^{k_3} S_u^{k_1 k_2}) du.$$

- $1 \leq k_1, k_2, k_3 \leq s-1$; p, k – even. In this case 3-point correlators are singular. The expression for them reduces to (we assume that $k_3 > k_1, k_2$)

$$\mathcal{Z}_{k_1 k_2 k_3} = -\frac{Y_u^{k_1} Y_u^{k_2} Y_u^{k_3}}{(Y_u^0)'}\Bigg|_{x=1} + \int_0^1 Y_u^{k_3} Y_u^{k_1 k_2} dx$$

Two other integral terms are absent. Thus to satisfy fusion rules one needs the following condition to hold

$$\int_0^1 Y_u^{k_3} Y_u^{k_1 k_2} dx = \begin{cases} 0, & \text{if } k_3 \leq k_1 + k_2 \\ \frac{1}{p}, & \text{if } k_3 > k_1 + k_2 \end{cases}$$

This determines

$$\boxed{Y_u^{k_1 k_2} = \frac{1}{p} \sum_{k=0}^{\frac{s-k_1-k_2-p_0-2}{2}} (6k + 4p_0 - 3) x^{2p_0-2} P_k^{(0, \frac{4p_0-6}{3})}, \quad 1 \leq k_{1,2} \leq s-1}$$

We arrive at the expression for nonzero three-point correlation numbers

$$\mathcal{Z}_{k_1 k_2 k_3} = -\frac{Y_u^{k_1} Y_u^{k_2} Y_u^{k_3}}{(Y_u^0)'}\Bigg|_{x=1} = -\frac{1}{p}$$

Now we can evaluate non-vanishing correlation numbers, when selection rules are satisfied. In this case the three-point correlation numbers are given by the first term in the formula above. So we have

$$Z_0 = \frac{p}{(p+3)(p-3)} \tag{75}$$

$$Z_{kk} = \begin{cases} \frac{1}{p-3(k+1)}, & 1 \leq k \leq s-1, \quad k \text{ even} \\ \frac{1}{3(k+1)-p}, & s \leq k \leq p-2, \quad k \text{ even} \end{cases} \tag{76}$$

$$Z_{k_1 k_2 k_3} = -\frac{1}{p}, \qquad 1 \leq k_i \leq p-2, \quad k \text{ even}. \tag{77}$$

The quantity that doesn't depend on the normalization of the operators and correlators is

$$\frac{(Z_{k_1 k_2 k_3})^2 Z_0}{\prod_{i=1}^3 Z_{k_i k_i}} = \frac{\prod_{i=1}^3 |p - k_i q|}{p(p+q)(p-q)}$$

where $p = 3s + p_0, q = 3$. It is in agreement with the direct calculation of the integrals over the moduli space in Minimal Gravity.

12. Conclusion

We have considered the Douglas approach to Liouville Minimal Gravity. We suggested a useful integral representation for the partition function of the Liouville Minimal Gravity perturbed by primary operators. We showed that there exists a special solution of the Douglas string equation and an appropriate resonance relations $t(\lambda)$ which allows to receive the correlation numbers satisfying conformal selection rules of Liouville Minimal Gravity. The three- and four-point correlation numbers derived from the Douglas string equation coincide with the direct calculations in $(q, p) = (3, 3s + p_0)$ MLG when the selection rules are satisfied.

References

1. A. Belavin and A. Zamolodchikov, On Correlation Numbers in 2D Minimal Gravity and Matrix Models, *J.Phys.A* **42** (2009) 304004, arXiv:0811.0450.
2. A. Belavin, B. Dubrovin, and B. Mukhametzhanov, Minimal Liouville Gravity correlation numbers from Douglas string equation, arXiv:1310.5659.
3. V. Knizhnik, A. M. Polyakov, and A. Zamolodchikov, Fractal Structure of 2D Quantum Gravity, *Mod.Phys.Lett.A* **3** (1988) 819.
4. A. Belavin, Al. Zamolodchikov, Integrals over moduli spaces, ground ring, and four-point function in minimal Liouville gravity, *Theor.Math.Phys.* **147** (2006) 729–754.
5. Al. Zamolodchikov, Higher equations of motion in Liouville field theory, *Int.J.Mod.Phys.A* **19S2** (2004) 510–523, hep-th/0312279.
6. V. Kazakov, A. A. Migdal, and I. Kostov, Critical Properties of Randomly Triangulated Planar Random Surfaces, *Phys.Lett.B* **157** (1985) 295–300.
7. V. Kazakov, Ising model on a dynamical planar random lattice: Exact solution, *Phys.Lett.A* **119** (1986) 140–144.
8. V. Kazakov, The Appearance of Matter Fields from Quantum Fluctuations of 2D Gravity, *Mod.Phys.Lett.A* **4** (1989) 2125.
9. M. Staudacher, The Yang-Lee edge singularity on a dynamical planar random surface, *Nucl.Phys.B* **336** (1990) 349.
10. E. Brezin and V. Kazakov, Exactly solvable field theories of closed strings, *Phys.Lett.B* **236** (1990) 144–150.

11. M. R. Douglas and S. H. Shenker, Strings in Less Than One-Dimension, *Nucl.Phys.B* **335** (1990) 635.
12. D. J. Gross and A. A. Migdal, Nonperturbative Two-Dimensional Quantum Gravity, *Phys.Rev.Lett.* **64** (1990) 127.
13. G. W. Moore, N. Seiberg, and M. Staudacher, From loops to states in 2-D quantum gravity, *Nucl.Phys.B* **362** (1991) 665–709.
    P. Di Francesco and D. Kutasov, Unitary Minimal Models coupled to 2d Quantum gravity, *Nucl.Phys.B* **342** (1990) 589–624.
    A.Zamolodchikov, Unpublished.
14. Al. Zamolodchikov, Three-point function in the minimal Liouville gravity, *Theor.Math.Phys.* **142** (2005) 183–196.
15. M. R. Douglas, Strings in less than one-dimension and the generalized KdV hierarchies, *Phys.Lett.B* **238** (1990) 176.
16. P. H. Ginsparg, M. Goulian, M. Plesser, and J. Zinn-Justin, (p, q) String actions, *Nucl.Phys.B* **342** (1990) 539–563.

MIRROR MATTER AND OTHER MODELS FOR DARK MATTER

S.I. BLINNIKOV

ITEP, B.Cheremushkinskaya, 25, 117218 Moscow,
SAI, Lomonosov MSU, 119991 Moscow
and Novosibirsk State University, Novosibirsk, 630090
E-mail: Sergei.Blinnikov@itep.ru

A brief review of the Mirror Matter (MM) model in modern pattern of the Dark Matter (DM) problem. This model was proposed by I.Yu.Kobzarev, L.B.Okun, and I.Ya.Pomeranchuk in 1966. Mirror Matter is the oldest but still viable candidate for the description of real DM, but probably not in the pure form. The difficulties of standard ΛCDM model with WIMPs (Weakly Interacting Massive Particles) are touched upon. It is shown where Mirror Matter could help in solving some of those problems. Mirror particles are a subclass of a wider set of ADM (Asymmetric DM, when DM particles do not annihilate contrary to supersymmetric particles in ΛCDM). The references on limits for cross-sections of bosons in ADM models accreting onto neutron stars are given. They can be much lower than experimental limits for WIMPs.

1. Introduction

Hypothesis of the presence of Dark Matter in the Universe (German: Dunkle Materie, DM hereafter) was put forward by Fritz Zwicky in 1933.[1] In this paper, Zwicky has discovered a virial paradox in Coma cluster of galaxies (in the constellation Coma – Coma Berenices) and suggested the presence of "dunkle Materie" in the cluster. Very soon, just after the paper by Einstein[2] on gravitational lensing (already, 12 years earlier, discussed by Chwolson[3] for ordinary stars), Zwicky has proposed that the presence of DM can be tested by the effect of gravitational lensing,[4,5] – this case we call now macrolensing.

For the investigations of DM it is very important to study the rotation curves of galaxies, which was also first proposed by Zwicky[6] and soon the missing mass was found in M31 (Andromeda) galaxy.[7]

But Zwicky's seminal ideas and papers were almost forgotten for half a century, and rediscovery of DM occurred only decades later, in 1970s.

At that time this idea was much more popular in USSR, than in the West, where it was accepted only after strong resistance, see the review by Einasto.[8] The problem of hidden or dark mass in galaxies became acute because of the mismatch of direct (kinematic) and indirect (photometric) estimates of the mass of galaxies (and their systems).[9] DM necessity became clear also after the discovery in X-rays of a huge amount of hot gas (with a temperature of the order of keV) in clusters of galaxies. The hot gas cannot be trapped by the gravity of visible baryons.

Fluctuations of the Cosmic Microwave Background (CMB) tell us that at the epoch of recombination the perturbations of baryonic matter were very small: for redshift $z = z_{\rm rec} \approx 10^3$ they had a level of $\sim 10^{-5}$. This means that to our epoch they could grow only as a scale factor, i.e. as $(1 + z_{\rm rec}) \sim 10^3$, and we would get now, for $z = 0$, not more than 1% of density fluctuations, which drastically contradicts the observed structure. If DM dominates in gravity, it may have at the time of recombination a larger amplitude of perturbations, and the "dark matter boost" is provided for structure formation in visible matter, see, e.g.[10,11] DM "gas" should be non-relativistic (cold), otherwise the structure will be smeared. The basic model for DM assumes the Cold Dark Matter (CDM).

The technique for investigating DM has been developed not only for the gravitational macrolensing but also for cases of "microlensing" and "weak lensing",[12–19] see the review.[20]

2. The first Dark Matter model – paper by Kobzarev, Okun, Pomeranchuk (1966)

The paper[21] (English translation[22]) has actually suggested a first model for DM. The seminal article by Kobzarev, Okun and Pomeranchuk[21] was published in 1966, when physicists could not even know that dark matter will soon become one of the most important problems in cosmology and elementary particle physics. Authors had other motives. They sought to show how the broken CP–symmetry can be restored.

In this article it was shown that if there are "mirror" particles, reducing CP–symmetry, they may interact with conventional particles only very weakly. This is exactly what is needed to explain dark matter.

Here are a few detailed quotes from this wonderful article.

Abstract of:[21]

"In connection with the discovery of CP violation in the decay $K_2^0 \to 2\pi$ the possibility is discussed, that "mirror" (R) particles exist in addition to the ordinary (L) particles. The introduction of these latter particles reestablishes the equivalence of left and right. It is shown that mirror particles cannot interact with ordinary particles strongly, semi-strongly, or electromagnetically. Weak interactions between L and R particles are possible, owing to the exchange of neutrinos. L and R particles must have the same gravitational interaction. The possibility of existence and detection of macroscopic bodies (stars) made up of R-matter is discussed."

The central point of the paper[21] is a proof of the weakness of common electromagnetic interaction of mirror and ordinary particles:

"We start with the case of a common electromagnetic interaction for the two kinds of particles. In this theory there should exist two kinds of π^0-mesons: π^0_L and π^0_R ($\pi^0_R = -CPA\pi^0_L$). In the presence of a common photon, the following transition is possible

$$\pi^0_R \rightleftarrows 2\gamma \rightleftarrows \pi^0_L .$$

This will give rise to two states

$$\pi^0_1 = \frac{1}{\sqrt{2}}(\pi^0_R - \pi^0_L), \quad \pi^0_2 = \frac{1}{\sqrt{2}}(\pi^0_R + \pi^0_L) ,$$

where π^0_1 is even under CPA (odd under CA), and π^0_2 is odd under CPA (even under CA). We shall assume that the electromagnetic interaction is even under CPA and P, and thus also under CA, but is not necessarily even under C [Bernstein J, Feinberg G, Lee T D *Phys. Rev. B* 139, 1650 (1965), Barshay S *Phys. Lett.* 17, 78 (1965)]. By assumption, a γ-(photon) transforms into itself under CA. Then the π^0_1 state, which has odd CA, cannot decay into two photons, real or virtual, but can decay into three photons. Since the width of such a decay is small, this π^0_1 meson would be long-lived, in contradiction with experiments."

It was also shown in the paper that common strong interaction is impossible. The conclusion was rather pessimistic:

"The main result of the present paper, consisting in the conclusion of the possibility of only a very weak interaction between mirror matter and ordinary matter, and that the upper bound for the concentration of mirror-matter in the solar system is small, does not add to the attractiveness of the hypothesis of the existence of mirror matter."

3. Problems which arose after the paper by Kobzarev, Okun and Pomeranchuk (1966)

In fact, Kobzarev, Okun and Pomeranchuk[21] suggested the first model for Dark Matter (DM), which may contain invisible stars. As I told above, even astronomers, to say nothing about physicists, were aware of the severity of the DM problem. Soon the problem became urgent in astronomy, while in physics predictions on supersymmetric particles and on axions appeared. Currently there is a large flow of papers on CDM, consisting of WIMPs. Not so much work is published on mirror matter. Nevertheless, the hypothesis of the existence of mirror matter is still attractive in the study of Dark Matter.

Below I discuss briefly some facts about DM and related issues.

3.1. Who introduced hidden sector into DM

It is often written that mirror particles were introduced by Lee and Yang in 1956,[23] see, e.g., a recent paper by Pavšič.[24] L.B.Okun himself was writing in his review of 1966[25,26] in Physics Uspekhi: "The hypothesis on existence of mirror particles was put forward in the paper[23] (see also Lewis Carroll 'Through the Looking-Glass, and What Alice Found There') and was discussed in detail in.[21]"

In reality, Lee and Yang[23] have introduced the concept of right-handed protons, but their "R-matter" was not hidden! – as some of recent papers claim, e.g.[27]

An exact citation from[23] reads:

"... the interaction between them is not necessarily weak. For example, p_R and p_L could interact with the same electromagnetic field and perhaps the same pion field. They could then be separately pair-produced, giving rise to interesting observational possibilities."

Thus Lee and Yang write explicitly about the possibility of direct electromagnetic and strong interactions between left- and right-handed protons, while Kobzarev, Okun and Pomeranchuk have shown that this is impossible. They demonstrated that the Mirror particles "live" in the hidden sector, within which the microphysics is the same as in the visible sector. The terms "mirror particles" and "Mirror Matter" were also first introduced in.[21]

One can mention here the article of their predecessors Nishijima and Saffouri,[28] who discussed the idea of a "shadow Universe". Okun and Pomeranchuk[29–31] have shown that the proposed model of the shadow universe is in sharp conflict with neutrino experiments. The existence of "shadow" K_1^0-mesons, which should have high penetrating power, were excluded by the data obtained in the neutrino experiment at CERN wherein behind the protection (25 m of iron) no abnormal particles were revealed with properties similar to the "shadow" K_1^0-mesons.

On other ways of coming to the concept of mirror matter from various considerations see.[32,33]

3.2. *Literature of last years on MM*

Detailed review of the MM problem status as of 2007 is given by L.B.Okun in.[34] Optimistic view of the problem is offered by Z.K.Silagadze.[35] Many papers are published by R.Foot, see, e.g.,[36,37] of the most recent publications one can point out.[38] The latter paper explains how galaxies formed structure taking into account the dissipation in MM.

Interesting development of the model in terms of fundamental physics is given in a series.[39–41]

Model of mirror particles is discussed in[42] in the context of asymmetric DM (ADM), – see more about ADM below in section 3.6.

P.Ciarcelluti published a review,[43] where he discusses the growth of perturbations, CMB and large-scale structure formation, taking into account the mirror matter as DM (see on this topic paper by Z.Berezhiani *et al.*[44])

On astrophysical aspects, in particular on the heating of neutron stars by MM-particles[45] see also my review.[46]

3.3. *Possible models for DM*

There are many candidates for DM particles and the most discussed in detail are the following.

- WIMPs - supersymmetric particles (neutralino for heavy or gravitino for lighter particles) and their clouds;
- The axion-like particles and objects;
- Mirror particles and objects – this model is not as popular now as the two above, but this is the oldest of all the proposed models and it is still surprisingly useful.

Until now no reliable signals are found in underground laboratories, aimed at detection of WIMPs! This stimulates the search for alternatives to WIMPs (see section 3.6), including MM.

Neutralinos and axions have electromagnetic interactions (on weak scale). Mirror matter $\equiv$ MM in its pure form is not involved in the electromagnetic interactions with ordinary matter $\equiv$ OM (or in other interactions of the standard model). We share only common gravity. MM has analogues of all "our" interactions in the mirror sector. In addition to these models for MM, there are options for models with a slightly broken mirror symmetry, which have a weak electromagnetic coupling with OM, see e.g.[37] Richness of other opportunities and oscillations is discussed in.[47]

Let us briefly list some predictions about the possible properties of MM in "pure" form, without the admixture of electromagnetic interactions or oscillations.

- Basically MM is able to replace the cold dark matter (CDM) in the cosmological evolution (P.Ciarcelluti[48–50]), if $T_{MM} < 0.3T_{OM}$ in the early universe.
- MM-baryons can form compact objects of stellar masses and sizes.
- Mirror stars should be invisible, but should give the effect of microlensing, see next subsection.
- Due to lower T_{MM} a larger fraction of mirror helium is produced in cosmological nucleosynthesis than in OM helium, hence the stellar evolution is faster, and the proportion of MM intergalactic gas may be lower than in OM.
- Possible life and intelligence in MM.[34]

3.4. Microlensing: end of MACHO era?

Invisible stars can be found by effect of Gravitational microlensing may be caused by both visible and invisible stars. Those objects are called now MACHOs. for "Massive Astrophysical Compact Halo Object". This phenomenon was first discussed in relation with MM by Berezhiani, Dolgov and Mohapatra,[51] and Blinnikov.[52]

3.4.1. MACHO, EROS, AGAPE, MEGA, OGLE – contradicting results

MACHO group[53] has revealed 13 – 17 microlensing events in the Large Magellanic Cloud (LMC), significantly higher number than expected from known stars. Not all DM in the halo may be in MACHO, only a fraction, usually denoted by f. MACHO group concluded that compact objects in the mass range $0.15M_\odot < M < 0.9M_\odot$ have a fraction f in DM halo on the level $0.08 < f < 0.50$ (95% CL).

Bennett[54] has concluded (based on the results of MACHO group) that MACHOs have been really found.

EROS (Expérience pour la Recherche d'Objets Sombres) collaboration has placed only an upper limit on the halo fraction, $f < 0.2$ (95% CL) for objects in this mass range, while EROS-2[55] gives $f < 0.1$ for $10^{-6}M_\odot < M < 1M_\odot$.

AGAPE collaboration,[56] working on microlensing in M31 (Andromeda) galaxy, finds the halo MACHO fraction in the range $0.2 < f < 0.9$, while MEGA group marginally conflicts with them finding an upper limit $f < 0.3$.[57]

Detailed analysis of the controversial situation with the results of different groups is given in.[58] Newer results[59] for EROS-2 and OGLE (Optical Gravitational Lensing Experiment) in the direction of the Small Magellanic Cloud claim: $f < 0.1$ obtained at 95% confidence level for the MACHO with a mass of $10^{-2}M_{\odot}$ and $f < 0.2$ for the MACHO with mass $0.5M_{\odot}$.

Recent data on the microlensing in other aspects are discussed in.[60]

3.4.2. *Destruction of wide pairs of visible stars*

Evans and Belokurov[61] criticize some papers in the series "End of MACHO era (1974–2004)". Paper,[62] which appeared in this series, asserts that wide pairs of visible stars must be destroyed by invisible MACHO flying near them. In addition to the criticism of,[61] one can point out that it is necessary to consider not only the process of destruction, but also a reverse process of creating pairs of visible stars from single individual stars not binded previously by mutual gravitation.

Probability of microlensing[14–16] is naturally measured by the so-called optical depth τ. Evans and Belokurov[61] confirmed lower number of compact objects in the direction to LMC, than obtained in the MACHO group, i.e. they got $\tau < 0.36 \times 10^{-7}$ in agreement with EROS results.[55] Later, a paper of the same Cambridge group[63] was published where, on the basis of studies of binary stars, arguments in favor of real existence of MACHO's and against the pessimistic conclusions in[62] were put forward.

Total DM halo mass in the Galaxy can not be explained by invisible stars, therefore, it is more likely that DM consists of combinations like CDM + MM or WDM + MM, where WDM (Warm dark matter) may consist, for example, of sterile neutrinos (with mass of a few keV) or gravitino, see, e.g.[64,65]

3.5. *Constraints on DM particles cross-sections from observations of clusters of galaxies*

3.5.1. *"Bullet" cluster 1E0657-56*

Cluster of galaxies 1E0657-558 ("Bullet" cluster) is formed as a result of a collision of two clusters. This is clearly seen in X-ray images. The observations of this cluster give important clues to the proof of reality of DM. It is difficult to explain the observations in alternatives to DM like modified gravity (MOND or MG). It is clear that the hot X-ray emitting gas which dominates the baryon mass in the cluster is shifted from the DM distribution in this cluster. The latter is traced by the effect of gravitational lensing and follows the distribution of stars and galaxies (which are effectively a collisionless gravitating gas) and.[66–68]

Many physicists tend to use this example as a direct proof of a small cross-section of self-interaction of DM particles. In reality it proves only what is observed and nothing more: DM behaves like ordinary stars which interact only by gravity and form a collisionless matter. But of course O-stars are made of strongly and electromagnetically interacting particles. The same may be true for the DM-component.

3.5.2. Cluster MACS J0025.4-1222

This cluster is very similar to the Bullet cluster. It is found there that the DM self-interaction cross-section obeys the limit $\sigma/m < 4$ cm^{2}g^{-1}.[69] For particle masses in GeV range this limit is lower than for collisions of ordinary atoms. But this limit would be true for DM particles only if they do not form bound objects! If in DM there is anything like our normal solid bodies, say, ice with density like 1 g cm^{-3} with the size r larger than a few cm, then this limit is very well satisfied (σ is $\propto r^2$, while mass is $\propto r^3$).

So if the properties of DM particles are similar to our particles, and they are able to form stars, planets, asteroids, etc. (e.g., like MM), then all observations of the merging clusters are reproduced. However, we have to squeeze a major fraction of MM particles into those compact objects, not leaving a large fraction in the form of gas like we have in OM in clusters of galaxies where the O-gas dominates in baryon component.

3.5.3. DM in Abell 520 cluster

Dark matter in the Abell 520 cluster is hard to explain in CDM with WIMPs. In contrast to the Bullet cluster, the lensing signal and the X-ray emission coincide here, and are both shifted away from galaxies, indicating that DM is collisional like in O-gas: "...a mass peak without galaxies cannot be easily explained within the current collisionless dark matter paradigm".[70]

We may get this behaviour leaving a major fraction of DM in form of gas in MM models. Mirror Matter models are richer than the WIMP based dark matter models, they "are more flexible and for them diverse behaviour of the dark matter is a natural expectation".[35]

3.6. Difficulties of CDM, limits from underground experiments and ADM models

Arguments in favor of the model, which is the most popular at the moment, that is ΛCDM (cold dark matter), are strong. For example, already Sandage[71] studied the influence of the local group of galaxies on the velocity field. According to current research, clearly visible traces of the DM-infall in the local group are found in the phase plane distance-speed[72-74] in accordance with the predictions of ΛCDM model.

Difficulties of ΛCDM model are discussed by Dolgov in this volume. In particular, one of the problems is the lack of dwarf satellite galaxies of large galaxies. CDM predicts an order of magnitude more satellites of galaxies than observed. CDM theory predicts, moreover, a singular distribution of matter in the centers galaxies, a sharp cusp. It is believed that in reality there is a shallow (cored) profile in the center, although the situation is not yet obvious in observations.[75,76]

Interesting results are obtained at SAO[77-79] on the voids in the surrounding space, and on the deficit of dark matter near our Local Group of galaxies. It is not easy to understand in the ΛCDM.

Many other problems in addition to those are discussed by P.Kroupa.[80] For example, satellites of our Galaxy and M31 tend to form disk distributions. Contrary to this observation, the CDM model would predict isotropic random distribution of orbits, while MM can resolve this problem, see the discussion in.[81]

3.6.1. Underground detectors

Currently there are several working installations aimed at direct detection of WIMPs, in particular CDMS, DAMA, XENON, LUX. A review of methods, facilities, and other references, see, for example, in.[82,83]

Recent results of the LUX detector[84] give the strongest limits on the DM cross-sections. At a confidence level of 90%, the minimum upper limit on the elastic scattering cross-section of WIMP on nucleon is 7.6×10^{-46} cm^2 at WIMP mass 33 GeV. This result has overrun all other limits and hints of WIMPs detection in other detectors. It is a powerful impetus for the development of models having particles very weakly interacting with ordinary nucleons. MM is exactly such a kind of model.

3.6.2. *Asymmetric Dark Matter, ADM*

Density of relic WIMPs in the conventional CDM models is explained by freeze-out in the expanding hot universe with a proper choice of their masses and annihilation cross sections.[85–87] Models of asymmetric dark matter (ADM) are based on a very different assumption: they assume that asymmetry in the density of DM particles and antiparticles is generated in the hot universe, similar to the baryon-antibaryon asymmetry of ordinary matter.[42,88] This class includes also MM – Mirror Matter, since mirror baryons should be as asymmetrical to mirror antibaryons as our ordinary baryons to ordinary antibaryons. In light of the strong constraints obtained in LUX experiment[84] the MM model becomes attractive because mirror baryons should not give any signals in underground detectors.

Other options for ADM models are also developed, for example, the boson ADM is being considered in detail.[42,88,89] There are non-zero cross-sections, but they may be well below the LUX limits. Strong constraints on the boson ADM are obtained when ADM bosons accreting onto a neutron star are considered, since the maximum mass for a cold boson star is m_{Pl}/m times smaller than the Chandrasekhar limit for fermion stars.[90] Here m_{Pl} is the Planck mass, and m is the particle mass, defining self-gravitation of the star. If the boson mass m if of order 1 GeV, the mass limit is by 19 orders of magnitude smaller than the mass of the Sun. The accumulation of such a small number of ADM bosons inside a neutron star may lead to black hole formation in its center, which will eventually "eat" the whole neutron star. This allows one to obtain strong limits on the cross-section of ADM bosons.[89]

A review on different DM models with new physics, see, e.g., in.[91]

4. Conclusions

Until now there are still several arguments motivating Mirror Matter search.

- WIMPs – weakly interacting massive particles are still not discovered.
- Properties of DM particles with respect to clustering on stellar-size and mass scale are unknown.
- Microlensing: deficit of normal stars to explain all MACHO events (although the total mass of DM-halo can not be explained by invisible stars).
- Clusters of galaxies like Abell 520 do exist which are hard to explain in pure CDM picture.[35]

- Other difficulties of CDM, which can be resolved in MM model.[81]
- New kinds of mysterious transients like reported by K. Barbary et al.,[92] may be explained by MM objects.[45]

Finally, I can draw several conclusions.

- Mirror matter (MM) is still a useful model of DM, which shows how rich can be the world structure in the Dark Matter sector.
- Observations of microlensing are not yet in full agreement, but it is clear that the Milky Way DM halo may contain no more than $\sim$ 20 % of its mass in invisible stars.
- Calculations of the perturbation growth at the linear stage based on MM have been carried out: MM can give dates for the growth of perturbations in the baryonic matter structures[48,49] – but this awaits an independent check.
- One can not use the limits on cross-sections of self-interacting DM individual particles, obtained from observations of the colliding clusters: such particles can form macroscopic bodies which behave as collisionless gas.

References

1. F. Zwicky, *Helvetica Phys. Acta* **6** (1933) 110.
2. A. Einstein, *Science* **84** (1936) 506.
3. O. Chwolson, *Astronom. Nachrichten* **221** (1924) 329.
4. F. Zwicky, *Phys. Rev.* **51** (1937) 290.
5. F. Zwicky, *Phys. Rev.* **51** (1937) 679.
6. F. Zwicky, *Astrophys. J.* **86** (1937) 217.
7. H. W. Babcock, *Lick Obs. Bull.* **19** (1939) 41.
8. J. Einasto, arXiv:0901.0632 [astro-ph.CO].
9. A. V. Zasov, Distribution of Dark Matter in Galaxies, in *Dark matter, dark energy and their detection, Conference at NSU*, 2013.
10. A. V. Gurevich and K. P. Zybin, *Soviet Physics Uspekhi* **165** (1995) 723.
11. A. V. Gurevich and K. P. Zybin, *Soviet Physics Uspekhi* **38** (1995) 687.
12. S. Refsdal, *Mon. Not. R. Astron. Soc.* **128** (1964) 295.
13. S. Liebes, *Phys. Rev.* **133** (1964) 835.
14. A. V. Byalko, *Astron. Zh.* **46** (1969) 998.
15. A. V. Byalko, *Sov. Astron.* **13** (1970) 784.
16. B. Paczynski, *Astrophys. J.* **304** (1986) 1.
17. A. V. Gurevich, K. P. Zybin and V. A. Sirota, *Physics Letters A* **214** (1996) 232.
18. A. V. Gurevich, K. P. Zybin and V. A. Sirota, *Soviet Physics Uspekhi* **167** (1997) 913.

19. A. V. Gurevich, K. P. Zybin and V. A. Sirota, *Soviet Physics Uspekhi* **40** (1997) 869.
20. A. F. Zakharov, *Phys. Part. Nucl.* **39** (2008) 1176.
21. I. Y. Kobzarev, L. B. Okun and I. Y. Pomeranchuk, *Yadernaya Fizika* **3** (1966) 1154.
22. I. Y. Kobzarev, L. B. Okun and I. Y. Pomeranchuk, *Sov. J. Nucl. Phys.* **3** (1966) 837.
23. T. D. Lee and C. N. Yang, *Phys. Rev.* **104** (1956) 254.
24. M. Pavšič, arXiv:1310.6566 [hep-th].
25. L. B. Okun', *Soviet Physics Uspekhi* **89** (1966) 603.
26. L. B. Okun', *Soviet Physics Uspekhi* **9** (1967) 574.
27. J. L. Feng, H. Tu and H.-B. Yu, *JCAP* **10** (2008) 43.
28. K. Nishijima and M. H. Saffouri, *Physical Review Letters* **14** (1965) 205.
29. L. B. Okun' and I. Y. Pomeranchuk, *JETP Lett.* **1** (1965) 28.
30. L. B. Okun' and I. Y. Pomeranchuk, *Soviet Journal of Experimental and Theoretical Physics Letters* **1** (1965) 167.
31. L. B. Okun' and I. Y. Pomeranchuk, *Physics Letters* **16** (1965) 337.
32. M. Pavšič, *Int. J. Theor. Phys.* **9** (1974) 229.
33. R. Foot, H. Lew and R. R. Volkas, *Physics Letters B* **272** (1991) 67.
34. L. B. Okun', *Soviet Physics Uspekhi* **50** (2007) 380.
35. Z. K. Silagadze, arXiv:0808.2595 [astro-ph].
36. R. Foot, *Int. J. Mod. Phys. D* **13** (2004) 2161.
37. R. Foot, *Phys. Rev. D* **78**, 043529 (2008).
38. R. Foot, *Phys. Rev. D* **88**, 023520 (2013).
39. C. R. Das, L. V. Laperashvili, H. B. Nielsen and A. Tureanu, *Phys. Rev. D* **84**, 063510 (2011).
40. C. R. Das, L. V. Laperashvili, H. B. Nielsen and A. Tureanu, *Physics Letters B* **696**, 138 (2011).
41. C. R. Das, L. V. Laperashvili and A. Tureanu, *Int. J. Mod. Phys. A* **28**, 50085 (2013).
42. K. Petraki and R. R. Volkas, *Int. J. Mod. Phys. A* **28**, 30028 (2013).
43. P. Ciarcelluti, *Int. J. Mod. Phys. D* **19**, 2151 (2010).
44. Z. Berezhiani, P. Ciarcelluti, D. Comelli and F. L. Villante, *Int. J. Mod. Phys. D* **14**, 107 (2005).
45. F. Sandin and P. Ciarcelluti, *Astroparticle Physics* **32**, 278 (2009).
46. S. I. Blinnikov, *Physics of Atomic Nuclei* **73**, 593 (2010).
47. M. Sarrazin and F. Petit, *European Physical Journal C* **72**, 2230 (2012).
48. P. Ciarcelluti, *Int. J. Mod. Phys. D* **14**, 187 (2005).
49. P. Ciarcelluti, *Int. J. Mod. Phys. D* **14**, 223 (2005).
50. P. Ciarcelluti, Early Universe cosmology with mirror dark matter, in *American Institute of Physics Conference Series*, American Institute of Physics Conference Series, Vol. 1241 (2010).
51. Z. G. Berezhiani, A. D. Dolgov and R. N. Mohapatra, *Physics Letters B* **375**, 26 (1996).
52. S. I. Blinnikov, arXiv:astro-ph/9801015.
53. C. Alcock, R. A. Allsman, D. R. Alves *et al.*, *Astrophys. J.* **542** (2000) 281.

54. D. P. Bennett, *Astrophys. J.* **633** (2005) 906.
55. P. Tisserand, L. Le Guillou, C. Afonso *et al.*, *Astron. Astrophys.* **469** (2007) 387.
56. A. Riffeser, S. Seitz and R. Bender, *Astrophys. J.* **684** (2008) 1093.
57. G. Ingrosso, S. Calchi Novati, F. de Paolis *et al.*, *Astron. Astrophys.* **462** (2007) 895.
58. M. Moniez, *General Relativity and Gravitation* **42** (2010) 2047.
59. S. Calchi Novati, S. Mirzoyan, P. Jetzer and G. Scarpetta, *Mon. Not. R. Astron. Soc.* **435** (2013) 1582.
60. S. Mao, *Research in Astronomy and Astrophysics* **12** (2012) 947.
61. N. W. Evans and V. Belokurov, *Mon. Not. R. Astron. Soc.* **374** (2007) 365.
62. J. Yoo, J. Chanamé and A. Gould, *Astrophys. J.* **601** (2004) 311.
63. D. P. Quinn, M. I. Wilkinson, M. J. Irwin *et al.*, arXiv:astro-ph/0603706.
64. D. Gorbunov, A. Khmelnitsky and V. Rubakov, *JHEP* **12** (2008) 55.
65. D. Gorbunov, A. Khmelnitsky and V. Rubakov, *JCAP* **10** (2008) 41.
66. D. Clowe, A. Gonzalez and M. Markevitch, *Astrophys. J.* **604** (2004) 596.
67. D. Clowe, M. Bradač, A. H. Gonzalez *et al.*, *Astrophys. J. Lett.* **648**, L109 (2006).
68. M. Bradač, D. Clowe, A. H. Gonzalez *et al.*, *Astrophys. J.* **652**, (2006) 937.
69. M. Bradač, S. W. Allen, T. Treu *et al.*, *Astrophys. J.* **687**, (2008) 959.
70. A. Mahdavi, H. Hoekstra, A. Babul *et al.*, *Astrophys. J.* **668**, (2007) 806.
71. A. Sandage, *Astrophys. J.* **307** (1986) 1.
72. P. Sikivie, I. I. Tkachev and Y. Wang, *Physical Review Letters* **75** (1995) 2911.
73. P. Sikivie, I. I. Tkachev and Y. Wang, *Phys. Rev. D* **56** (1997) 1863.
74. K. Dolag, A. D. Dolgov and I. I. Tkachev, *Soviet Journal of Experimental and Theoretical Physics Letters* **96** (2013) 754.
75. G. A. Shchelkanova, S. I. Blinnikov, A. S. Saburova and A. D. Dolgov, *Pisma v Astron. Zh.* **39** (2013) 743.
76. G. A. Shchelkanova, S. I. Blinnikov, A. S. Saburova and A. D. Dolgov, *Astronomy Letters* **39** (2013) 665.
77. I. D. Karachentsev, *Astrophysical Bulletin* **67** (2013) 123.
78. I. D. Karachentsev, V. E. Karachentseva, O. V. Melnyk, A. A. Elyiv and D. I. Makarov, *Astrophysical Bulletin* **67** (2012) 353.
79. A. A. Elyiv, I. D. Karachentsev, V. E. Karachentseva, O. V. Melnyk and D. I. Makarov, *Astrophysical Bulletin* **68** (2013) 1.
80. P. Kroupa, *Publ. Astr. Soc. Australia* **29** (2012) 395.
81. R. Foot and Z. K. Silagadze, *Physics of the Dark Universe* **2** (2013) 163.
82. D. Akimov, *Nuclear Instruments and Methods in Physics Research A* **628** (2011) 50.
83. M. Pospelov, *Phys. Rev. D* **84**, 085008 (2011).
84. LUX Collaboration: D. S. Akerib, H. M. Araujo, X. Bai *et al.*, arXiv:1310.8214 [astro-ph.CO].
85. B. W. Lee and S. Weinberg, *Physical Review Letters* **39** (1977) 165.
86. M. I. Vysotsky, Y. B. Zeldovich, A. D. Dolgov, *JETP Lett.* **26** (1977) 200.

87. M. I. Vysotsky, A. D. Dolgov and Y. B. Zeldovich, *Soviet Journal of Experimental and Theoretical Physics Letters* **26** (1977) 188.

88. K. M. Zurek, arXiv:1308.0338 [hep-ph].

89. J. Kumar, arXiv:1308.4513 [hep-ph].

90. I. Goldman and S. Nussinov, *Phys. Rev. D* **40** (1989) 3221.

91. S. Gardner and G. M. Fuller, *Progress in Particle and Nuclear Physics* **71** (2013) 167.

92. K. Barbary, K. S. Dawson, K. Tokita *et al.*, *Astrophys. J.* **690** (2009) 1358.

COSMOLOGY:
FROM POMERANCHUK TO THE PRESENT DAY

A.D. DOLGOV

Novosibirsk State University, Novosibirsk, 630090, Russia
ITEP, Bol. Cheremushkinsaya ul., 25, 113259 Moscow, Russia
Dipartimento di Fisica e Scienze della Terra, Università degli Studi di Ferrara, Polo
Scientifico e Tecnologico - Edificio C, Via Saragat 1, 44122 Ferrara, Italy
Istituto Nazionale di Fisica Nucleare (INFN), Sezione di Ferrara, Polo Scientifico e
Tecnologico - Edificio C, Via Saragat 1, 44122 Ferrara, Italy

A short and, due to lack of time (half a century of the science progress in half an hour), rather superficial review on the development of cosmology for scientists working on particle and nuclear physics is presented. The introductory historical part is mostly dedicated to the fundamental works done in Russia (USSR), but not always well known outside the country. Next, the key papers on theory and on astronomical observations, which determined the progress in cosmology during the last half-century or posed crucial problem, are discussed. Among them there are inflation, baryogenesis, dark matter, dark energy, the vacuum energy problem, modification of gravity at large scales, and angular fluctuations of the cosmic microwave background radiation. The presentation is probably biased towards cosmology as it is seen from Russia (or from ITEP) and reflects personal prejudice of the author.

1. Some pieces of history

Slightly more than half a century ago cosmology was an outcast in respectable scientific family. The wide-spread attitude was well characterized by an ironic quotation from L. Landau: "always in error but never in doubt", which the author if this review heard from Ya. Zeldovich. Nevertheless just at that time many outstanding works have been done which enter the fund of gold of the modern cosmology. Of course in connection with the hot universe model (or big bang cosmology) one could not help mentioning the names of Friedmann,[1] who on the basis of the Einstein equations predicted the universe expansion, later discovered by Lemaitre[2] and Hubble,[3] and surely Gamow,[4] who is justly the father of big bang cosmology.

Returning to more recent epoch, we have to recall the pioneering papers by Zeldovich[5] of 1965, where the freezing (in Russian literature "quenching") of massive stable particles in cosmology was first calculated on the basis of the derived there fundamental equation, which at the present time is a cornerstone for calculations of the density of dark matter particles in the universe. Twelve years later this equation was applied to the calculations of the cosmological density of heavy neutral leptons[6,7] and not really fair obtained the name Lee-Weinberg equation.

The pioneering work by Kobzarev, Okun, and Pomeranchuk[8] about possible mirror particle dark matter was created at approximately the same time in 1966 long before dark matter became one of the popular themes in cosmology and particle physics.

In the same year, 1966, the paper by Gershtein and Zeldovich on the cosmological bound on neutrino mass was published. It has to be admitted that the attitude of the scientific establishment to this work was quite skeptical in the spirit of the Landau utterance mentioned above. However, now it is commonly accepted that the most precise device to weigh neutrinos is telescope, see below. The work of Gershtein and Zeldovich was essentially repeated 6 years later by Cowsick and McClelland,[10] however, the authors made two essential mistakes which resulted in approximately 7 times stronger limit. Nevertheless the cosmological bound on the neutrino mass for a long time was called the Cowsick-McClelland bound, which was absolutely unjust.

A year later, in 1967, A. Sakharov[11] suggested a remarkable mechanism of explanation of the baryon asymmetry of the universe, i.e. of the observed predominance of matter over antimatter. He formulated three sufficient conditions for dynamical generation of the asymmetry: 1) breaking of C and CP invariance; 2) deviation from thermal equilibrium; 3) non-conservation of the baryonic number. Even at that "ancient" time there were no doubts about the first two conditions, but nobody believed that the hypothesis of non-conservation of baryons might be true. Nowadays it is practically experimental fact proved by astronomers.

Until 1965 among Russian cosmologists the cold universe model was the commanding one, due to a dominant influence of Zeldovich. The situation radically changed after discover of the cosmic microwave background radiation (CMB), made by Penzias and Wilson,[12] who came upon it accidentally testing an antenna designed for detecting a weak microwave radiation. Several years before this discovery, in 1957, T.A. Ter Shamaonov[13] registered similar radiation, though with lower accuracy, when he worked on calibration of the antenna for the constructed giant radio-telescope RATAN-600. The importance of this observation was not understood at that time due to

unwelcome hot universe model in the Soviet Union at that time. Still half
a year before the Penzias and Wilson discovery and despite hostile attitude
of the establishment to big bang cosmology a paper by Doroshkevich and
Novikov[14] on the possibility of observation of the cosmic microwave background radiation was published. In this work the most favorable frequency
band for the observation was specified. In this connection we should again
mention Gamow, who predicted existence of the cosmic microwave background arriving to us from the early hot stage of the universe evolution.

The measured frequency spectrum of the microwave background is very
accurately given (with precision about 10^{-4}) by the equilibrium Bose-
Einstein spectrum:

$$f = [\exp(\omega/T) - 1]^{-1} \tag{1}$$

with temperature $T = 2.7260 \pm 0.0013$ K and vanishing chemical potential
$\mu/T < 10^{-4}$. The temperature of radiation arriving from different patches
in the sky is almost ideally the same. Before 1980s such equality of the
temperature all over the sky was considered as one greatest cosmological
mysteries, since celestial points separated more than by one degree never
knew about each other in the usual Friedmann cosmology (see discussion
about inflationary paradigm below).

However, this cheerless perfection of VMB is a little bit broken: very
small temperature fluctuations, δT, must exist and they are indeed observed. The point is that the universe is noticeably inhomogeneous. There
exist stars, galaxies, clusters and superclusters of galaxies. All that cannot evolve from perfectly homogeneous world. Hence there must be small
but non-vanishing density perturbations and such perturbations would be
imprinted in the angular fluctuations of the CMB temperature.

The first measurements of δT have been done at the beginning of 90s.[15]
According to them $\delta T/T = 10^{-4} - 10^{-5}$ in different frequency and angular scales. In an earlier work done of the Russian satellite "Relict",[16] only
the quadrupole anisotropy has been detected and, unfortunately, with the
rather poor precision, $6.6 \times 10^{-6} < \delta T_2/T < 3.3 \times 10^{-5}$.

It is interesting that at the time when the detectors for measurements
of the angular fluctuations of CMB were discussed, the magnitude of δT
was expected to be much larger than it was subsequently observed. Larger
temperature fluctuations should exist in the universe without dark matter (DM). Dominating presence of DM in the universe allowed for much
smaller $\delta T/T$. If it were known in advance, then probably the search for

temperature fluctuations would be postponed to better times. Thus, sometimes incorrect theoretical expectations stimulate successful experiments.

2. Contemporary universe

To the present time a huge amount of precise measurements of $\delta T/T$ is accumulated, done on balloons and the satellite detectors, in particularly, on WMAP, see Fig. 1. Recently new results from Planck mission have been published which greatly exceed all others by accuracy, see Fig. 2. Analysis of the spectrum of the angular fluctuations of CMB allows to determine cosmological parameters with unprecedented precision. The measurements of CMB temperature fluctuations definitively brought cosmology into the exact science "club".

Returning to the precision of observations in the 60s, recall that Hubble parameter was known inside the factor of two: $H = 50 - 100$ km/sec/Mpc. The baryon-to-photon ratio was known even worse: $N_B/N_\gamma = 10^{-9\pm1}$. In this connection I recall an acid comment of a well known physicist during my talk on cosmological baryon asymmetry at an ITEP seminar: "What kind of science is it, where the error enters into the exponent?"

At this ancient times people believed that the usual baryonic matter made practically 100% of the universe mass. The first seriously perceived observations, which shook this believe, appeared only in 1974,[17] though already 40 years earlier there were serious indications that the universe is not so simple.[18]

Nowadays practically nobody doubts that baryons contribute petty 5% fraction of the whole cosmological matter, while the dominant 95% remains unknown. This unknown part in turn consists of two parts: dark matter (DM) whose contribution is around 25%, and dark energy (DE), contributing approximately 70%. There is a rich zoo of possible forms of DM. They may be stable not yet known elementary particles, compact stellar-like objects or black holes, but in all the cases they have normal gravitational interactions. As for the dark energy, it is a mysterious substance which drives accelerated cosmological expansion and in this sense it antigravitates. Mainly the following two sources of the cosmic acceleration are considered, either a light scalar field or a modification of gravity at large scales. We discuss them in what follows.

The fraction of energy density of one or other form of matter, ρ_j, in cosmology is expressed through the dimensional parameter $\Omega_j = \rho_j/\rho_c$, where ρ_c is the so called critical energy density, determined from the first

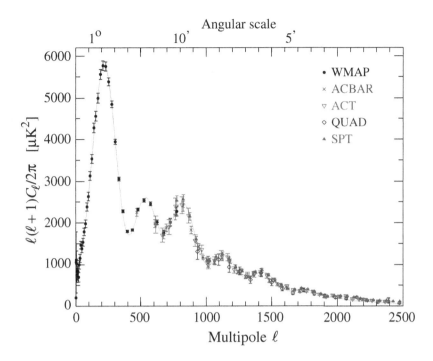

Fig. 1. Combined data of temperature fluctuations before Planck. Here C_l is the amplitude of the fluctuation squared for the multipole l.

Friedmann equation with $k = 0$:

$$H^2 \equiv \left(\frac{\dot{a}}{a}\right)^2 = \frac{8\pi \, \rho_{tot}}{3m_{Pl}^2} - \frac{k}{a^2}, \tag{2}$$

where $a(t)$ is the cosmological scale factor which characterizes the universe expansion, ρ is the average cosmological energy density, and k is a constant usually taken as $k = +1$ (closed universe), $k = -1$ (open "curved") and $k = 0$ (open flat).

According to contemporary combined astronomical data the Hubble parameter is known with the accuracy about 2%, $H = 67.3 \pm 1.2$ km/sec/Mpc. The precision in measurement of H based only on the Planck data is sightly less precise, see below. Knowing H we can find the value of the critical energy density:

$$\rho_c = \frac{3H^2 m_{Pl}^2}{8\pi} = 0.85 \cdot 10^{-29} \, \text{g/cm}^3. \tag{3}$$

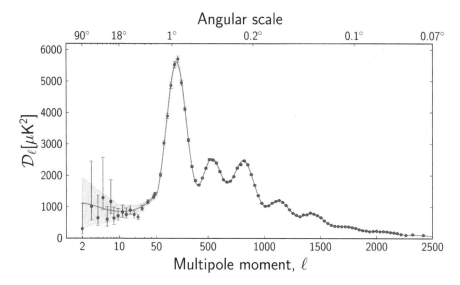

Fig. 2. Temperature fluctuations of CMB according to Planck measurements. The notations are the same as in Fig. 1, with $D_l = C_l l(l+1)/(2\pi)$.

As we mentioned above, the ratio $\Omega_{tot} = \rho_{tot}/\rho_c$, determines 3-dimensional geometry of the universe. If $\Omega_{tot} > 1$, the universe is closed and have geometry of three-dimensional sphere. For $\Omega_{tot} < 1$ the universe is open and have geometry of three-dimensional hyperboloid. In the special and seemingly quite improbable case of $\Omega_{tot} = 1$ the universe geometry is the flat Euclidean one. Surprisingly just this special case is realized in nature. According to the existing data $\Omega_{tot} = 1 \pm 0.02$. Inflationary theory predicts that $\Omega_{tot} = 1$ with much better precision, at the level about 10^{-4}.

Not so long ago it was believed that the final destiny of the universe was uniquely connected with its geometry: an open universe (including a flat one) will expand forever, while for a closed universe expansion will ultimately turn into contraction. Indeed, since the energy density of the usual matter drops down faster than $1/a^2$, it follows from Eq. (2) that for a closed universe H will turn to zero in finite time, while for an open universe H will always be positive, tending to zero asymptotically. However, dark energy breaks this connection and independently on geometry the universe will expand forever if only the equation of state of DE will not change in some distant future.

Usually a simple linear equation of state is taken:

$$P = w\rho, \tag{4}$$

where P is the pressure density, w is a constant parameter, different for different forms of matter. For example for non-relativistic matter, for which $P \ll \rho$ it is assumed that $w_{nr} = 0$; for relativistic matter $w_{rel} = 1/3$. For dark energy, according to the observations, $w_{DE} = -1.13^{+0.13}_{-0.10}$.

If we look at the second Friedmann equation

$$\frac{\ddot{a}}{a} = -\frac{4\pi\, G_N}{3}\,(\rho + 3P), \tag{5}$$

we can conclude that for $w < -1/3$ cosmological acceleration becomes positive, as is indicated by the contemporary astronomical data. It is practically established that the universe expanded with acceleration at the very early stage during the so called inflation. This primordial thrust led to creation of our large suitable for life universe out of microscopically small initial state. Antigravity created by negative pressure is necessary for life. In the world governed by the Newtonian theory, in which the source of gravitational force is a positive definite mass, life would be impossible

The following two comments are in order here. First, even in General relativity (GR) antigravity may appear only for infinitely large systems. Any finite object in canonical GR always gravitates. Second, if gravity is modified at large scales to create cosmological acceleration, gravitational repulsion may emerge even in finite systems.[19]

The Planck measurements result in the following values of the basic cosmological parameteres;

$$H = 67.9 \pm 1.5 \ (71 \pm 2.5), \tag{6}$$

$$\Omega_B = 4.9\% \ (4.5\%), \tag{7}$$

$$\Omega_{DM} = 26.8\% \ (22.7\%), \tag{8}$$

$$\Omega_{DE} = 68.3\% \ (72.8\%). \tag{9}$$

The numbers in brackets indicate the corresponding parameter values which were accepted before publication of the Planck data.

Close magnitudes of parameters Ω_j for absolutely different forms of matter bring with them the so called problem of cosmic conspiracy. Indeed, according to the present understanding the density of baryons, dark matter, and dark energy are not connected to each other and could be different by several (many) orders of magnitude. A natural explanation of their proximity is not yet known.

In not so distant past the most accurate way to determine the total amount of baryons in the universe was the analysis of big bang nucleosynthesis (see below). Presently CMB gives a compatible value for the baryonic density but much higher precision. It is intriguing that less than 10% of all baryons are observed directly and where the rest of them is hidden is not yet known.

In addition to baryons, dark matter, and dark energy, the universe is populated by the photons of CMB with $\Omega_{CMB} = 4.8 \cdot 10^{-5}$ and neutrinos for whose density only upper, $\Omega_\nu < 5 \cdot 10^{-3}$, and lower, $\Omega_\nu > 10^{-3}$, bounds are known. The last limit does not follow from astronomy but from the lower limit on the neutrino mass obtained from the analysis of neutrino oscillations, $\Delta m^2 = 0.0024$ eV^2, while cosmology allows to calculate the number density of neutrinos today: $n_\nu = 112$ cm^{-3}, as it was done by Gershtein and Zeldovich.[9]

One of the most impressive results of Planck is a very strong upper bound on the neutrino mass:

$$\sum m_\nu < 0.23 \text{ eV}, \tag{10}$$

where the sum is taken over three known neutrino types. Keeping in mind their very small mass differences we conclude that the mass of any neutrino does not exceed 0.08 eV. For comparison the best bound on the neutrino mass obtained in direct experiments is about 2 eV. It is spectacular that the best device to weigh neutrinos is telescope.

Apart from that, the Planck measurements lead to an upper bound on the effective number of neutrino species: $N_\nu^{(eff)} = 3.30 \pm 0.27$, which is comparable and even a little stronger than the bound obtained from big bang nucleosynthesis. Probably it worthwhile to explain that $N_\nu^{(eff)}$ is the number of light or massless particle species with the energy density equal to that of the usual equilibrium neutrinos. For example $N_\nu^{(eff)} = 3.1$ means that there are some new particles in plasma with the energy density equal to 0.1 of the energy density of the usual neutrinos, which are in thermal equilibrium with photons and electron-positron pairs. The standard theory (without any new particles) predicts $N_\nu^{(eff)} = 3.046$. The excess over 3 by 0.012 comes from diminishing of equilibrium photon density due to plasma corrections,[20] because the neutrino density is normalized to the density of photons in CMB. The remaining 0.034 comes from neutrino heating by the annihilation of hotter electron-positron pairs in the primeval plasma.[21]

Note in conclusion that the shape of the spectrum of the angular fluctuations of the CMB temperature perfectly well confirms qualitative features of the standard cosmological model and allows to precisely measure the basic cosmological parameters. However, there are unexplained anomalies at large angles: namely too small amplitudes of multipoles with low l and very large left-right asymmetry with respect to ecliptics.

3. Inflation

The Friedmann cosmology very well described the universe starting from some mysterious singularity point up to the present time, explaining baryon asymmetry of the universe, observed abundances of light elements, which were synthesized during a few first minutes from the universe creation, equilibrium microwave background radiation, and large scale structure formation given the spectrum of the initial density perturbations, see e.g. the review.[22]

However, many important features of the Friedmann model were inexplicable, such as:

(1) Equality of the CMB temperature over whole sky, despite the fact that two points in the sky separated more than by one degree never knew anything about each other.

(2) Quasi-homegeneity of the universe at large scales.

(3) Proximity of Ω_{tot} to unity, even with a very low precision of the order of unity. To realize such close to unity value, Ω must be precisely fine-tuned to unity in the early universe, e.g. during BBN there should be $|\Omega_{tot} - 1| < 10^{-16}$ and at the Planck epoch the fine-tuning must be really astonishing, $|\Omega_{tot} - 1| < 10^{-60}$. During this epoch the space-time curvature was of the order of m_{Pl}^2 or even higher, so the classical space-time did not exist. It was the earliest period in the universe history down to which we could extrapolate back in time our knowledge of the universe evolution and the classical gravity.

(4) Existence of small density perturbations at astronomically large scales, which were necessary for large scale structure formation. No reasonable mechanism for generation of such perturbations was known.

(5) And last but not the least, the driving force causing the universe expansion uniformly at the distances much larger than the cosmological horizon.

All these problems were solved in economical and beautiful way by the hypothesis that at a pre-big-bang stage the universe expanded exponentially (or quasi-exponentially). To understand it better it would be instructive to present a few simple equations.

According to the law of covariant conservation of the energy-momentum tensor for homogeneous matter distribution:

$$\dot{\rho} + 3H(\rho + P) = 0. \tag{11}$$

For the vacuum-like equation of state for which $P = -\rho$ (or $w = -1$), the energy density does not change during cosmological expansion. Correspondingly the Hubble parameter also remains constant, see Eq. (2), and hence the expansion proceeds exponentially due to powerful antigravity of vacuum-like energy. Such space-time is called De Sitter space.

If by one or other reason the equation of state with $w = -1$, was realized in the early universe, even in a microscopically small patch of space, then due to exponential expansion of this tiny volume our huge universe could be formed out of practically nothing. The conditions in almost all such universe must be the same, except for the far-away boundary regions, because the universe came from a huge stretching of the causally connected bubble. This solves the problems 1 and 2 indicated above.

It is astonishing that the mass of matter in the initial bubble could be microscopically tiny, $M_{in} \sim \rho_{in} l_{in}^3$, where ρ_{in} and l_{in} are respectively the initial energy density and the size of the bubble. It is incomparable with the total mass of the universe inflated from this bubble but the energy conservation law is not violated.

According to Eq. (2) the role of the last term in the r.h.s. exponentially drops down during inflation and this effect can easily create the necessary fine-tuning described in point 3 above.

The wave length of quantum fluctuations exponentially rises together with the scale factor and from microscopically small it turns into astronomically large. Apart from that the fluctuation amplitude becomes somewhat larger. This mechanism naturally creates density perturbations required for the large scale structure formation, i.e for creation of galaxies and their clusters. The spectrum of perturbations in this scenario can be calculated and well agrees with astronomical observations. Nowadays this is a commonly accepted mechanism of creation of perturbations at astronomically large scales.

The first paper, where inflationary hypothesis was deployed for the solution of some problems indicated above, was published by Kazanas.[23] A little later a more detailed and better known paper by Guth[24] appeared. The generation of density perturbations and their spectrum were first calculated by Mukhanov and Chibisov[25] in the frameworks of Starobinsky[26] R^2-model (see below). According to their results the spectrum has the power law form with the spectral index $n = 0.96$. This result well agrees with the recent Planck data and is considered as a strong support in favor of the Starobinsky model. However, the analysis is model dependent and e.g. in a model with millicharged particles[27] the spectral index can be higher. In particular, the flat Harrison-Zeldovich spectrum[28] with $n = 1$ may remain viable. Two possible sources of inflation are discussed in the literature: quasi-homogeneous scalar field, inflaton, and gravity modification at large curvature. The most beautiful of the inflaton models is probably the chaotic inflation suggested by Linde.[29] According to this model, any scalar field ϕ, homogeneous at the Hubble scale and with slowly varying potential such that $U''(\phi) < H^2$, would lead to an exponential expansion in the early universe, which quite naturally would last sufficiently long, so all the visible part of the universe and even the parts far beyond would be created by such colossally inflated initially quasi-homogeneous region.

Gravity modification at large curvature are usually based on the introduction to the standard GR action an additional non-linear term: $R \to R - R^2/m^2$, where m is a constant parameter with dimension of mass, as it was suggested by Starobinsky. Such terms can be generated by radiative corrections to the usual GR action.[26] However, radiative corrections generate not only the terms which depends on R but more complicated ones as well, e.g. the terms proportional the square of the Ricci tensor, $R_{\mu\nu}R^{\mu\nu}$. The latter lead to some undesirable properties of the theory, such as emergence of ghosts and tachyons.

Note that an introduction of the quadratic in curvature terms into the action has a limited applicability because for very large curvatures the effective action becomes more complicated. Roughly speaking instead of being proportional to R^2 the corrections go into denominator, e.g. due to a rise of particle effective masses in strong gravitational fields.

The successful solution of the fundamental problems of the Friedmann cosmology and quantitative prediction of the form of the spectrum of inflationary density perturbations permits to consider inflation as an experimentally or better to say observationally established fact. An additional strong argument in favor of inflation would be registration of long gravitational waves generated during inflation.[30] However, the intensity of such waves is model dependent and may be quite low. So their absence would not kill inflationary scenario and their detection would be one more proof of its validity.

According to inflationary picture, the universe initially looked as dark expanding emptiness until the inflaton field dropped down so that the

Hubble parameter became smaller than the inflaton mass (or the potential curvature became larger than H). After that ϕ started to oscillated simultaneously and everywhere, producing elementary particles mostly with masses smaller than m_ϕ. This moment of explosive particle production would be proper to call "big bang". There is an amazing similarity of this process to that described in the Bible: "... the earth was without form, and void; and darkness was upon the face of the deep" and then all of a sudden: "let there be light". However, in contrast to the Bible, the inflationary picture is described by well defined mathematical equations and permits to make quantitative prediction which agree with astronomical observations.

Let us mention in conclusion an important fact that inflation is possible only if baryonic number is not conserved.[31] Indeed, for sufficiently long inflation the cosmological energy density must be (approximately) constant. However, if the baryonic number is conserved the necessary constancy of ρ is impossible. Since inflation seems to be the only way to create our suitable for life universe, then from our existence follows that baryons are not conserved and thus there is a nonzero probability to discover the proton decay or neutron-antineutron oscillations. It's amusing that half a century ago the statement was exactly opposite: "Our existence proves that baryonic number is not conserved", which, as we know now, is not correct.

4. The Standard Cosmological Model

As a result of an impressive development of theory and a fantastic progress in astronomical observations of all possible messengers from the universe the Standard Cosmological Model (SCM) was established. The model will hardly change in its essential features at phenomenological level. The SCM successfully describes cosmological history starting from inflation to the present days. However, there are some clouds in this sky, maybe not so serious as two famous clouds of Lord Kelvin, namely the ultraviolet catastrophe and the Michelson-Morley problem, out of which heavy rains of quantum mechanics and relativity theory poured. On the other hand, we cannot exclude that the cosmological clouds will also lead to similar grave thunderstorms. In any case modern cosmology surely demands new physics beyond the minimal standard model of particle physics.

The theoretical foundation of SCM is General relativity, though its validity is constantly questioned, especially at recent times, both for large and small curvatures, as e.g. the mentioned above R^2-theory and $F(R)$ theories discussed below in Sec. 6. In addition to GR one needs to specify the chemical content of the cosmological matter, i.e. a set of fundamental particles and fields and their equations of state. It is worth noticing that the state of the system not always can be described by equation of state, or in other words pressure density is not always can be expressed as a function of energy density locally, i.e. at the same time moment and in the same space point. In this case dynamical equations of motion can serve instead of equation of state.

The main epochs of the cosmological history in crude approximation are the following:

I. Inflation, that is the initial inflationary stage when due to sufficiently long exponential expansion our astronomically large universe was created. To create our huge universe almost homogeneous at the scale of the present day horizon and with the same temperature of CMB over all the sky the necessary rise of the scale factor during inflationary period should be at least 60–70 Hubble times, so that $a(t)$ rises not less than by $\exp(60 - 70)$ $= 10^{26} - 10^{30}$. As it was argued in Sec. 3, inflation is practically an "experimental" fact.

Strictly speaking we know nothing about the epoch before inflation and it is even unclear if the conception of time existed before inflation. In accordance with Linde's ideas[29] it is quite probable that classical space-time did not exist before inflation and "before" is not a well defined notion. So this period is still terra incognita. Nevertheless there are some heroic attempts to describe the universe at preinflationary epoch, formulating the so called pre-big-bang cosmology, though it may be proper to call the end of inflation as big bang.

II. Big bang. At some moment the energy density of the inflaton dropped down to a value at which the Hubble parameter, $H \sim \sqrt{\rho_{infl}}/m_{Pl}$, became smaller than the inflaton mass (if $m_\phi = 0$, we can speak about the properly defined slope of the potential). From this moment on a slow roll-down of the inflaton towards the minimum of the potential turns into oscillations around it. These oscillations resulted in production of elementary particles which had non-negligible coupling to inflaton. This moment, when the cold expanding emptiness turned into hot cosmological plasma is natural to call the big bang. A list of first papers on particle production by inflaton and the universe heating is presented in ref.[32]

III. Baryogenesis and possible antimatter creation. The observed in the universe excess of matter over antimatter is simply and beautifully explained by the Sakharov's mechanisms,[40] but only in principle. There exist myriads of different scenarios of baryogenesis,[33] but it is not known which of them took place in reality. Is is even possible that several different mechanisms had operated. Moreover the models of baryogenesis are possible, where any of three Sakharov's conditions is not obligatory, see the second paper in ref.[33]

Very important for understanding of the mechanism of baryogenesis would be an observation of primordial cosmic antimatter. The search of it is actively performed now at several apparatuses: BESS, PAMELA, AMS, and more sensitive detectors are in project. All existing devices are dedicated to search of anti-helium nuclei but up to now not a single anti-helium nucleus has been detected.

All observed to the present time antiprotons and positrons almost surely are secondary produced in cosmic ray interactions or as a result of catastrophic stellar processes. However, even over this dull background there are two phenomena of unknown origin. First, there is an intensive source of photons with energy 0.511 MeV in the Galactic center.[34] This energy is exactly equal to the photon energy from the two-photon annihilation of e^+e^--pairs at rest. It would be exciting if there is a clump of antimatter in the Galactic center but an absence of higher energy quanta from $\bar{p}p$-annihilation seemingly excludes such possibility. The second mysterious fact is an excess of high energy positrons with $E \sim 100$ GeV in cosmic rays discovered by PAMELA[35] and confirmed by Fermi[36] and AMS[37] space missions.

IV. Primordial or big bang nucleosynthesis (BBN). It was always considered as one of the cornerstones of big bang cosmology, since the abundances of light elements: ^4He, ^2H, ^3He, which have been synthesized in the very early universe at temperatures in the interval from 1 MeV down to 60–70 keV, when the universe age was from 1 to 200 seconds, very well agree with observations and there are no other known way to create them. At the present time there is only one piece of data, where a disagreement between theory and observations is found, namely the observed amount of ^7Li is factor three smaller than theoretically predicted. It may indicate some new physical phenomena, e.g. existence of new elementary particles or, what is more probable, could be a result of an erroneous interpretation of the data.

The observed abundances of ^4He and deuterium during many years were used to derive bounds on the concentration of unknown light particles in the primeval plasma. These bounds are presented in the form of effective neutrino species, see discussion after Eq. (10). At different times this bounds varied from $N_\nu^{(eff)} < 4$ down to $N_\nu^{(eff)} < 3.1$. Surprisingly recent observations indicate that $N_\nu^{(eff)} > 3$, though do not contradict the canonical number 3.046 inside the existing error bars. For example, the latest measurement of the primordial abundance of ^4He[38] leads to conclusion $N_\nu^{(eff)} = 3.51 \pm 0.35$ (68% CL) and the data on deuterium[39] are best described with $N_\nu^{(eff)} = 3.28 \pm 0.28$. The future will tell, if these results are an evidence of a discovery of new light elementary particles, which are now called dark radiation, or the data will finally converge to the canonical value $N_\nu^{(eff)} = 3.046$, but it will be wonderful to discover new elementary particles just looking up the sky, e.g. a sterile neutrino, which might explain anomalies observed in neutrino oscillations.

Let us note finally that the abundance of the produced deuterium strongly depends upon the total baryon density in the universe. That's why primordial deuterium is called the baryometer. Now the cosmic microwave background can claim this title as well.

V. Hydrogen recombination and cosmic microwave background radiation. Due to very large magnitude of the elastic photon-electron scattering the mean free path of photons in the primeval plasma in the early universe was much shorter than the cosmological horizon. Hence the photons slowly diffuse in this medium. The situation crucially changed after the hydrogen recombination at temperature $T = 3000$ K, i.e. at the redshift $z_{rec} \approx 1100$, when electrons and protons formed neutral atoms, for which the cross-section of scattering of long wave photons sharply dropped down. After that the photons started to freely propagate in the universe bringing us information about their temperature at the recombination epoch all over the sky. As we mentioned above, the measured angular temperature fluctuations allow quite accurately determine fundamental cosmological parameters.

VI. Formation of the large scale structure of the universe. Small density perturbations, created during inflation, stayed with practically constant amplitude for a large part of the universe history. To be more precise, the relative density contrast $\Delta = \delta\rho/\rho$ did not change during the radiation dominated stage. In a static world the gravitational attraction of the regions with excessive density would lead to an exponential rise of Δ, as it was shown at the beginning of the last century by Jeans.[40] Cosmological expansion slows down this rise and it turns at most into a power law one. In fact, it can be shown that at the cosmological stage when relativistic matter dominated the relative density contrast rose not faster than the logarithm of the scale factor, so they may be considered as practically constant. The essential rise of density perturbations started at the stage of dominance of non-relativistic matter or, as is usually said, matter dominated (MD) stage, which began at redshift $z_{eq} \approx 10^4$. Theory says that at this stage Δ rises as the cosmological scale factor, i.e. from the onset of MD stage it may increase by four orders of magnitude*. This is exactly what is necessary for the initial density perturbations equal to $\Delta \sim 10^{-4}$, which are known from the measurements of the temperature fluctuations of CMB. After the density contrast reached unity the perturbations started to rise very quickly and that's how the systems with huge density contrast $\Delta \gg 1$ can be created.

5. Dark matter

The simple considerations presented in the previous section already permit to conclude that the dominant part of matter in the universe must be invisible, i.e. the matter which does not interact with electromagnetic radiation and in particular with light. It is called dark matter (DM), though a more precise word would be "invisible", because "dark" means that the matter absorbs light, while invisible one simply does not interact with it. Without dark matter the formation of structure in the usual matter could not start at $z \approx 10^4$, because the strong light pressure would inhibit the rise of perturbations. This rise could start only after hydrogen recombination at $z \approx 10^3$. It means in turn that the perturbations would increase at most by factor 10^3 and could not reach unity to the present time. On the other hand, the light pressure does not prevent from perturbation rise in dark matter and thus they could start rising already at $z_{eq} \approx 10^4$. Later after recombination the usual matter would be able to fall down into the pre-prepared by dark matter potential wells. So without dark matter life in the universe would not appear to the present time and maybe never.

Apart from the presented above qualitative arguments there are a lot of precise data which unambiguously prove the existence of dark matter and precisely measure the amount of it. Among them are the following:

---

*Pioneering works on evolution of density perturbations in cosmology belong to Lifshits. The up-to-date presentation of the theory can be found e.g. in the books.[42]

(1) Flat rotational curves. The velocities of gas particles and galaxy-satellites around a larger galaxy do not fall down as $v \sim 1/\sqrt{r}$ with rising distance, as expected, but tend to a constant value. It shows that there is an invisible matter around the galaxy with density dropping as $1/r^2$.

(2) Gravitational lensing of distant light sources permits to estimate amount of matter on the way from the source to the Earth and shows that the amount of invisible matter is approximately 5 times larger than that of visible.

(3) Equilibrium of hot gas in rich galactic clusters also demands approximately 5 times more matter than it is directly observed.

(4) The quantitative analysis of large scale structure formation, including the so called baryon acoustic oscillations (BAO) also lead to conclusion about dominant role of dark matter.

(5) Analysis of the spectrum of angular fluctuations of CMB.

All these different pieces of data well agree between each other giving for the total cosmological fraction of dark and baryonic matter the value $\Omega_m \approx 0.3$, out of which the baryonic matter makes $\Omega_B \approx 0.05$. More precise values are presented above in Eqs. (7) and (8).

From the astronomical point of view dark matter can be separated into three classes: hot (HDM), warm (WDM), and cold (CDM) depending upon the characteristic damping scale. This scale is essentially the free streaming length, l_{FS}, of the DM particles till they stop due to the cosmological redshift. For HDM the mass inside their free streaming length is larger than the galactic mass, $M_{FS} > M_{gal} \sim 10^{12} M_\odot$. An example of WDM particle is neutrino, for which $M_{FS} \sim m_{Pl}^3/m_\nu^2 \sim 10^{17} M_\odot (1\text{eV}/m_\nu)^2$. For warm dark matter $M_{FS} \sim M_{gal}$.

Warm dark matter particles may consist of sterile neutrinos with the mass in keV range or pseudo-goldstone bosons with similar properties.

Popular candidates for CDM particles are the lightest supersymmetric particles (LSP) with masses in the interval 100 – 1000 GeV or some other weakly interacting massive particles (WIMPs). Recently CDM made of LSPs somewhat lost its attractiveness because such particles were not discovered at LHC. Nevertheless, LSPs with higher masses (above the LHC threshold) remain in the game but at the expense of some modification of the standard cosmological evolution.

Another natural candidate for CDM particle is axion despite its vanishingly small mass, $m_a \lesssim 10^{-5}$ eV. Axions are cold particles because they are produced at rest and hence their free streaming length is much shorter than the galactic size.

A separate group of possible bearers of CDM consists of the so called massive astrophysical compact halo objects (MACHOs). This group may consist of dwarf stars of low luminosity with mass comparable to the solar mass or primordial black holes with $M \geq 10^{16}$ g.

A very interesting group of all kind particles of dark matter may make mirror particles from the whole mirror world, similar but not identical to ours.

Apart from that some more exotic possibilities might be realized, e.g. quark nuggets, topological or non-topological solitons or something new not included into this list.

The canonical most popular model at the present time is the ΛCDM model i.e. the model where the dominant form of matter in the universe is cold dark matter plus dark energy in the form of Lambda-term or, in other words, vacuum energy. In more detail vacuum energy is discussed in the next section. This model well describes gross features of the observed large scale structure but some details do not fit into the model. Among them there are:

1. Missing galactic satellites. CDM model predicts an order of magnitude more galactic satellites (i.e. smaller galaxies gravitationally bound to the host large galaxy) than observed.

2. Destruction of the galactic disk. Even if the number of the satellites is reduced by star formation winds, many smaller tightly bound DM systems would survive and destroy galactic disk by gravitational heating.

3. Cusps in galactic centres. Theory predicts singular matter distribution in galactic centers, $\rho_{DM} \sim r^{-\kappa}$ with $\kappa = 1$–2, while a smooth profile is observed.

4. Excessive angular momentum of galaxies. CDM model predicts galactic angular moment several time smaller than the observed one.

Possibly these problems arise because of insufficient precision of numerical simulation or due to neglect of essential physical effects e.g. of the role of baryonic mater. More revolutionary explanation could an introduction of another form of dark matter with different properties, e.g. self-interacting dark matter such as mirror matter. However, this way has its own serious obstacles. In view of these problems hypothesis about warm dark matter gains more and more popularity. The best choice would be a mixture of CDM and WDM but this assumption makes deeper the cosmic conspiracy problem because in this case four forms of matter should have energy densities of comparable magnitude and not "good old" three: baryonic, CDM and DE.

6. Dark energy and/or modified gravity

Dark energy is an unknown substance which forces the universe to expand with acceleration. According to observations it has the equation of state $P = w\rho$ with $w = -1.13^{+0.13}_{-0.10}$. If $w < -1$, the density of dark energy would rise with time and reach infinite value during a finite interval. Such a disaster may be avoided if w is larger or equal to (-1) or w depends upon time and will shift to safe $w \geq -1$ value in the future. Eq. (5) shows that for such w acceleration is positive as demanded. However, in the future w may shift to non-negative value and so we return to the old decelerated cosmology.

During the last two decades different kinds of data in favor of the accelerated expansion were accumulated:

1. In the 90s there arose the universe age crisis. For the Hubble parameter exceeding 70 km/sec/Mps the calculated universe age happened to be shorter than 10 billion years, especially if the total energy density was equal to the critical one, as predicted by inflationary theory. On the other hand, nuclear chronology and the age of old stellar clusters demanded $t_U \geq 13$ billion years. The universe age can be expressed through the present day values of the Hubble parameter and densities of different forms of matter as:

$$t_U = \frac{1}{H} \int_0^1 \frac{dx}{\sqrt{1 - \Omega_{tot} + (\Omega_m/x) + (\Omega_r/x^2) + x^2\Omega_v}}, \qquad (12)$$

where Ω_{tot} is the total relative cosmological energy density and $\Omega_{r,m,v}$ are contributions from relativistic, nonrelativistic, and vacuum-like energies respectively. Evidently for $\Omega_m = 0.3$ and $\Omega_v = 0.7$ the universe age would be close to 13 billion years in agreement with the data.

2. As it is mentioned above the set of many independent kinds of measurements gives the value $\Omega_m = 0.3$, while the spectrum of angular fluctuations of CMB (in particular, the position of the first acoustic peak) demands $\Omega_{tot} = 1$ in agreement with inflationary prediction. In addition the analysis of large scale structure not only showed that $\Omega_m = 0.3$, but also demonstrated a suppression of structure formation at large scales. The latter can be explained by the accelerated expansion.

3. A direct evidence of cosmic acceleration was presented by the discovery of dimming of distant supernova at redshifts of the order of unity. There

are strong arguments in favor that such supernovae, SN1a, are the standard candles and so their diminished brightness indicates that they may be at a larger distance than expected and hence the universe should expand faster than it would take place with the usual Friedmann expansion regime. For the works where this phenomenon was observed three astronomers S. Perlmutter, B.P. Schmidt, and A.G. Riess got the Nobel Prize of 2011 as it is stated "for the discovery of the accelerating expansion of the Universe through observations of distant supernovae". It is noteworthy that an alternative explanation of the effect by the attenuation of light from distant supernovae in the interstellar and intergalactic media can be excluded because the effect is not monotonic as a function of the redshift. First it rises with z and then drops down. It is in agreement with a dark energy interpretation because ρ_m rises with z, while ρ_v remains constant.

As a simplest agent which could induce accelerate expansion a light scalar field is considered with an almost constant amplitude. The energy-momentum tensor of such a field has the form:

$$T_{\mu\nu} = \partial_\mu \phi \partial_\mu \phi - (1/2)g_{\mu\nu} \left[\partial_\alpha \phi \partial^\alpha \phi - U(\phi) \right]. \tag{13}$$

For quasi-homogeneous field this tensor becomes (approximately) proportional to metric tensor, $T_{\mu\nu} \sim g_{\mu\nu}$, realizing vacuum-like equation of state $w \approx -1$, which leads to cosmic acceleration.

Another possible mechanism to create accelerated expansion is the modification of gravity at large distances by introduction an additional nonlinear in curvature term into the gravitational action:

$$S = \frac{m_{Pl}^2}{16\pi} \int d^4x \sqrt{-g}[R + F(R)]. \tag{14}$$

The first such models[43] with $F(R) = -\mu^4/R$, where μ is a constant parameter with dimension of mass, though could induce an accelerated expansion but led to a strong instability of gravitational equation inside celestial bodies.[44] To fix this problem a further modifications of modified gravity were suggested, as e.g. the following:[45]

$$F(R) = \lambda R_0 \left[\left(1 + \frac{R^2}{R_0^2} \right)^{-n} - 1 \right] - \frac{R^2}{6m^2}, \tag{15}$$

or similar to it, where the function $F(R)$ is chosen in such a way that the corresponding equations of motion in vacuum have the solution $R = const$, describing De Sitter space-time.

The problem of dark energy is closely related to the problem of vacuum energy or, what is the same, with the cosmological constant problem (or Lambda-term). In an attempt to apply the GR equations to cosmology Einstein found that there are no stationary solutions and suggested to correct this "flaw" by an addition of cosmological constant Λ[46] into the gravitational equations:

$$R_{\mu\nu} - \frac{1}{2}g_{\mu\nu}R - \Lambda g_{\mu\nu} = \frac{8\pi}{m_{Pl}^2}T_{\mu\nu}\,. \tag{16}$$

Later it became clear that cosmological constant is equivalent to vacuum energy with the energy-momentum tensor:

$$T_{\mu\nu}^{(vac)} = g_{\mu\nu}\rho^{(vac)} \tag{17}$$

and equations of state $P = -\rho$.

It might be natural to assume that dark energy coincides with vacuum energy but theoretical evaluation of different contributions into vacuum energy exceeds the observed value by 50–100 orders of magnitude. Especially impressive is the contribution to vacuum energy from the effects of quantum chromodynamics (QCD). It is established that the QCD vacuum is not empty but filled with the condensates of gluon and quark fields[47] with negative energy density which is 45 orders of magnitude larger by the absolute value than the cosmological energy density. Such condensates are absolutely necessary to obtain the correct value of the proton mass, which grossly exceeds the sum of masses of constituent quarks. Inside proton the condensate is destroyed by quarks and so the proton mass is not equal to the sum of quark masses minus the binding energy but it also includes the mass of the condensate inside the proton volume $V_p \sim 1/(100\text{ MeV})^3$:

$$m_p = 2m_u + m_d + |\rho_{vac}^{QCD}|V_p \approx 1\text{ GeV}, \tag{18}$$

where $m_u \sim m_d \sim 5\text{MeV}$ and

$$\rho_{vac}^{QCD} \approx -10^{45}\rho_c. \tag{19}$$

One of the deepest mysteries of Nature is what compensates negative ρ_{vac}^{QCD}, so a positive but 10^{45} smaller by magnitude vacuum energy emerges as a result. So it seems evident that something "lives" in vacuum to achieve the necessary compensation. Of course it is impossible to formally exclude a boring possibility that the QCD and other contributions into vacuum energy are precisely compensated by fantastically fine-tuned subtraction constant. Together with anthropic principle and the assumption of almost infinite number of worlds with different values of the subtraction constant such a solution of the vacuum energy problem even does not look too unnatural. Anyhow it seems that the solution of the dark energy problem is impossible without understanding of the closely related vacuum energy problem.

7. Conclusion

Instead of conclusion let us make a list, surely non-complete, of unsolved cosmological and astrophysical problems. Some of them have been mentioned in the body of the talk, while the others which were omitted due to lack of space and time, are mentioned now simply because of being important and interesting. So, here are the questions for which we yet do not have answers:

(1) What are the particles of dark matter?
(2) Which is the mechanism of the cosmological acceleration? Scalar field, modified gravity, none of the above?
(3) What is the explanation of the cosmic conspiracy, which leads to comparable values of different forms of energy, $\Omega_B \sim \Omega_{DM} \sim \Omega_{DE}$?
(4) Which of many scenarios of baryogenesis is realized in nature? Or there are several ones?
(5) Is there primordial cosmic antimatter? Will the search of antinuclei at the detectors PAMELA, BESS, and AMS has a chance to be successful or we will have to wait for new generation of detectors?
(6) What is the mechanism of intensive emission of 0.511 line from the Galactic center?
(7) How the excess of positrons in cosmic rays with energies about 100 GeV, observed by PAMELA, Fermi, and AMS, can be explained?
(8) What is the mechanism of production of ultrahigh energy cosmic rays? Is there Greissen-Zatsepin-Kuznim cutoff?
(9) How the supermassive black holes, observed in galactic centers, were created? Did they appear in already formed galaxies or were produced earlier and served as seeds of galaxy formation?
(10) What is the mechanism of creation of quasars at high redshifts and why the metallicity in their neighborhood is so high?
(11) What is the mechanism of gamma-bursters emission?
(12) How galactic and intergalactic magnetic fields were generated?
(13) What is the origin of CMB anomalies at large angles or low multipoles?
(14) How the discrepancy between theory and observation of ^7Li will be resolved? What is it, an error in observations (or their interpretation) or new physics?
(15) Does dark radiation, indicated by BBN and by the spectrum of angular fluctuations of CMB exist?
(16) And last but surely not the least, what compensates experimentally known contribution of QCD condensates into vacuum energy with 10^{-45} precision?

This is surely not the complete list of problems, but even it demonstrates the tremendous progress of cosmology during the last half century and how exciting became this field of science. After it became an exact science, cosmology solved a lot of old problems but discovered on the way a variety of new ones.

I thank E. Arbuzova for helpful comments. The work was supported by the Grant of the Government of the Russian Federation N 11.G34.31.0047.

References

1. A. Friedmann, Z. Phys. 10 (1922) 377;
   A. Friedmann, Z. Phys. 21 (1924) 326.
2. G. Lemaitre, Annales de la Socété Scientifique de Bruxelles, Série A, 47 (1927), 49; English translation in MNRAS, 91 (1931) 483;
   see also H. Nussbaumer, L. Bieri, arXiv:1107.2281v2 [physics.hist-ph].
3. E. Hubble, PNAS, 15 (1929) 168.
4. G. Gamow, Phys. Rev. 70 (1946) 572.
5. Ya. B. Zeldovich, Adv. Astron. Ap. 3 (1965) 42;
   Ya. B. ZelâĂŹdovich, L. B. Okun, and S. B. Pikelner, Usp. Fiz. Nauk 84 (1965) 113.
6. B. W. Lee, S. Weinberg, 1977, Phys. Rev. Lett. 39 (1977) 165.
7. M. I. Vysotsky, Ya. B. Zeldovich, and A. D. Dolgov, JETP Lett. 26 (1977) 200-202.
8. I. Yu. Kobzarev, I. Yu, L. B. Okun, and I. Ya Pomeranchuk, Yadern. Fiz. 3 (1966).
9. S. S. Gershtein, and Ya B. Zeldovich. "Rest mass of muonic neutrino and cosmology." JETP Lett. 4 (1966) 120–122.
10. R. Cowsick, J. McClelland, Phys. Rev. Lett. 29 (1972) 669.
11. A. D. Sakharov, JETP Lett. 5 (1967) 24.
12. A. A. Penzias, R. Wilson, Astorphys. J. 142 (1965) 419.
13. T. A. Ter-Shmaonov, Instruments and Experimental Techniques, 1 (1957) 83.
14. A. G. Doroshkevich, and I. D. Novikov, Soviet Physics Doklady, Vol. 9. 1964.
15. G. F. Smoot *et al.*, Astrophys. J. 396 (1992) L1.
16. I. Strukov *et al.*, "PisâĂŹ ma A. Zh. 18 (1992) 387." Month. Not. Roy. Astron. Soc 258 (1992) 37.
17. J. Einasto, A. Kaasik, E. Saar, Nature 250 (1974) 309;
    J. P. Ostriker, P. J. E. Peebles, A. Yahil, Astrophys. J. 193 (1974) L1.
18. J. H. Oort, Bull. Astron. Inst. Netherlands 6 (1932) 249;
    F. Zwicky, Helv. Phys. Acta 6 (1933) 110.
19. E. V. Arbuzova, A. D. Dolgov, L. Reverberi, Astroparticle Physics 54C (2014) 44; arXiv:1306.5694 [gr-qc].
20. R. E. Lopez, S. Dodelson, A. Heckler and M. S. Turner, Phys. Rev. Lett. 82 (1999) 3952.

21. A. D. Dolgov, M. Fukugita, JETP Lett. 56 (1992) 123; 56 (1992) 129; Phys. Rev. D 46 (1992) 5378;
    A. D. Dolgov, S. H. Hansen and D. V. Semikoz, Nucl. Phys. B 503 (1997) 426; Nucl. Phys. B 543 (1999) 269;
    A. D. Dolgov, Phys. Repts. 370 (2002) 333.

22. A. D. Dolgov, Ya. B. Zeldovich, Usp. Fiz. Nauk 130 (1980) 559;
    A. D. Dolgov, Ya. B. Zeldovich, Rev. Mod. Phys. 53 (1981) 1.

23. D. Kazanas, Ap. J. 241 (1980) L59.

24. A. Guth, Phys. Rev. D 23 (1981) 347.

25. V. F. Mukhanov, and G. V. Chibisov, JETP Lett. 33 (1981) 532–535.

26. A. A. Starobinsky, JETP Lett. 30 (1979) 682; V. Ts. Gurovich, A. A. Starobinsky, Sov. Phys. JETP 50 (1979) 844; Zh. Eksp. Teor. Fiz. 77 (1979) 1683.

27. A. D. Dolgov, S. L. Dubovsky, G. I. Rubtsov, I. I. Tkachev, arXiv:1310.2376 [hep-ph].

28. E. R. Harrison, Phys. Rev. D 1 (1970) 2726;
    Ya. B. Zeldovich, MNRAS. 160 (1972) 1.

29. A. D. Linde, Phys. Lett. B 129 (1983) 177.

30. V. A. Rubakov, M. V. Sazhin, A. V. Veryaskin, Phys. Lett. B 115 (1982) 189.

31. A. Dolgov, Phys. Rept. 222 (1992) 309; Surv. High Energy Phys. 13 (1998) 83.

32. A. D. Dolgov and A. D. Linde, Phys. Lett. B 116 (1982) 329;
    L. F. Abbott, E. Farhi and M. B. Wise, Phys. Lett. B 117 (1982) 29;
    A. J. Albrecht, P. J. Steinhardt, M. S. Turner, F. Wilczek, Phys. Rev. Lett. 48 (1982) 1437;
    A. D. Dolgov, D. P. Kirillova, Sov. J. Nucl. Phys. 51, 172 (1990);
    J. H. Traschen, R. H. Brandenberger, Phys. Rev. D 42 , 2491 (1990);
    L. Kofman, A. D. Linde and A. A. Starobinsky, Phys. Rev. Lett. 73 (1994) 3195; Phys. Rev. D56 (1997) 3258.

33. A. D. Dolgov, Phys. Rep. 222 (1982) 309 (1992);
    A. D. Dolgov, Surv. High Energy Phys. 13 (1998) 83;
    V. A. Rubakov, M. E. Shaposhnikov, Uspekhi Fiz. Nauk, 166 (1996) 493;
    A. Riotto, M. Trodden, Annu. Rev. Nucl. Part. Sci. 49 (1999) 35;
    M. Dine, A. Kusenko, Rev. Mod. Phys. 76 (2004) 1.

34. W. N. Johnson, F. R. Harnden, R. C. Haymes, Astrophys. J. 172 (1972) L1;
    M. Leventhal, C. J. MacCallum, P. D. Stang, Astrophys. J. 225 (1978) L11;
    W. R. Purcell et al., Astrophys. J. 491 (1997) 725;
    P. A. Milne, J. D. Kurfess, R. L. Kinzer, M. D. Leising, New Astron. Rev. 46 (2002) 553 (2002);
    J. Knodlseder et al., Astron. Astrophys. 441 (2005) 513;
    P. Jean et al., Astron. Astrophys. 445 (2006) 579;
    G. Weidenspointner et al., Astron. Astrophys. 450 (2006) 1013.

35. O. Adriani, G. C. Barbarino, G. A. Bazilevskaya et al., Nature 458 (2009) 607.

36. M. Ackermann et al., Phys. Rev. Lett. 108 (2012) 011103.

37. AMS Collaboration (M. Aguilar et al.) Phys. Rev. Lett. 110 (2013) 14, 141102.

38. Y. I. Izotov, G. Stasinska, N. G. Guseva, arXiv:1308.2100.
39. R. Cooke, M. Pettini, R. A. Jorgenson *et al.*, arXiv:1308.3240.
40. J. H. Jeans, Philosophical Transactions of the Royal Society A 199 (1902) 1.
41. E. M. Lifshitz, JETP 16 (1946) 587;
    E. M. Lifshitz, and I. M. Khalatnikov, Soviet Phys. Uspekhi 6 (1964): 495.
42. V. Mukhanov, *Physical Foundations of Cosmology*, Cambridge University Press, 2005;
    D. S. Gorbunov, and V. A. Rubakov, *Introduction to the Theory of the Early Universe: Cosmological Perturbations and Inflationary Theory*, World Scientific, 2011.
43. S. Capozziello, S. Carloni, A. Troisi, Recent Res. Dev. Astron. Astrophys. 1 (2003) 625;
    S. M. Carroll, V. Duvvuri, M. Trodden, M. S. Turner, Phys. Rev. D 70 (2004) 043528.
44. A. D. Dolgov, M. Kawasaki, Phys. Lett. B 573 (2003) 1.
45. A. A. Starobinsky, JETP Lett. 86 (2007) 157;
    W. Hu, I. Sawicki, Phys. Rev. D 76 (2007) 064004;
    A. Appleby, R. Battye, Phys. Lett. B 654 (2007) 7.
46. A. Einstein, Sitzgsber. Preuss. Acad. Wiss. 1 (1918) 142.
47. M. Gell Mann, R. J. Oakes, B. Renner, Phys. Rev. 175 (1968) 2195;
    M. A. Shifman, A. I. Vainshtein, V. I. Zakharov, Nucl. Phys. B147 (1978) 385.

HADRON STRUCTURE AND ELASTIC SCATTERING

I. DREMIN

Lebedev Physical Institute, Moscow 119991, Russia
E-mail: dremin@lpi.ru

When colliding, the high energy hadrons can either produce new particles or scatter elastically without change of their quantum numbers and other particles produced. Namely elastic scatterings of hadrons are considered in this paper. The general machinery of their theoretical treatment is described. Some new experimental data are presented and confronted to phenomenological approaches. The internal structure of hadrons is the main subject of these studies. Its impact on properties of their interactions is reviewed. It is shown that protons become larger and darker with increase of their collision energy and reveal some substructure. The violation of the geometric scaling in the diffraction cone and new problems of description of differential cross sections outside it are described.

Prelude. It was in September 1954 when I met Pomeranchuk. He lectured Theory of Relativity. Next year he became a tutor of our "theoretical group". In some time he suggested me to study the book of Flügge on nuclear physics. After struggling with difficult subject written in German I asked him some questions. He liked it and one by one proposed several problems to be solved. In parallel, he insisted on passing through the series of famous Landau exams. It was a good school. Its lessons I described in the book of reminiscences about Pomeranchuk. As a supervisor, he advised my diploma paper to be published in JETP Letters and recommended me to Prof. Tamm as a PhD student. Soon I proposed the one-pion exchange model which was later extended to multiperipheral and multireggeon models. Pomeranchuk got interested in it and asked me to come for discussion. He was deeply interested in properties of hadron collisions. Our contacts lasted till his death.

In this paper I describe some new findings about elastic scattering of hadrons studied now up to LHC energies. I briefly reviewed this at the Pomeranchuk centennial seminar at ITEP. This would be extremely interesting subject for him. Let me just mention Pomeron and famous Pomeranchuk theorem to remind you his basic contributions in this field. I dedicate the paper to the memory of my teacher Isaak Pomeranchuk.

1. Introduction

Hadron interactions are strong and, in principle, should be described by quantum chromodynamics (QCD). However, experimental data show that their main features originate from the non-perturbative sector of QCD. Only the comparatively rare processes with high transferred momenta can be treated theoretically rather successfully by the perturbative methods due to the well-known property of the asymptotical freedom of QCD. Thus, in absence of methods for rigorous solution of QCD equations, our understanding of the dynamics of the main bulk of strong interactions is severely limited by the model building or some rare rigorous relations. In fact, our approach to high energy hadronic processes at present is at best still in its infancy.

From experiment we have learned, at least, about five subregions of the elastic scattering differential cross section. Here, we discuss only two of them: the diffraction cone and the Orear regime. The diffraction cone at small angles reminds the semiclassical effects with Gaussian decrease in angles. The region at larger angles with exponential decrease of the cross section called Orear region became noticeable only at energies of colliding particles above several GeV. It persists till present LHC energies of 7 and 8 TeV.

The extended review was earlier published in.[1] The sections 2 and 3 are the abbreviated versions of its corresponding parts. Others present some results obtained later.

2. The main relations

The measurement of the differential cross section is the only source of the experimental information about their elastic scattering. Therefore, the main characteristics of hadron interactions directly related to the elastic scattering amplitude such as the total cross section, the elastic scattering cross section, the ratio of the real to imaginary part of the amplitude, the slope of the diffraction cone are obtained. The first two of them are functions of the total energy only, while others depend on two variables, the total energy and the transferred momentum (or the scattering angle).

The dimensionless elastic scattering amplitude A defines the differential cross section in a following way:

$$\frac{d\sigma(s)}{dt} = \frac{1}{16\pi s^2}|A|^2 = \frac{1}{16\pi s^2}(\text{Im}A(s,t))^2(1 + \rho^2(s,t)). \qquad (1)$$

Here, the ratio of the real and imaginary parts of the amplitude has been defined

$$\rho(s,t) = \frac{\mathrm{Re}A(s,t)}{\mathrm{Im}A(s,t)}. \tag{2}$$

In what follows, we consider the very high energy processes. Therefore, the masses of the colliding particles may be neglected, and one uses the expression $s = 4E^2 \approx 4p^2$, where E and p are the energy and the momentum in the center of mass system. The four-momentum transfer squared is

$$-t = 2p^2(1 - \cos\theta) \approx p^2\theta^2 \approx p_t^2 \quad (\theta \ll 1) \tag{3}$$

with θ denoting the scattering angle in the center of mass system and p_t the transverse momentum.

The elastic scattering cross section is given by the integral of the differential cross section (1) over all transferred momenta:

$$\sigma_{el}(s) = \int_{t_{min}}^{0} dt \frac{d\sigma(s)}{dt}. \tag{4}$$

The total cross section σ_t is related by the optical theorem with the imaginary part of the forward scattering amplitude at high energy s as

$$\sigma_t(s) = \frac{\mathrm{Im}A(p, \theta = 0)}{s}. \tag{5}$$

Elastically scattered hadrons escape from the interaction region declining mostly at quite small angles within the so-called diffraction cone*. Therefore the main attention has been paid to this region. As known from experiment, the diffraction peak has a Gaussian shape in the scattering angles or exponentially decreasing as the function of the transferred momentum squared:

$$\frac{d\sigma}{dt} \bigg/ \left(\frac{d\sigma}{dt}\right)_{t=0} = e^{Bt} \approx e^{-Bp^2\theta^2}. \tag{6}$$

In view of the relations (4), (5), (6), any successful theoretical description of the differential distribution must succeed in fits of the energy dependence of the total and elastic cross sections as well.

The diffraction cone slope B is given by

$$B(s,t) \approx \frac{d}{dt}\left[\ln\frac{d\sigma(s,t)}{dt}\right]. \tag{7}$$

---

*The tiny region of the interference of the Coulomb and nuclear amplitudes at extremely small angles does not contribute to the elastic scattering cross section and we discard it.

In experiment, the slope B depends slightly on t at the given energy s. E.g., at the LHC, its value changes by about 10% within the cone for $|\Delta t| \approx 0.3$ GeV^2. We neglect it in a first approximation.

The normalization factor in Eq. (6) is

$$\left(\frac{d\sigma}{dt}\right)_{t=0} = \frac{\sigma_t^2(s)(1 + \rho_0^2(s))}{16\pi},\tag{8}$$

where $\rho_0 = \rho(s, 0)$. Eq. (8) follows from Eqs. (1) and (5) at $t = 0$.

According to the dispersion relations which connect the real and imaginary parts of the amplitude and the optical theorem Eq. (5), the value ρ_0 may be expressed as an integral of the total cross section over the whole energy range. In practice ρ_0 is mainly sensitive to the local derivative of the total cross section. Then to a first approximation the result of the dispersion relation may be written in a form[2–4]

$$\rho_0(s) \approx \frac{1}{\sigma_t}\left[\tan\left(\frac{\pi}{2}\frac{d}{d\ln s}\right)\right]\sigma_t = \frac{1}{\sigma_t}\left[\frac{\pi}{2}\frac{d}{d\ln s} + \frac{1}{3}\left(\frac{\pi}{2}\right)^3\frac{d^3}{d\ln s^3} + \dots\right]\sigma_t.\tag{9}$$

At high energies $\rho_0(s)$ is mainly determined by the derivative of the logarithm of the total cross section with respect to the logarithm of energy.

The bold extension of the first term in this series to non-zero transferred momenta would look like

$$\rho(s, t) \approx \frac{\pi}{2}\left[\frac{d\ln \mathrm{Im}A(s, t)}{d\ln s} - 1\right].\tag{10}$$

If one neglects the high-$|t|$ tail of the differential cross section, which is several decades lower than the optical point, and integrates in Eq. (4) using Eq. (6) with constant B, then one gets the approximate relation between the total cross section, the elastic cross section and the slope

$$\frac{\sigma_t^2(1 + \rho_0^2)}{16\pi B\sigma_{el}} \approx 1.\tag{11}$$

The elastic scattering amplitude must satisfy the general properties of analiticity, crossing symmetry and unitarity. The unitarity of the S-matrix $SS^+ = 1$ imposes definite requirements on it. In the s-channel it looks like

$$\mathrm{Im}A(p, \theta) = I_2(p, \theta) + F(p, \theta) =$$
$$\frac{1}{32\pi^2}\int\int d\theta_1 d\theta_2 \frac{\sin\theta_1 \sin\theta_2 A(p, \theta_1)A^*(p, \theta_2)}{\sqrt{[\cos\theta - \cos(\theta_1 + \theta_2)][\cos(\theta_1 - \theta_2) - \cos\theta]}} +$$
$$F(p, \theta).\tag{12}$$

The region of integration in Eq. (12) is given by the conditions

$$|\theta_1 - \theta_2| \le \theta, \quad \theta \le \theta_1 + \theta_2 \le 2\pi - \theta. \tag{13}$$

The integral term represents the two-particle intermediate states of the incoming particles. The function $F(p, \theta)$ represents the shadowing contribution of the inelastic processes to the elastic scattering amplitude. Following[5] it is called the overlap function. It determines the shape of the diffraction peak and is completely non-perturbative. Only some phenomenological models pretend to describe it.

In the forward direction $\theta = 0$ this relation in combination with the optical theorem (5) reduces to the general statement that the total cross section is the sum of cross sections of elastic and inelastic processes:

$$\sigma_t = \sigma_{el} + \sigma_{inel}. \tag{14}$$

The unitarity relation (12) has been successfully used[6–9] for the model-independent description of the Orear region between the diffraction cone and hard parton scattering which became the crucial test for phenomenological models.

Experimentally, all characteristics of elastic scattering are measured as functions of energy s and transferred momentum t. However, it is appealing to get knowledge about the geometrical structure of scattered particles and the role of different space regions in the scattering process. Then one should use the Fourier-Bessel transform to get correspondence between the transferred momenta and these space regions. The transverse distance between the centers of colliding particles called as the impact parameter **b** determines the effective transferred momenta t. The amplitudes in the corresponding representations are related as

$$h(s, b) = \frac{1}{16\pi s} \int_{t_{min}=-s}^{0} dt A(s, t) J_0(b\sqrt{-t}). \tag{15}$$

Peripheral collisions at large b lead to small transferred momenta $|t|$.

The amplitude $A(s, t)$ may be connected to the eikonal phase $\delta(s, \mathbf{b})$ and to the opaqueness (or blackness) $\Omega(s, \mathbf{b})$ at the impact parameter **b** by the Fourier-Bessel transformation

$$A(s, t = -q^2) = \frac{2s}{i} \int d^2 b e^{i\mathbf{q}\mathbf{b}} (e^{2i\delta(s,\mathbf{b})} - 1) = 2is \int d^2 b e^{i\mathbf{q}\mathbf{b}} (1 - e^{-\Omega(s,\mathbf{b})}). \tag{16}$$

Assuming $\Omega(s, \mathbf{b})$ to be real and using Eq. (5) one gets

$$\sigma_t = 4\pi \int_0^\infty (1 - e^{-\Omega(s,\mathbf{b})}) b db. \tag{17}$$

Also

$$\sigma_{el} = 2\pi \int_0^\infty (1 - e^{-\Omega(s,\mathbf{b})})^2 b db, \tag{18}$$

and

$$B = \frac{\int_0^\infty (1 - e^{-\Omega(s,\mathbf{b})}) b^3 db}{2 \int_0^\infty (1 - e^{-\Omega(s,\mathbf{b})}) b db}. \tag{19}$$

To apply the inverse transformation one must know the amplitude $A(s,t)$ at all transferred momenta. Therefore, it is necessary to continue it analytically to the unphysical region of t.[10] This may be done.[11] Correspondingly, the mathematically consistent inverse formulae contain, in general, the sum of contributions from the physical and unphysical parts of the amplitude $A(s,t)$. The amplitude in (12) enters only in the physical region. Only this part of its Fourier-Bessel transform is important in the unitarity relation for the impact parameter representation as well. It is written as

$$\mathrm{Im}h(s,b) = |h(s,b)|^2 + F(s,b), \tag{20}$$

where $h(s,b)$ and $F(s,b)$ are obtained by the direct transformation of $A(s,t)$ and $F(s,t)$ integrated only over the physical transferred momenta from t_{min} to 0. They show the dependence of the intensity of elastic and inelastic interactions on the mutual impact parameter of the colliding particles. The integrals over all impact parameter values in this relation represent analogously to the relation (14) the total, elastic and inelastic cross-sections, respectively.

However, the accuracy of the unitarity condition in b-representation (20) is still under discussion (see, e.g.,[12]) since some corrections due to unphysical region enter there even though their role may be negligible.

3. Where do we stand now?

First, let us discuss what we can say about asymptotic properties of such fundamental characteristics as the total cross section σ_t, the elastic cross section σ_{el}, the ratio of the real part to the imaginary part of the elastic amplitude ρ and the width of diffraction peak B at infinite energies. Then we compare this with some trends of present experimental data.

More than half a century ago it was claimed[13,14] that according to the general principles of the field theory and ideas about hadron interactions the total cross section can not increase with energy faster than $\ln^2 s$. The upper bound was recently improved[15] with the coefficient in front of the logarithm shown to be twice smaller than in the earlier limit:

$$\sigma_t \leq \frac{\pi}{2m_\pi^2} \ln^2(s/s_0), \qquad (21)$$

where m_π is a pion mass.

If estimated at present energies, this bound is still much higher than the experimentally measured values of the cross sections with $s_0 = 1$ GeV^2 chosen as a "natural" scale. Therefore it is of the functional significance. It forbids extremely fast growth of the total cross section exceeding asymptotically the above limits. Both the coefficient in front of logarithm in Eq. (21) and the constancy of s_0 are often questioned. In particular, some possible dependence of s_0 on energy s is proclaimed (see, e.g.[16]).

The Heisenberg uncertainty relation points out that such a regime favors the exponentially bounded space profile of the distribution of matter density $D(r)$ in colliding particles of the type $D(r) \propto \exp(-mr)$. Since the energy density is $ED(r)$ and there should be at least one created particle with mass m in the overlap region, then the condition $ED(r) = m$ gives rise to $r \leq \frac{1}{m} \ln(s/m^2)$ and, consequently, to the functional dependence of Eq. (21).

It was namely Heisenberg who first proposed earlier just such a behavior of total cross sections.[17] He considered the pion production processes in proton-proton collisions as a shock wave problem governed by some non-linear field-theoretical equations.

To study asymptotics, some theoretical arguments based on general principles of field theory and analogy of strong interactions to massive quantum electrodynamics[18] were promoted. The property that the limits $s \to \infty$ and $M \to 0$ (where M denotes the photon mass) commute has been used,[19] i.e. the asymptotics of strong interactions coincides with the massless limit of quantum electrodynamics. These studies led to the general geometrical picture of the two hadrons interacting as Lorentz-contracted black disks at asymptotically high energies (see also the review paper[20]). In what follows, we discuss some other possibilities as well. However, as a starting point for further reference, we describe the predictions of this proposal.

The main conclusions are:

(1) For black ($\Omega(s, \mathbf{b}) \to \infty$) and logarithmically expanding disks with finite radii R ($R = R_0 \ln s$, R_0=const) one gets from Eq. (17) that σ_t approaches infinity at asymptotics as

$$\sigma_t(s) = 2\pi R^2 + O(\ln s); \quad R = R_0 \ln s; \quad R_0 = \text{const.} \qquad (22)$$

(2) The elastic and inelastic processes should contribute on equal footing

$$\frac{\sigma_{el}(s)}{\sigma_t(s)} = \frac{\sigma_{in}(s)}{\sigma_t(s)} = \frac{1}{2} \mp O(\ln^{-1} s). \tag{23}$$

This quantum-mechanical result differs from "intuitive" classical predictions.

(3) The width of the diffraction peak $B^{-1}(s)$ should shrink because its slope increases as

$$B(s) = \frac{R^2}{4} + O(\ln s) \qquad (\text{see also}^{21}). \tag{24}$$

(4) The forward ratio of the real part to the imaginary part of the amplitude ρ_0 must vanish asymptotically as

$$\rho_0 = \frac{\pi}{\ln s} + O(\ln^{-2} s). \tag{25}$$

This result follows directly from Eq. (9) for $\sigma_t \propto \ln^2 s$.

(5) The differential cross section has the shape reminding the classical diffraction of light on the disk

$$\frac{d\sigma}{dt} = \pi R^4 \left[\frac{J_1(qR)}{qR} \right]^2, \tag{26}$$

where $q^2 = -t$.

(6) The product of σ_t with the value γ of $|t|$ at which the first dip in the differential elastic cross section occurs is a constant independent of the energy

$$\gamma \sigma_t = 2\pi^3 \beta_1^2 + O(\ln^{-1} s) = 35.92 \text{ mb} \cdot \text{GeV}^2, \tag{27}$$

where $\beta_1 = 1.2197$ is the first zero of $J_1(\beta\pi)$.

These are merely a few conclusions among many model dependent ones. None of these asymptotical predictions were yet observed in experiment.

Surely, there is another more realistic at present energies possibility that the black disk model is too extreme and the gray fringe always exists. It opens the way to the numerous speculations with many new parameters about the particle shape and opacity (see the list of references in[1]).

In the Table 1 we show the predictions of the gray disk model with the steep rigid edge described by the Heaviside step-function and the Gaussian disk model. $\Gamma(s, b)$ is the diffraction profile function.

The slope B is completely determined by the size of the interaction region R. Other characteristics are sensitive to the blackness of disks α.

In particular, the ratio X is proportional to α. The ratio Z plays an important role for fits at larger angles. It is inverse proportional to α. The corresponding formulae are given by Eq. (17), (18) and (19). The black disk limit follows from the gray disk model at $\alpha = 1$.

Table 1. The gray and Gaussian disks models ($X = \sigma_{el}/\sigma_t$, $Z = 4\pi B/\sigma_t$)

Model	$1 - e^{-\Omega} = \Gamma(s,b)$	σ_t	B	X	Z	$\frac{X}{Z}$	XZ
Gray	$\alpha\theta(R-b);\quad 0 \leq \alpha < 1$	$2\pi\alpha R^2$	$\frac{R^2}{4}$	$\frac{\alpha}{2}$	$\frac{1}{2\alpha}$	α^2	$\frac{1}{4}$
Gauss	$\alpha e^{-b^2/R^2};\, 0 \leq \alpha \leq 1$	$2\pi\alpha R^2$	$\frac{R^2}{2}$	$\frac{\alpha}{4}$	$\frac{1}{\alpha}$	$\frac{\alpha^2}{4}$	$\frac{1}{4}$

The parameter XZ is constant in these models and does not depend on the nucleon transparency. On the contrary, the parameter X/Z is very sensitive to it being proportional to α^2. Therefore, it would be extremely instructive to get some knowledge about them from experimental data.

In the Table 2 we show how the above ratios evolve with energy according to experimental data. Most primary entries there except the last two are taken from Refs.[22,23] with simple recalculation $Z = 1/4Y$. The data at Tevatron and LHC energies are taken from Refs.[24–26] All results are for pp-scattering except those at 546 and 1800 GeV for $p\bar{p}$ processes which should be close to pp at these energies.

Table 2. The energy behavior of various characteristics of elastic scattering.

$\sqrt{s}$, GeV	2.70	4.11	4.74	6.27	7.62	13.8	62.5	546	1800	7000
X	0.42	0.28	0.27	0.24	0.22	0.18	0.17	0.21	0.23	0.25
Z	0.64	1.02	1.09	1.26	1.34	1.45	1.50	1.20	1.08	1.00
$\frac{X}{Z}$	0.66	0.27	0.25	0.21	0.17	0.16	0.11	0.18	0.21	0.25
XZ	0.27	0.28	0.29	0.30	0.30	0.26	0.25	0.26	0.25	0.25

The most interesting feature of the experimental results is the minimum of the blackness parameter α at the ISR energies. It is clearly seen in the minima of X and X/Z and in the maximum of Z at $\sqrt{s} = 62.5$ GeV. The steady decrease of ratios X proportional to α and X/Z proportional to α^2 till the ISR energies and their increase at S$p\bar{p}$S, Tevatron and LHC energies means that the nucleons become more transparent till the ISR energies and more black to 7 TeV. The same conclusion follows from the behavior of Z. The value of Z approaches fast its limit for the Gaussian distribution of matter in the disk. This shows that the above crude models are not very bad for qualitative estimates in a first approximation.

We briefly comment on some of the important general trends of high-energy data observed in experiment.

1. Total cross sections increase with energy. At present energies, the power-like approximation is the most preferable one.

2. The ratio σ_{el}/σ_t decreases from low energies to those of ISR where it becomes approximately equal to 0.17 and then strongly increases up to 0.25 at the LHC energies. However, it is still pretty far from the asymptotical value 0.5 corresponding to the black disk limit.

3. The diffraction peak shrinks about twice from energies about $\sqrt{s} \approx 6$ GeV where $B \approx 10$ GeV^{-2} to the LHC energy where $B \approx 20$ GeV^{-2}.

4. A dip and subsequent maximum appear just at the end of the diffraction cone.

5. As regards the behavior of the differential cross section in the function of the transverse momentum behind the dip, the t-exponential of the diffraction peak is replaced, according to experimental data, by the $-\sqrt{|t|} \approx -p_t$-exponential at the intermediate angles:

$$d\sigma/dt \propto e^{-2a\sqrt{|t|}}, \quad a \approx \sqrt{B}. \tag{28}$$

The slope $2a$ in this region increases with energy and it shifts to lower $|t|$.

6. As a function of energy, the ratio ρ_0 increases from negative values at comparatively low energies, crosses zero in the region of hundreds GeV and becomes positive at higher energies where it passes through the maximum of about 0.14 and becomes smaller (about 0.11) at LHC energies.

7. The product $\gamma\sigma_t$ changes from 39.5 mb·GeV^2 at $\sqrt{s} = 6.2$ GeV to 51.9 mb·GeV^2 at $\sqrt{s} = 7$ TeV and deviates from the predicted asymptotic value (27). The total cross section σ_t increases faster than γ decreases.

From the geometrical point of view the general picture is that protons become blacker, edgier and larger (BEL).[27] We discuss it later. Thus even though the qualitative trends may be considered as satisfactory ones, we are still pretty far from asymptotics even at LHC energies.

4. LHC data and phenomenology

Here, we limit ourselves by the latest results of the TOTEM collaboration at the highest LHC energies 7 and 8 TeV.[25,26] The discussion of theoretical models is also concentrated near these data.

The total and elastic cross sections at 7 TeV are respectively estimated as 98.3 mb and 24.8 mb. The cross section shape in the region of the diffraction cone[25] is shown in Fig. 1. The t-exponential behavior with $B \approx 20.1$ GeV^{-2}

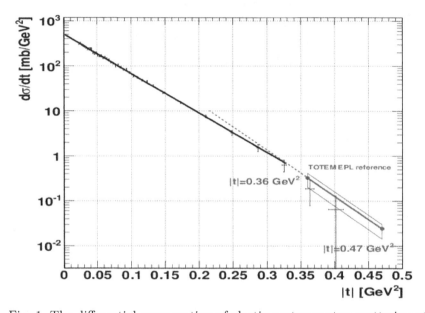

Fig. 1. The differential cross section of elastic proton-proton scattering at $\sqrt{s} = 7$ TeV measured by the TOTEM collaboration (Fig. 4 in[26]). The region of the diffraction cone with the $|t|$-exponential decrease is shown.

is clearly seen at $|t| < 0.3$ GeV^2. The peak steepens at the end of the diffraction cone so that in the $|t|$ interval of $(0.36–0.47)$ GeV^2 its slope becomes approximately equal to 23.6 GeV^{-2}. The results at somewhat larger angles[26] in the Orear region are presented in Fig. 2. The dip at $|t| \approx 0.53$ GeV^2 with subsequent maximum at $|t| \approx 0.7$ GeV^2 and the $\sqrt{|t|}$-exponential behavior are demonstrated. Some curves according to different model predictions[28–32] are also drawn there. All of them fail to describe the data. We conclude that namely this region becomes the Occam razor for all models.

As we see in the figures, various theoretical approaches have been attempted for description of different regions of the differential cross section. One by one we should name:

1. Purely geometrical approach with reference to the internal geometry

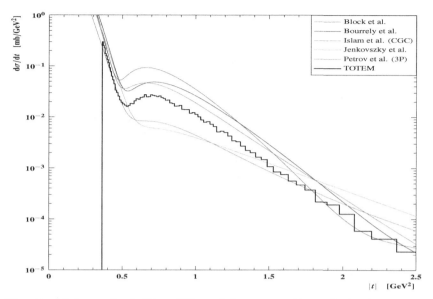

Fig. 2. (Color online) The differential cross section of elastic proton-proton scattering at $\sqrt{s} = 7$ TeV measured by the TOTEM collaboration (Fig. 4 in[25]).
The region beyond the diffraction peak is shown. The predictions of five models are demonstrated.

of colliding hadrons.

2. The analogy to the Fraunhofer diffraction.

3. The appeal to the electromagnetic and matter density distributions inside hadrons.

4. The dynamical picture of Reggeon exchanges which is the most popular one with Pomeron playing the distinguished role.

5. The OCD-inspired models.

Their detailed review is given in.[1] Here, we restrict ourselves by several recent developments not included or discussed there very briefly.

5. Proton structure

Since old days, it was clear that hadrons possess some internal structure. Namely this we discussed with Pomeranchuk after I proposed the one-pion exchange model of inelastic interactions.[33] It was treated as describing the peripheral interactions of hadrons. The deeper inside the hadron, the more pions should be exchanged and more dense populated should be the proton.

Since then, the principal features have not been changed with pions replaced by partons (quarks and gluons) even though pions as a chiral anomaly play the distinguished role.

Early attempts to consider elastic scattering of hadrons also stemmed from the analogous simple geometrical treatment of their internal structure.[34-37] Starting from simplified pictures, one tried to fit the elastic differential cross section. However, recent fits of LHC data failed outside the diffraction cone as demonstrated above. Thus these hypotheses were not precise enough.

At the same time, one can get the direct knowledge about the proton structure important for *inelasic* collisions using elastic scattering data. The impact of the proton structure on inelastic processes can be viewed from the overlap function defined by the unitarity condition in b-representation (20) for elastic scattering amplitudes. It has been directly computed[38] from experimental data of the TOTEM collaboration for pp-scattering at 7 TeV and shown in Fig. 3. The similar shape was obtained (see Fig. 4 in[39]) assum-

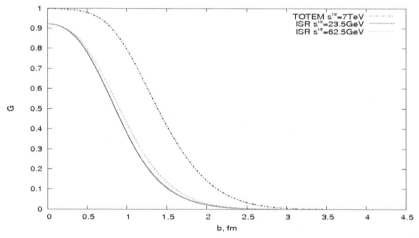

Fig. 3. The overlap function $G(s, b) = 4F(s, b) \leq 1$ at 7 TeV (upper curve)[38] compared to those at ISR energies 23.5 GeV and 62.5 GeV.

ing the Gaussian profile of the elastic contribution $h(s, b)$. Both shapes show the pattern with rather flat shoulder close to 1 (i.e. to complete blackness) at small impact parameters b and subsequent quite steep fall-off. Attempts to fit it by a single Gaussian fail.

In view of the supposed proton substructure with a darker and stepwise kernel surrounded by a more transparent cloud of partons it is reasonable to attempt the fit with a stepwise behavior of the Gaussian exponential like

$$\ln \frac{F(s,0)}{F(s,b)} = \frac{b^2}{a[1 - \frac{2}{\pi} \arctan \frac{b-b_0}{\lambda}]}. \tag{29}$$

The fit reveals quite strong separation of the two regions at $b_0 \approx 0.3$ fm with a width of the transition region $\lambda \approx 0.1$ fm. According to Eq. (5) the exponential becomes three times larger in the narrow strip between the borders of the transition region $b_0 - \lambda$ and $b_0 + \lambda$. The region $b < b_0$ is completely black while at $b > b_0$ it becomes more transparent. However when compared to ISR results[38] its blackness increases with energy, especially at rather large impact parameters about 1 fm as seen in Fig. 4. These pecu-

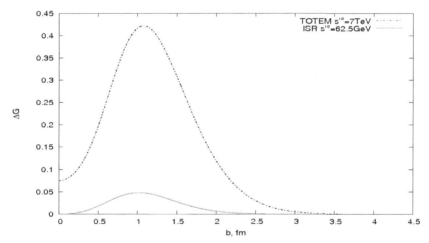

Fig. 4. The difference between the overlap functions. Dash-dotted curve is for 7 TeV and 23.5 GeV energies, solid curve is for 62.5 GeV and 23.5 GeV energies. Conclusion: The parton density at the periphery increases strongly with energy increase!

liar features have been used for description[40] of the CMS data[41] about jet production at high multiplicities. The increased role of central pp-collisions with small impact parameters for events triggered by jets was demonstrated. The important conclusion of this analysis is that the perturbative QCD can be applied to these processes only at high enough transverse momenta of jets exceeding 7–8 GeV.

6. Scaling laws

Since long ago, it was discussed[42,43] a possibility that the differential cross sections might be described as functions of a single scaling variable representing a definite combination of energy and transferred momentum. No rigorous proof of this assumption has been proposed. This property was recently obtained[44] from the solution of the partial differential equation for the imaginary part $\mathrm{Im}A(s,t)$ of the elastic scattering amplitude. The equation has been derived by equating the two expressions for the ratio of the real to imaginary parts of the amplitude $\rho(s,t)$. They were known from the local dispersion relations $(10)^{2-4}$ with the s-derivative and from the linear t-approximation[42,45] with the t-derivative (for more details see[44]).

Therefrom the following partial differential equation is valid

$$p - f(x)q = 1 + f(x), \tag{30}$$

where $p = \partial u/\partial x$; $q = \partial u/\partial y$; $u = \ln \mathrm{Im}A(s,t)$; $f(x) = 2\rho(s,0)/\pi \approx d\ln\sigma_t/dx$; $x = \ln s$; $y = \ln|t|$; σ_t is the total cross section. The variables s and $|t|$ should be considered as scaled by the corresponding constant factors s_0^{-1} and $|t_0|^{-1}$.

The general solution of Eq. (30) reveals the scaling law

$$\frac{t}{s}\mathrm{Im}A(s,t) = \phi(t\sigma_t). \tag{31}$$

For the differential cross section it looks like

$$t^2 d\sigma/dt = \phi^2(t\sigma_t), \tag{32}$$

if the real part of the amplitude is neglected compared to the imaginary part. Thus the scaling law is predicted not for the differential cross section itself but for its product to t^2. Let us note that the often used ratio (see, e.g.,[46]) of $d\sigma/dt$ to $d\sigma/dt|_{t=0} \propto \sigma_t^2$ is also a scaling function. However, the expression (32) is more suitable for comparison with experiment.

The scaling law with the $t\sigma_t$-scale is known as the geometrical scaling.

In Fig. 5, we plot $t^2 d\sigma/dt$ for pp-scattering at energies $\sqrt{s}$ from 4.4 GeV to 7 TeV as functions of $t\sigma_t$ with σ_t provided by the corresponding experiment. Rather reasonable scaling is observed in the diffraction cone except the TOTEM data at 7 TeV. Thus the simple geometrical scaling is not fulfilled at high energies even at low transferred momenta. To restore some approximate scaling one should replace the variable $t\sigma_t$ by $t^a\sigma_t$ and plot $t^{2a}d\sigma/dt$ with $a \approx 1.2$ as shown in.[47] However, this is not a simple geometric scaling anymore.

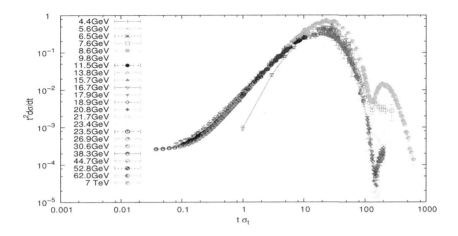

Fig. 5. (Color online) Violation of geometrical scaling at LHC energies.

7. Orear region and real part at any t

The theoretical explanation of the new regime of exponential decrease of the differential cross section beyond the diffraction cone with angles based on consequences of the unitarity condition in the s-channel has been proposed in Refs.[6,7] The careful fit to experimental data showed good quantitative agreement with experiment.[8] Nowadays the same approach helped explain the TOTEM findings.[9]

We consider the left-hand side and the integral term I_2 in the unitarity condition (12) at the angles θ outside the diffraction peak. Because of the sharp fall off of the amplitude with angle, the principal contribution to the integral arises from a narrow region around the line $\theta_1 + \theta_2 \approx \theta$. Therefore one of the amplitudes should be inserted at small angles within the cone as a Gaussian while another one is kept at angles outside it. Integrating over one of the angles the linear integral equation is obtained:

$$\mathrm{Im}A(p,\theta) = \frac{p\sigma_t}{4\pi\sqrt{2\pi B}} \int_{-\infty}^{+\infty} d\theta_1 e^{-Bp^2(\theta-\theta_1)^2/2} f_\rho \mathrm{Im}A(p,\theta_1) + F(p,\theta), \quad (33)$$

where $f_\rho = 1 + \rho_0\rho(\theta_1)$.

It can be solved analytically (for more details see[6,7]) with two assumptions that the role of the overlap function $F(p,\theta)$ is negligible outside the diffraction cone and the function f_ρ may be approximated by a constant, i.e. $\rho(\theta_1) = \rho_l = \mathrm{const}$.

It is easy to check that the solution of this equation is

$$\mathrm{Im}A(p,\theta) = C_0 \exp\left(-\sqrt{2B\ln\frac{Z}{f_\rho}}p\theta\right) + \sum_{n=1}^{\infty} C_n e^{-(\mathrm{Re}b_n)p\theta}\cos(|\mathrm{Im}b_n|p\theta - \phi_n)$$
(34)

with

$$b_n \approx \sqrt{2\pi B|n|}(1 + i\,\mathrm{sign}n) \qquad n = \pm1, \pm2, \ldots \qquad (35)$$

This expression contains the exponentially decreasing with θ (or $\sqrt{|t|}$) term (Orear regime!) with imposed on it oscillations strongly damped by their own exponential factors. These oscillating terms are responsible for the dip. The exponential in Eq. (34) is well defined. It contains the value of Z in the logarithm becoming very sensitive to ρ when Z approaches 1 (see Table 2). The comparison with LHC data has shown that this ratio must be negative and quite large (about -2) in this region. Most of the widely used models do not predict such values. Moreover many of them get it positive. This follows from the equal numbers of zeros of real and imaginary parts. Only those models with odd sum of this number can succeed in getting negative $\rho(s,t)$. The unitarity condition does not ask for a zero of the imaginary part to fit the dip as the models do but ascribes it to the damped oscillations contained in the solution of the equation. This discrepancy is not resolved yet. No correspondence between the two approaches has yet been established. Some equations for the ratio have been derived and solved. They favor the variable sign of it. However the problem asks for further investigation.

8. Conclusions

There are new exciting findings at LHC. They pose serious problems for theoreticians. I am sure that all of them would be of great interest to I.Ya. Pomeranchuk.

References

1. I.M. Dremin, *Physics-Uspekhi* **56** 3 (2013)
2. V.N. Gribov, A.A. Migdal, *Sov. J. Nucl. Phys.* **8** 583 (1969)
3. J.B. Bronzan, G.L. Kane, U.P. Sukhatme, *Phys. Lett. B* **49** 272 (1974)
4. J. Fischer, P. Kolar, *Phys. Lett. B* **64** 45 (1976); *Phys. Rev. D* **17** 2168 (1978)
5. L. Van Hove, *Nuovo Cimento* **28** 798 (1963)
6. I.V. Andreev, I.M. Dremin, *ZhETF Pis'ma* **6** 810 (1967)
7. I.V. Andreev, I.M. Dremin, *Sov. J. Nucl. Phys.* **8** 473 (1969)
8. I.V. Andreev, I.M. Dremin, I.M. Gramenitskii, *Nucl. Phys.* **10** 137 (1969)
9. I.M. Dremin, V.A. Nechitailo, *Phys. Rev. D* **85** 074009 (2012)

10. M. Adachi, T. Kotani, *Prog. Theor. Phys.* **35** 485 (1966); **39** 785 (1968)
11. M. Islam, *Nucl. Phys. B* **104** 511 (1976)
12. V. Kundrat, J. Kaspar, M. Lokajicek, arXiv:0912.1188 (2009)
13. M. Froissart, *Phys. Rev.* **123** 1053 (1961)
14. A. Martin, *Il Nuovo Cimento A* **42** 930 (1966)
15. A. Martin, *Phys. Rev. D* **80** 065013 (2009)
16. Ya.I. Azimov, *Phys. Rev. D* **84** 056012 (2011); arXiv:1204.0984; 1208.4304
17. W. Heisenberg, *Z. Phys.* **133** 65 (1952)
18. H. Cheng, T.T. Wu, *Phys. Rev. Lett.* **24** 1456 (1970)
19. H. Cheng, T.T. Wu, *Phys. Lett. B* **34** 647 (1971)
20. M. Block, R. Cahn, *Rev. Mod. Phys.* **57** 563 (1985)
21. T. Kinoshita, *Phys. Rev. D* **3** 2346 (1970)
22. T.T. Chou, C.N. Yang, *Phys. Lett. B* **128** 457 (1983)
23. A.W. Chao, C.N. Yang, *Phys. Rev. D* **8** 2063 (1973)
24. N.A. Amos *et al.*, *Phys. Lett. B* **247** 127 (1990)
25. G. Antchev *et al.*, TOTEM Collaboration, *Europhys. Lett.* **95** 41001 (2011)
26. G. Antchev *et al.*, TOTEM Collaboration, *Europhys. Lett.* **96** 21002 (2011)
27. R. Henzi, P. Valin, *Phys. Lett. B* **132** 443 (1983)
28. M. Block, F. Halzen, *Phys. Rev. D* **83** 077901 (2011); arXiv:1205.5514; 1208.4086
29. C. Bourrely, J. Soffer, T.T. Wu, *Eur. Phys. J. C* **28** 97 (2003)
30. M. Islam, J. Kaspar, R. Luddy, *Mod. Phys. Lett. A* **24** 485 (2009)
31. L. Jenkovsky, A. Lengyel, D. Lontkovskyi, *Int. J. Mod. Phys. A* (2012)
32. V.A. Petrov, E. Predazzi, A.V. Prokudin, *Eur. Phys. J. C* **28** 525 (2003)
33. I.M. Dremin, D.S. Chernavsky, *ZhETP* **38** 229 (1960)
34. A.D. Krisch, *Phys. Rev.* **135** B1456 (1964)
35. T.T. Chou, C.N. Yang, *Phys. Rev.* **170** 1591 (1966); *Phys. Rev. Lett.* **20** 1615 (1968)
36. M. Islam, *Nuovo Cim. A* **48** 251 (1967)
37. H. Cheng, T.T. Wu, *Phys. Rev. Lett.* **22** 666 (1969)
38. I.M. Dremin, V.A. Nechitailo, *Nucl. Phys. A* **916** 241 (2013), arXiv:1306.5348.
39. L. Frankfurt, M. Strikman, C. Weiss, arXiv:1009.2559.
40. A.Yu. Azarkin, I.M. Dremin, M. Strikman, to be published.
41. CMS Collaboration, *Eur. Phys. J. C* (2013); arXiv:1310.4554.
42. G. Auberson, T. Kinoshita, A. Martin, *Phys. Rev. D* **3** 3185 (1971)
43. J. Dias de Deus, *Nucl. Phys. B* **59** 231 (1973)
44. I.M. Dremin, A.A. Radovskaya, *Europhys. Lett.* **100** 61001 (2012); arXiv:1209.1935
45. A. Martin, *Lett. Nuovo Cim.* **7** 811 (1973)
46. P. Brogueira, J. Dias de Deus, *Eur. Phys. J. C* **37** 075006 (2010)
47. I.M. Dremin, V.A. Nechitailo, *Phys. Lett. B* **720** 177 (2013), arXiv:1212.3313.

RG LIMIT CYCLES

K. BULYCHEVA

Institute of Theoretical and Experimental Physics, Moscow 117218, Russia
E-mail: bulycheva@itep.ru

A. GORSKY

Institute of Theoretical and Experimental Physics, Moscow 117218, Russia
Moscow Institute of Physics and Technology, Dolgoprudny 141700, Russia
E-mail: gorsky@itep.ru

In this review we consider the concept of limit cycles in the renormalization group flows. The examples of this phenomena in the quantum mechanics and field theory will be presented.

1. Generalities

It is usually assumed that the RG flow connects fixed points, starting at a UV repelling point and terminating at a IR attracting point. However it turned out that this open RG trajectory does not exhaust all possibilities and the clear-cut quantum mechanical example of the nontrivial RG limit cycle has been found in[1] confirming the earlier expectations. This example triggered the search for patterns of this phenomena which was quite successful. They have been identified both in the systems with finite number of degrees of freedom[2-5] and in the field theory framework.[6-8] Now the cyclic RG takes its prominent place in the world of RG phenomena however the subject certainly deserves much more study.

The appearance of critical points corresponds to phase transitions of the second kind, hence there exists a natural question concerning the connection of RG cycles with phase transitions. The very phenomenon of the cyclic RG flow has been interpreted in the important paper[7] as a kind of generalization of the BKT phase transitions in two dimensions. One can start from a usual example of an RG flow connecting UV and IR fixed points and then consider a motion in a parameter space which results in a merging of the fixed points. In[7] it was argued that when the parameter goes into the complex region the cyclic behaviour of the RG flow gets manifested and a gap in the spectrum arises. This happens similar to the BKT transition case when a deconfinement of vortices occurs at the critical temperature and the conformal symmetry is restored at lower temperatures. The appearance of the RG cycles can be also interpreted as the peculiar anomaly in the classical conformal group.[9] This anomaly has the origin in some "falling to the center" UV phenomena which could have quite different reincarnations. We would like to emphasize one more generic feature of the phenomena — the cyclic RG usually occurs in the system with at least two couplings. One of them undergoes the RG cyclic flow while the second determines the period of the cycle.

The collision of the UV and IR fixed points can be illustrated in a quite general manner as follows. Assume that there are two couplings (α, g) in the theory and we focus at the renormalization of the coupling which enjoys the following β-function

$$\beta_g = (\alpha - \alpha_0) - (g - g_0)^2, \tag{1}$$

which vanishes at the hypersurface in the parameter space

$$g = g_0 \pm \sqrt{\alpha - \alpha_0}. \tag{2}$$

It was argued in[7] that the collision of two roots at $\alpha = \alpha_0$ can be interpreted as the collision of UV and IR fixed points. Upon the collision the points move into the complex g plane and an RG cycle emerges. The period of the cycle can be immediately estimated as

$$T \propto \int_{g_{UV}}^{g_{IR}} \frac{dg}{\beta(\alpha; g)} \propto \frac{1}{\sqrt{\alpha - \alpha_0}}. \tag{3}$$

The phenomena is believed to be generic once the beta–function has the form (1). Note that is was shown that the RG cycles are consistent with the c-theorem.[10]

Breaking of the conformal symmetry results in the generation of the mass scale which has non-perturbative nature. Due to the RG cycles the scale is not unique and the whole tower with the Efimov-like scaling gets manifested

$$E_{n+1} = \lambda E_n, \tag{4}$$

where λ is fixed by the period of RG cycle.

In the examples available we could attempt to trace the physical picture behind. It turns out that the origin of two couplings is quite general. One coupling does not break the conformal symmetry which is exact in some subspace of the parameter space. The second coupling plays the role of UV regularization which can be imposed in one or another manner. It breaks the conformal symmetry however some discrete version of the scale symmetry survives which is manifested in the cycle structure. The UV regulator will have different reincarnations in the examples considered: the account of the finite size of the nuclei, contact interaction in the model of superconductivity or the brane splitting in the supersymmetric models.

Historically the first example of this phenomena has been found long time ago by Efimov[11] in the context of nuclear physics. He considered the three-body system when two particles are near threshold and have attractive potential with the third particle. It was shown that two-particle bound states are absent in the spectrum, but there is a tower of the three-particle bound states with the geometrical scaling corresponding to $\lambda \approx 22,7$. The review on the RG interpretation of the Efimov phenomena can be found in.[12]

When considering the system with finite number of degrees of freedom the meaning of the RG flows has to be clarified. To this aim some UV cutoff should be introduced. In the first example in[1] the step of the RG corresponds to the integrating out the highest energy level taking into account its correlation with the rest of the spectrum. This approach has a lot in common with the renormalization procedure in the matrix models considered in.[13] The same UV cutoff for formulation of RG procedure has been used for the Russian Doll (RD) model describing the restricted BCS model of superconductivity.[14] In that case the coupling providing the Cooper pairing undergoes the RG cycle while the CP-violating parameter defines the period.

In the second class of the examples the UV cutoff is introduced not at the high energy scale but at small distances. The RG cycles have been found in the non-relativistic Calogero-like models with $\frac{1}{r^2}$ potential which enjoys the naive conformal symmetry.[3–5] The RG flow is formulated in terms of the short distance regularization of the model. It is assumed that the wave function with $E = 0$ at large r does not depend on the UV cutoff at small r. This condition yields the equation for the parameter of a cutoff in the regularization potential. This equation has multiple solutions which can be interpreted as the manifestation of the tower of shallow bound states with the Efimov scaling in the regularized Calogero model with attraction. The scaling factor in the tower is determined by the Calogero coupling constant which reflects the remnant of the conformal group upon the regularization.

The list of the field theory examples in different dimensions with the cyclic RG flows is short but quite representative. In two dimensions the explicit example with the RG cycle has been found in some range of parameters in the sin-Gordon model. The cycle manifests itself in the pole structure of the S-matrix. Efimov-like tower of states corresponds to the specific poles with the Regge-like behavior of the resonance masses[6]

$$m_n = m_s e^{\frac{n\pi}{h}},\qquad(5)$$

where h is a certain parameter of the model. Moreover it was argued that the S-matrix behaves universally under the cyclic RG flows. The tower of Efimov states scales in the same manner as in the quantum mechanical case.

The origination of the cyclic RG behavior in the sin–Gordon model is not surprising. Indeed it was argued in[7] that the famous Berezinsky–Kosterlitz–Thouless (BKT) phase transition in XY system belongs to this universality class. On the other hand one can map the XY system at the T temperature into the sin-Gordon theory with the parameters:

$$L_{SG} = T(\partial\phi)^2 - 4z\cos\phi,\qquad(6)$$

and look at the renormalization of the interaction coupling. The β-functions read as

$$\beta_u = -2v^2,\qquad \beta_v = -2uv,\qquad(7)$$

where

$$u = 1 - \frac{1}{8\pi T},\qquad v = \frac{2z}{T\Lambda^2},\qquad(8)$$

and Λ is the UV cutoff introduced to regularize the vortex core. The β-function implies the existence of the limit cycle with the following expression for the correlation length:

$$\xi_{BKT}\Lambda \propto \exp\left(\frac{c}{\sqrt{|T - T_c|}}\right),\qquad(9)$$

above the phase transition. This RG behavior gets mapped into the RG cycle in the sine-Gordon model.

The example of the Efimov tower in 2+1 dimensions has been found in[15] in the holographic representation. The model is based on the $D3 - D5$ brane configuration and corresponds to the large N 3d gauge theory with fundamentals enjoying $\mathcal{N} = 4$ supersymmetry. In addition the magnetic field and the finite density of conserved charge are present. At strong coupling the gauge theory is described in terms of the probe N_f flavor branes in the nontrivial $AdS_5 \times S^5$ geometry when the $U(1)$ bulk gauge field is added providing the magnetic field in the boundary theory.

The generation of the tower of the Efimov states happens at some value of the "filling fraction" ν in external magnetic field. The phase transition corresponds to the change of the minimal embedding of the probe $D5$ branes in the bulk geometry with the BKT critical behavior of the order parameter. In that case the order parameter gets identified with the condensate σ which behaves as:

$$\sigma \propto \exp\left(-\frac{1}{\nu}\right). \tag{10}$$

Above the phase transition the embedding gets changed and the brane becomes extended in one more coordinate. The scale associated with this extension into new dimension is nothing but the nonperturbative scale amounting to the mass gap. The phenomena of the cyclic RG flow in this case has the Breitenlohner–Freedman instability as the gravitational counterpart.

In four dimensions the most famous example of the Efimov tower is the so-called Miransky scaling for the condensate in the magnetic field. In[16] was argued that the chiral condensate is generated in the external magnetic field in the abelian theory with the following behavior:

$$\langle \bar{\Psi}\Psi \rangle \propto \Lambda^3 \exp(-\frac{c}{\sqrt{\alpha - \alpha_{crit}}}), \tag{11}$$

where α is the fine-structure constant, and c is some parameter of the model.

More recent example[17] of the Efimov tower in four dimensions concerns the Veneziano limit of QCD when $N_f, N_c \to \infty$ while the ratio $x = \frac{N_f}{N_c}$ is fixed. It turns out that this parameter can be considered as the variable in the RG flow which reminds the finite-dimensional examples. At some value of RG scale the tower of condensates gets generated with geometrical Efimov scaling. The period of the RG cycle reads as:

$$T \propto \frac{\kappa}{\sqrt{x_c - x}}, \tag{12}$$

where x_c is the critical value of the x parameter. Finally the 4d example with the RG cycle has been found in the $\mathcal{N} = 2$ SUSY gauge theory in the

Ω-background.[8] In that case the gauge coupling undergoes the RG cycle whose period is determined by the parameter of the Ω-background,

$$T \propto \epsilon^{-1}. \tag{13}$$

The appearance of the RG cycle in this model can be traced from its relation with the quantum integrable systems of the spin chain type.

In this review we provide the reader with the examples of this phenomenon. The list of the systems with finite number of degrees of freedom involves the Calogero model and the relativistic model with the classical conformal symmetry describing the external charge in graphene. Another finite-dimensional example concerns the RD model of the restricted BCS superconductivity. The field theory examples concern the 3d and 4d theories in external fields. We shall focus on their brane representations and use heavily their relations to the finite dimensional integrable systems.

2. RG cycles in non-relativistic quantum mechanics

In this Section we consider the example of the limit cycle in RG in the non-relativistic system with the inverse-square potential, or the Calogero system:

$$H = \frac{\partial^2}{\partial r^2} - \frac{\mu(\mu - 1)}{r^2}. \tag{14}$$

The distinctive feature of the system described by the Hamiltonian (14) is its conformality. Namely, the operators (H, D, K), where D is the dilation generator and K is the conformal boost, generate the conformal SL(2) algebra (see Section 4):

The eigenfunctions of (14) having finite energy immediately break this symmetry; more non-trivial is the fact that even the ground state breaks conformality. Namely, the solution to the $H\psi = 0$ equation is the following:

$$\psi_0 = c_+ r^\mu + c_- r^{1-\mu}. \tag{15}$$

This solution is scale-invariant only if one of the coefficients $c_\pm$ is zero. If both the coefficients are present, they define an intrinsic length scale $L = (c_+/c_-)^{1/(-2\mu+1)}$. Requiring that the quantity c_+/c_- which describes the ground-state solution be invariant under the change of scale,

$$\frac{c_+}{c_-} = -r_0^{-2\mu+1} \frac{\gamma - \mu + 1}{\gamma + \mu}, \tag{16}$$

we arrive at the beta-function for the γ parameter,

$$\beta_\gamma = \frac{\partial \gamma}{\partial \log r_0} = -(\gamma + \mu)(\gamma - \mu + 1) = \left(\mu - \frac{1}{2}\right)^2 - \left(\gamma - \frac{1}{2}\right)^2, \quad (17)$$

where r_0 is the RG scale. We can identify $\gamma = \mu - 1, \gamma = -\mu$ points, i.e. solutions with $c_+ = 0$, $c_- = 0$, with UV and IR attractive points of the renormalization group flow.[7]

If $\mu = i\nu$ is imaginary, i.e. the potential is attracting, then the equation (17) allows us to determine the period of the renormalization group:

$$T = -\int_{-\nu+1}^{\nu} \frac{d\gamma}{\beta_\gamma} = \frac{\pi}{\nu - \frac{1}{2}}. \quad (18)$$

This means that an infinite number of scales is generated, differing by a factor of $\exp\left(-\frac{\pi}{\nu - \frac{1}{2}}\right)$. To see this explicitly, we find the solutions to the Schrödinger equation at finite energies.

In the attracting potential the solution (15) can be written as:

$$\psi_0 \propto \sqrt{r} \sin\left(\left(\nu - \frac{1}{2}\right)\log\left(\frac{r}{r_0}\right) + \alpha\right). \quad (19)$$

We see that this solution oscillates indeterminately in the vicinity of the origin and there is no way to fix the α constant. To regularize this behaviour, we can break the scale invariance at the level of the Hamiltonian and introduce a regularizing potential. Two most popular regularizations involve the square-well potential[4,5] or the delta-shell potential.[3] One more choice is to introduce a δ-function at the origin.[7]

Choosing the square-well regularization,

$$V(r) = \begin{cases} -\frac{\nu(\nu-1)}{r^2}, r > R, \\ -\frac{\lambda}{R^2}, r \leq R, \end{cases} \quad (20)$$

we require that the action of the dilatation operator on the wavefunction inside the well and outside it be equal at $r = R$. This condition amounts to the equation on λ,

$$\sqrt{\lambda}\cot\sqrt{\lambda} = \frac{1}{2} + \nu\cot\left(\nu\log\left(\frac{R}{r_0}\right)\right). \quad (21)$$

The multivalued function $\lambda(R)$ can be chosen to be continuous.[5]

The wavefunction regular at infinity is given as a combination of the Bessel functions,[5]

$$\psi\left(r, \kappa_m\right) = \sqrt{r}(-1)^m \left(ie^{-i\nu\frac{\pi}{2}} J_{i\nu}\left(\kappa_m r\right) - ie^{i\nu\frac{\pi}{2}} J_{-i\nu}\left(\kappa_m r\right)\right), \qquad (22)$$

where κ is the energy of the state. The spectrum consists of infinitely many shallow bound states with adjacent energies differing by an exponential factor,

$$\frac{\kappa_{m+1}}{\kappa_m} = e^{-\frac{\pi}{\nu}}. \qquad (23)$$

Note that the coordinate enters the wavefunction (22) only in combination with energy, and the spectrum is generated by the dilation operator:

$$\psi_{m+1} = \exp\left(-\frac{\pi}{\nu} r \partial_r\right) \psi_m. \qquad (24)$$

One can think of that relation as that the action of the dilatation operator shifts zeroes of the wave function from the area of $r < R$ to the area with the inverse square potential, and one step of (24) evolution corresponds to elimination of a single zero in the area with the square-well potential. Since the wave function oscillates infinitely at the origin, the elimination of all the zeroes would require an infinite number of steps, and in this way a whole tower of states gets generated.

3. RG cycle in graphene

In this Section we shall consider the similar problem in 2+1 dimensions which physically corresponds to the external charge in the planar graphene layer. The problem has the classical conformal symmetry and is the relativistic analogue of the 3+1 conformal non-relativistic Calogero-like system. Due to conformal symmetry we could expect the RG cycles and Efimov-like states in this problem upon imposing the short distance cutoff. The issue of the charge in the graphene plane has been discussed theoretically[19–21] and experimentally.[22,23] It was argued that indeed there is the tower of "quasi-Rydberg" states with the exponential scaling.[24] The situation can be interpreted as an atomic collapse phenomena similar to the instability of $Z > 137$ superheavy atoms in QED.[25]

Turn now to the consideration of an electron in graphene which interacts with an external charge. The two-dimensional Hamiltonian reads as,

$$H_D = v_F \sigma_i p^i + V(r), \qquad i = 1, 2. \qquad (25)$$

The external charge creates a Coulomb potential,

$$V(r) = -\frac{\alpha}{r}, \qquad r \geq R. \tag{26}$$

As we shall see, the solution in presence of the potential (26) oscillates indefinitely at the origin and needs to be regularized by some cutoff R. Hence close enough to the origin $r \leq R$ the potential (26) gets replaced by some constant potential $V_{reg}(r, \lambda(R))$. The renormalization condition for the λ parameter is that the zero-energy wave function is not dependent on the short-distance regularization. This condition is chosen similarly to that of the renormalization of the Calogero system (see Section 2). Hence our primary task is to find the zero-energy solution to the Dirac equation,

$$H_D \psi_0 = 0. \tag{27}$$

Since the Hamiltonian commutes with the J_3 operator,

$$J_3 = i\frac{\partial}{\partial\varphi} + \sigma_3, \qquad [H_D, J_3] = 0, \tag{28}$$

we can look for the solutions of (27) in the form:

$$\psi_0 = \begin{pmatrix} \chi_0(r) \\ \xi_0(r)e^{i\varphi} \end{pmatrix}, \qquad J_3\psi_0 = \psi_0. \tag{29}$$

Then in polar coordinates the equation (27) reads as:

$$\begin{cases} -i\hbar v_F \left(\partial_r + \frac{1}{r}\right)\xi_0 = -V(r)\chi_0, \\ -i\hbar v_F \partial_r \chi_0 = -V(r)\xi_0, \end{cases} \tag{30}$$

which is equivalent to:

$$\begin{cases} \xi_0(r) = i\hbar v_F (V(r))^{-1} \partial_r \chi_0, \\ \partial_r^2 \chi_0 + \left(\frac{1}{r} - \frac{V'(r)}{V(r)}\right)\partial_r \chi_0 + \frac{V^2(r)}{\hbar^2 v_F^2}\chi_0 = 0. \end{cases} \tag{31}$$

For the potential $V = -\frac{\alpha}{r}$ we get the following equation on $\chi_0(r)$:

$$\partial_r^2 \chi_0 + \frac{2}{r}\partial_r \chi_0 + \frac{\beta^2}{r^2}\chi_0 = 0, \qquad \beta = \frac{\alpha}{\hbar v_F}. \tag{32}$$

Supposing that $\beta^2 = \frac{1}{4} + \nu^2$ we write the solution as:

$$\chi_0 = \sqrt{r}\left(c_-\left(\frac{r}{r_0}\right)^{-i\nu} + c_+\left(\frac{r}{r_0}\right)^{i\nu}\right) \propto \sqrt{r}\sin\left(\nu\log\frac{r}{r_0} + \varphi\right). \quad (33)$$

We see that this solution shares the properties of the ground-state Calogero wavefunction (15), namely at nonzero $c_\pm$ it generates its own intrinsic length scale and it oscillates indeterminately at the origin. In order to fix the φ constant we need to introduce a cut-off potential. Hence we consider the solution in the potential:

$$V(r) = \begin{cases} -\frac{\alpha}{r}, r > R, \\ V_{reg} = -\hbar v_F \frac{\lambda}{R}, r \leq R. \end{cases} \quad (34)$$

The dilatation operator acts on χ as following:

$$r\partial_r\chi_0 = \left(\frac{1}{2} + \nu\cot\left(\nu\log\frac{r}{r_0}\right)\right)\chi_0. \quad (35)$$

For the constant potential V_{reg} we get from (31):

$$\partial_r^2\chi_0^{reg} + \frac{1}{r}\partial_r\chi_0^{reg} + \frac{\lambda^2}{R^2}\chi_0^{reg} = 0. \quad (36)$$

Choosing the solution of (36) which is regular at the origin we obtain,

$$\chi_0^{reg} \propto J_0\left(\lambda\frac{r}{R}\right). \quad (37)$$

Computing the action of the dilation operator on the solution in the area of constant potential and equating it to the action of the dilation operator (35) we get the equation on the λ regulator parameter:

$$\frac{1}{2} + \nu\cot\left(\nu\log\left(\frac{R}{r_0}\right)\right) = -\lambda\frac{J_1(\lambda)}{J_0(\lambda)}. \quad (38)$$

The equation (38) defines λ as a multi-valued function of R. The period of the RG flow corresponds to jump from one branch of the $\lambda(R)$ function to another.

Now we proceed to find the bound states in the (26) potential. We consider again the Dirac equation,

$$H_D\psi_\kappa = -\hbar v_F \kappa\psi_\kappa. \quad (39)$$

Then the equation on χ analogous to (31) is as following:

$$\partial_r^2 \chi_\kappa + \frac{2\beta - \kappa r}{\beta - \kappa r}\frac{1}{r}\partial_r \chi_\kappa + \left(\frac{\beta}{r} - \kappa\right)^2 \chi_\kappa = 0. \tag{40}$$

Asymptotically when $r \gg \frac{\beta}{\kappa}$ the solution of (40) regular at infinity is given by the Hankel function,

$$\chi_\kappa \propto H_0^{(1)}(i\kappa r). \tag{41}$$

At small $r \ll \frac{\beta}{\kappa}$ the solution is not regular at the origin,

$$\chi_\kappa \propto \sqrt{r}\sin\left(\nu \log \frac{r}{r_0}\right), \tag{42}$$

and we are again in need for the regulator potential. Solving again the Dirac equation (39) in presence of the constant potential V_{reg} and computing the action of the dilatation operator,

$$r\partial_r \chi_k^{reg} = -(\lambda - \kappa R)\frac{J_1(\lambda - \kappa R)}{J_0(\lambda - \kappa R)}\chi_\kappa^{reg}, \tag{43}$$

we can equate (43) to the action of the dilatation operator on (42) and get the equation on the spectrum of the bound states,

$$\frac{1}{2} + \nu \cot(\nu \log(\kappa R)) = -(\lambda - \kappa R)\frac{J_1(\lambda - \kappa R)}{J_0(\lambda - \kappa R)}. \tag{44}$$

This condition gives the spectrum of infinitely many shallow bound states,

$$\kappa_n = \kappa_* \exp\left(-\frac{\pi n}{\nu}\right), \qquad \kappa \to \infty. \tag{45}$$

4. Anomaly in the SO(2,1) algebra

Let us make some comments on the algebraic counterpart of the phenomena considered following.[9] As we have mentioned the conformal symmetry is the main player since Hamiltonians under consideration are scale invariant before regularization. Actually this group can be thought of as the example of spectrum generating algebra when the Hamiltonian is one of the generators or is expressed in terms of the generators in a simple manner. This is familiar from the exactly or quasi-exactly solvable problems when the dimension of the representation selects the size of the algebraic part of the spectrum.

Let us introduce the generators of the SO(2,1) conformal algebra J_1, J_2, J_3: the Calogero Hamiltonian,

$$J_1 = H = p^2 + V(r), \tag{46}$$

the dilatation generator,

$$J_2 = D = tH - \frac{1}{4}(pr + rp), \tag{47}$$

and the generator of special conformal transformation,

$$J_3 = K = t^2 H - \frac{t}{2}(pr + rp) + \frac{1}{2}r^2. \tag{48}$$

They satisfy the relations of the SO(2,1) algebra:

$$[J_2, J_1] = -iJ_1, \qquad [J_3, J_1] = -2iJ_2, \qquad [J_2, J_3] = iJ_3. \tag{49}$$

The singular behavior of the potential at the origin amounts to a kind of anomaly in the SO(2,1) algebra,

$$A(r) = -i[D, H] + H, \tag{50}$$

which in d space dimensions can be presented in the following form:

$$A(r) = -\frac{d-2}{2}V(r) + (r^i \nabla_i)V(r). \tag{51}$$

The simple arguments imply the following relation

$$\frac{d}{dt}\langle D \rangle_{\text{ground}} = E_{\text{ground}}, \tag{52}$$

where the matrix element is taken over the ground state.

It turns out that (52) is fulfilled for the singular potentials in Calogero-like model or in models with contact potential, $V(r) = g\delta(r)$. The expression for anomaly does not depend on the regularization chosen. Moreover more detailed analysis demonstrates that the anomaly is proportional to the β-function of the coupling providing the UV regularization as can be expected.

A similar calculation of the anomaly for the graphene case can be performed for arbitrary state,

$$\left\langle \frac{dD}{dt} \right\rangle_\psi = \langle \Xi \rangle_\psi = -\int d^2x \psi^*(V(x) + x_i\partial_i V(x))\psi, \tag{53}$$

which yields using square-well regularization:

$$\langle \Xi \rangle_\psi = \hbar v_F \frac{\lambda(R)}{R} \frac{\int\limits_0^R r|\psi|^2 dr}{\int\limits_0^\infty r|\psi|^2 dr}. \tag{54}$$

It is convenient to use the two-dimensional identity in (51),

$$\nabla \frac{\vec{r}}{r} = 2\pi\delta(\vec{r}), \tag{55}$$

which simplifies the calculation of the anomaly for any normalized bound state,

$$\frac{d}{dt}\langle D \rangle_\Psi = -g\pi \int d^2 r \delta(r)|\Psi(r)|^2. \tag{56}$$

5. RG cycles in models of superconductivity

In this Section we explain how the cyclic RG flows emerge in truncated models of superconductivity. To this aim we shall first describe the Richardson model and then consider its generalization to the RD model which enjoys the cyclic RG flow. These models are distinguished by the finiteness of the number of fermionic levels. The relation with the integrable many-body systems proves to be quite useful.

5.1. *Richardson model versus Gaudin model*

Let us recall the truncated BCS-like Richardson model of superconductivity[26] with some number of doubly degenerated fermionic levels with the energies $\epsilon_{j\sigma}, j = 1, \ldots, N$. It describes the system of a fixed number of the Cooper pairs. It is assumed that several energy levels are populated by Cooper pairs while levels with the single fermions are blocked. The Hamiltonian reads as

$$H_{BCS} = \sum_{j,\sigma=\pm}^N \epsilon_{j\sigma} c_{j\sigma}^+ c_{j\sigma} - G \sum_{jk} c_{j+}^\dagger c_{j-}^\dagger c_{k-} c_{k+}, \tag{57}$$

where $c_{j\sigma}$ are the fermion operators and G is the coupling constant providing the attraction leading to the formation of the Cooper pairs. In terms of the hard-core boson operators it reads as

$$H_{BCS} = \sum_j \epsilon_j b_j^\dagger b_j - G \sum_{jk} b_j^\dagger b_k, \tag{58}$$

where

$$[b_j^\dagger, b_k] = \delta_{jk}(2N_j - 1), \qquad b_j = c_{j-}c_{j+}, \qquad N_j = b_j^\dagger b_j. \tag{59}$$

The eigenfunctions of the Hamiltonian can be written as,

$$|M\rangle = \prod_i^M B_i(E_i)|\text{vac}\rangle, \qquad B_i = \sum_j^N \frac{1}{\epsilon_j - E_i} b_j^\dagger, \tag{60}$$

provided the Bethe ansatz equations are fulfilled,

$$G^{-1} = -\sum_j^N \frac{1}{\epsilon_j - E_i} + \sum_j^M \frac{2}{E_j - E_i}. \tag{61}$$

The energy of the corresponding states reads as:

$$E(M) = \sum_i E_i. \tag{62}$$

It was shown in[27] that the Richardson model is exactly solvable and closely related to the particular generalization of the Gaudin model. To describe this relation it is convenient to introduce the so-called pseudospin SL(2) algebra in terms of the creation-annihilation operators for the Cooper pairs,

$$t_j^- = b_j, \qquad t_j^+ = b_j^\dagger, \qquad t_j^0 = N_j - 1/2. \tag{63}$$

The Richardson Hamiltonian commutes with the set of operators R_i,

$$R_i = -t_i^0 - 2G\sum_{j\neq i}^N \frac{t_i t_j}{\epsilon_i - \epsilon_j}, \tag{64}$$

which are identified as the Gaudin Hamiltonians,

$$[H_{BCS}, R_j] = [R_i, R_j] = 0. \tag{65}$$

Moreover the Richardson Hamiltonian itself can be expressed in terms of the operators R_i as:

$$H_{BCS} = \sum_i \epsilon_i R_i + G\left(\sum R_i\right)^2 + \text{const.} \tag{66}$$

The number N of the fermionic levels coincides with the number of sites in the Gaudin model and the coupling constant in the Richardson Hamiltonian corresponds to the "twisted boundary condition" in the Gaudin model. The Bethe ansatz equations for the Richardson model (61) exactly coincide with the ones for the generalized Gaudin model. It was argued in[2] that the Bethe roots correspond to the excited Cooper pairs that is natural to think about the solution to the Baxter equation as the wave function of the condensate. In terms of the conformal field theory Cooper pairs correspond to the screening operators.[28]

For the nontrivial degeneracies of the energy levels d_j the BA equations read as:

$$G^{-1} = -\sum_j^N \frac{d_j}{\epsilon_j - E_i} + \sum_{j \neq i}^M \frac{2}{E_j - E_i}. \tag{67}$$

5.2. Russian Doll model of superconductivity and twisted XXX spin chains

The important generalization of the Richardson model describing superconductivity is the so-called RD model.[2] It involves the additional dimensionless parameter α and the RD Hamiltonian reads as:

$$H_{RD} = 2\sum_i^N (\epsilon_i - G)N_i - \bar{G}\sum_{j<k} (e^{i\alpha} b_k^+ b_j + e^{-i\alpha} b_j^+ b_k), \tag{68}$$

with two dimensionful parameters G, η and $\bar{G} = \sqrt{G^2 + \eta^2}$. In terms of these variables the dimensionless parameter α has the following form:

$$\alpha = \arctan\left(\frac{\eta}{G}\right). \tag{69}$$

It is also useful to consider two dimensionless parameters g, θ defined as $G = gd$ and $\eta = \theta d$ where d is the level spacing. The RD model reduces to the Richardson model in the limit $\eta \to 0$.

The RD model turns out to be integrable as well. Now instead of the Gaudin model the proper counterpart is the generic quantum twisted XXX spin chain.[29] The transfer matrix of such spin chain model $t(u)$ commutes with the H_{RD} which itself can be expressed in terms of the spin chain Hamiltonians.

The equation defining the spectrum of the RD model reads as:

$$e^{2i\alpha} \prod_{l=1}^N \frac{E_i - \varepsilon_l + i\eta}{E_i - \varepsilon_l - i\eta} = \prod_{j \neq i}^M \frac{E_i - E_j + 2i\eta}{E_i - E_j - 2i\eta}, \tag{70}$$

and coincides with the BA equations for the spin chain.

Taking the logarithm of the both sides of the equation (70) we obtain:

$$\alpha + \pi Q_i + \sum_{l=1}^{N} \arctan\left(\frac{\eta}{E_i - \varepsilon_l}\right) - \sum_{j=1}^{M} \arctan\left(\frac{2\eta}{E_i - E_j}\right) = 0. \quad (71)$$

Note that here we have added an arbitrary integer term to account for generically multivalued arctan function.

The RG step amounts to integrating out the N-th degree of freedom in the RD model, or equivalently to integrating out the N-th inhomogeneity in the XXX chain. This results into renormalization of the twist. From (71) it is easy to see that:

$$\arctan\left(\frac{\eta}{G_N}\right) - \arctan\left(\frac{\eta}{G_{N-1}}\right) = \sum_{i=1}^{M} \arctan\left(\frac{2\eta}{E_i - \varepsilon_N}\right). \quad (72)$$

When $M = 1$ it implies that:

$$G_{N-1} - G_N = \frac{G_N^2 + \eta^2}{\varepsilon_N - G_N - E}, \quad (73)$$

which is a discrete version of the (1) equation. Of course the same relation can be derived from the RD Hamiltonian (68). If we consider the wavefunction $\psi = \sum_i^N \psi_i b_i^\dagger |0\rangle$, the Schrödinger equation for a state with one Cooper pair amounts to:

$$(\varepsilon_i - G - E)\,\psi_i = (G + i\eta) \sum_{j=1}^{i-1} \psi_j + (G - i\eta) \sum_{j=i+1}^{N} \psi_j. \quad (74)$$

Integration out the N-th degree of freedom amounts to expressing ψ_M in terms of the other modes,

$$\psi_N = \frac{G + i\eta}{\varepsilon_N - G - E} \sum_{j=1}^{N-1} \psi_j, \quad (75)$$

and substituting it back into the Schrödinger equation (74). The G_{N-1} constant in the resulting equation will differ from the initial G_N value as in (73).

The key feature of the RD model is the multiple solutions to the gap equation. The gaps are parameterized as follows:

$$\Delta_n = \frac{\omega}{\sinh t_n}, \qquad t_n = t_0 + \frac{\pi n}{\theta}, \qquad n = 0, 1, \ldots, \quad (76)$$

where t_0 is the solution to the following equation:

$$\tan(\theta t_0) = \frac{\theta}{g}, \qquad 0 < t_0 < \frac{\pi}{\theta}. \tag{77}$$

and $\omega = dN$ for equal level spacing. Here $E^2 = \varepsilon^2 + |\Delta|^2$. This behavior can be derived via the mean field approximation.[14] The gap with minimal energy defines the ground state, and the other values of the gap describe excitations. In the limit $\theta \to 0$ the gaps $\Delta_{n>0} \to 0$ and

$$t_0 = \frac{1}{g}, \qquad \Delta_0 = 2\omega \exp\left(-\frac{1}{g}\right), \tag{78}$$

therefore the standard BCS expression for the gap is recovered. At the weak coupling limit the gaps behave as:

$$\Delta_n \propto \Delta_0 e^{-\frac{n\pi}{\theta}} \tag{79}$$

In terms of the solutions to the BA equations the multiple gaps correspond to the choices of the different branches of the logarithms, i.e. to different choices of the integer Q parameter in (70).

If the degeneracy of the levels is d_n then the RD model gets modified a little bit and is related to the higher spin XXX spin chain. The local spins s_i are determined by the corresponding higher pair degeneracy d_i of the i-th level,

$$s_i = d_i/2, \tag{80}$$

and the corresponding BA equations read as:

$$e^{2i\alpha} \prod_{l=1}^{N} \frac{E_i - \varepsilon_l + id_l + i\eta}{E_i - \varepsilon_l - id_l - i\eta} = \prod_{j \neq i}^{M} \frac{E_i - E_j + 2i\eta}{E_i - E_j - 2i\eta}. \tag{81}$$

5.3. Cyclic RG flows in the RD model

The RD model of truncated superconductivity enjoys the cyclic RG behavior.[2] The RG flows can be treated as the integrating out the highest fermionic level with appropriate scaling of the parameters using the procedure developed in.[1] The RG equations read as (73):

$$g_{N-1} = g_N + \frac{1}{N}(g_N^2 + \theta^2), \qquad \theta_{N-1} = \theta_N. \tag{82}$$

At large N limit the natural RG variable is identified with $s = \log(N/N_0)$ and the solution to the RG equation is:

$$g(s) = \theta \tan\left(\theta s + \tan^{-1}\left(\frac{g_0}{\theta}\right)\right). \tag{83}$$

Hence the running coupling is cyclic,

$$g(s + \lambda) = g(s), \qquad g(e^{-\lambda}N) = g(N), \tag{84}$$

with the RG period,

$$\lambda = \frac{\pi}{\theta}, \tag{85}$$

and the total number of the independent gaps in the model is:

$$N_{cond} \propto \frac{\theta}{\pi} \log N. \tag{86}$$

The multiple gaps are the manifestations of the Efimov-like states. The sizes of the Cooper pairs in the N-th condensates also have the RD scaling. The cyclic RG can be derived even for the single Cooper pair.

What is going on with the spectrum of the model during the period? It was shown in[14] that it gets reorganized. The RG flow experiences discontinuities from $g = +\infty$ to $g = -\infty$ when a new cycle gets started. At each jump the lowest condensate disappears from the spectrum,

$$\Delta_{N+1}(g = +\infty) = \Delta_N(g = -\infty), \tag{87}$$

indicating that the $(N + 1)$-th state wave function plays the role of N-th state wave function at the next cycle (see (75)).

The same behavior can be derived from the BA equation.[14] To identify the multiple gaps it is necessary to remind that the solutions to the BA equations are classified by the integers $Q_i, i = 1, \ldots, M$ parameterizing the branches of the logarithms. If one assumes that $Q_i = Q$ for all Bethe roots then this quantum number gets shifted by one at each RG cycle and was identified with the integer parameterizing the solution to the gap equations,

$$\Delta_Q \propto \Delta_0 \exp^{-\lambda Q}. \tag{88}$$

At the large N limit the BA equations of the RD model reduce to the BA equation of the Richardson-Gaudin model with the rescaled coupling,

$$G_Q^{-1} = \eta^{-1}(\alpha + \pi Q), \tag{89}$$

which can be treated as the shifted boundary condition in the generalized Gaudin model parameterized by an integer. Let us emphasize that the unusual cyclic RG behavior is due to the presence of two couplings in the RD model.

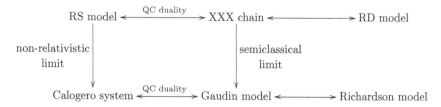

Fig. 1. Besides the triality shown on the picture, a bispectral duality acts on RS/Calogero and XXX/Gaudin sides of the correspondence. Being originated from three-dimensional mirror-symmetry,[30] this duality interchanges coordinates with Lax eigenvalues in the classical systems, and inhomogeneities with twists in quantum ones.

6. Triality in the integrable models and RG cycles

In this Section we summarize several dualities between the integrable models and consider the realizations of the cyclic RG flows in these systems. The question is motivated by the close relationship between the restricted BCS models and spin chains. Actually there are three different families of models related with each other by the particular identifications of phase spaces and parameters. The first family concerns the system of fermions (Richardson-Russian Doll) which develop superconducting gap. The second family involves the spin systems of twisted inhomogeneous Gaudin-XXX-XXZ type and their generalizations. The third family involves the Calogero-Ruijsenaars (CR) chain of the integrable many body systems.

We look for the answers on the following questions

- What is the condition yielding the RG equation for some coupling in each family?
- What is the RG variable?
- What determines the period of the cycle?

In the superconducting system at RG step one decouples the highest energy level and look at the renormalization of the interaction coupling constant. The RG time is identified with the number of energy levels $t = \log N$. The period of RG is defined by the T-asymmetric parameter of RD model.

In the spin chain model the RG step corresponds to the "integrating out" one "highest" inhomogeneity with the corresponding renormalization of the twist. The period of the RG flow is fixed by the Planck constant in the quantum spin chain. In the bispectral dual spin chain[33] one now "integrates out" one of the twists and "renormalizes" the inhomogeneity. Since the Planck constant gets inverted upon bispectrality $\hbar_{spin} \to \hbar_{spin}^{-1}$ the period of the RG cycle gets inverted as well. Note that the RG equation in the superconducting model can be mapped into BAE in the spin chain.[14] The condition yealding the RG equation corresponds to the independence of the Bethe root on the RG step.

For two-body system with attractive rational potential one can define the RG condition as the continuity of the zero-energy wave function under the changing the cutoff scale at small r. This condition imposes the RG equation at the cutoff UV coupling constant. This RG equation has the cycle with the period

$$T_{Cal} = \frac{\pi}{\nu - \frac{1}{2}}. \tag{90}$$

as was shown in Section 2.

The Quantum-Classical (QC) duality[30,31] relates the quantum spin chain systems and the classical Calogero–type systems. Through the QC correspondence, the rational Gaudin model can be linked with the rational Calogero system spin chain inhomogeneities being the Calogero coordinates, and the twist in the spin chain (which is a single variable in our case) being the Lax matrix eigenvalue. It is also possible to make a bispectrality transformation of rational Calogero model, which interchanges Lax eigenvalues with coordinates. This means that now the Calogero coordinates correspond to the twists at the spin chain side. In this case the Calogero coupling gets inverted which means that the period of the RG cycle gets inverted as well.

To consider the mapping of RG cycles in the Calogero system and the spin chain we need the generalization of QC duality to the quantum-quantum case. The spectral problem in Calogero model has been identified with the KZ equation involving the Gaudin Hamiltonian,

$$\frac{d}{dz_i}\Psi = H_{gaud}\Psi + \lambda\Psi. \tag{91}$$

Since we formulate RG condition on the Calogero side for the $E = 0$ state, the inhomogeneous term in the KZ equation is absent. The simplest test of the mapping of the RG cycles under QQ duality concerns the identifications of the periods. On the spin chain side it is identified with the Planck constant while at the Calogero side the period is defined by the coupling constant. The following identification holds for QC duality:[31]

$$\hbar_{spin} = \nu, \tag{92}$$

which implies that the periods of the cycles at the Calogero and spin chain sides match.

The Efimov-like tower in these families have the following interpretations. In the superconducting system it corresponds to the family of the gaps Δ_n with the Efimov scaling responsible of the scale symmetry broken down to the discrete subgroup. In the spin chain it corresponds to the different branches of the solutions to the BAE which can be also interpreted in terms of the allowed set of twists. Finally in the CR family it corresponds to the family of the shallow bound states near the continuum threshold.

7. RG cycles in Ω-deformed SUSY gauge theories

In this Section we shall explain how the RG flows in Ω-deformed SUSY gauge theories can be reformulated in terms of the brane moves. Why the very RG cycles could be expected in the deformed gauge theories? The answer is based on the identification of the quantum spin chains in one or another context in the SUSY gauge theory. Once such quantum spin chain has been found we can apply the results of the previous sections where the place of the RG cycles in the spin chain framework has been clarified.

First, we shall briefly review the Ω-deformation of the SUSY gauge theories. Then we make some general comments concerning the realization of the gauge theories as the worldvolume theories on D-branes to explain how the parameters of the gauge theory are identified with the brane coordinates.

7.1. *Four-dimensional Ω-deformed gauge theory*

The Bethe ansatz equations can be encountered not only in the models of superconductivity, but also in gauge theories. The quantum XXX spin chain governs the moduli space of vacua of an Ω-deformed four-dimensional theory in the Nekrasov–Shatashvili limit, i.e. when one of the deformation is chosen to be zero: $\epsilon_2 = 0, \epsilon_1 = \epsilon$.[35] Since the quantum XXX spin chain displays a cyclic RG behaviour, as we have seen in the Section 5, it is interesting to identify this phenomenon in the four-dimensional gauge theory.

Consider a four-dimensional $\mathcal{N} = 2$ theory with matter hypermultiplet, which has a vanishing β-function, i.e. when $N_f = 2N_c$. This theory is dual to a classical inhomogeneous twisted XXX chain, in a sense that the Seiberg-Witten curve for the gauge theory coincides with the spectral curve for the spin chain. The twist of the spin chain is identified with the modular parameter of the curve and with the complexified coupling of the gauge theory, the inhomogeneities of the spin chain get mapped into masses of the hypermultiplets. For more information on the correspondence between classical integrable systems and gauge theories the reader can consult.[36]

The Ω-deformation is introduced to regularize the instanton divergence in the partition function of the gauge theory.[37] We can consider the four-dimensional theory as a reduction of the six-dimensional $\mathcal{N} = 1$ theory with metric:

$$ds^2 = 2dzd\bar{z} + \left(dx^m + \Omega^{mn}x_nd\bar{z} + \bar{\Omega}^{mn}x_ndz\right)^2, \qquad m = 1,\ldots,4, \quad (93)$$

i.e. we can consider the theory on a four-dimensional space, fibered over a two-dimensional torus. One can imagine the $\epsilon_{1,2}$ deformation parameters as chemical potentials for the rotations in two orthogonal planes in four-dimensional Euclidean space. One can also think that the Euclidean $\mathbb{R}^4$ space gets substituted by a sphere S^4 with finite volume.

The non-trivial Ω-deformation modifies the correspondence between gauge theories and integrable systems. Namely, in the Nekrasov-Shatashvili limit the Ω-deformed gauge theory corresponds to a quantum XXX spin chain with ϵ playing the role of the Planck constant.[35] This deformed gauge theory also appears to be dual to the two-dimensional effective theory on a worldsheet of a non-abelian string.[39]

Consider Ω-deformed $\mathcal{N} = 2$ SQCD with $SU(L)$ gauge group, L fundamental hypermultiplets with masses m_i^f and L antifundamental hypermultiplets with masses m_i^{af}. Let us denote the set of the eigenvalues of the adjoint scalar in the vector multiplet by $\vec{a}$. We can expand the deformed partition function around $\epsilon = 0$ to identify the prepotential and effective twisted superpotential,

$$\log \mathcal{Z}\left(\vec{a},\epsilon_1,\epsilon_2\right) \sim \frac{1}{\epsilon_1\epsilon_2}\mathcal{F}(\vec{a},\epsilon) + \frac{1}{\epsilon_2}\mathcal{W}(\vec{a},\epsilon). \qquad (94)$$

The effective twisted superpotential is a multivalued function, with the branch fixed by the set of integers $\vec{k}$:

$$\mathcal{W}(\vec{a},\epsilon) = \frac{1}{\epsilon}\mathcal{F}(\vec{a},\epsilon) - 2\pi i\vec{k}\cdot\vec{a}, \qquad \vec{k} \in \mathbb{Z}^L. \qquad (95)$$

The equation on vacua,

$$\frac{\partial\mathcal{W}(\vec{a},\epsilon)}{\partial\vec{a}} = \vec{n}, \qquad \vec{n} \in \mathbb{Z}^L, \qquad (96)$$

provides the condition on $\vec{a}$,

$$\vec{a} = \vec{m}^f - \epsilon \vec{n}. \tag{97}$$

This theory admits the existence of non-abelian strings probing the four-dimensional space-time. The two-dimensional worldsheet theory of the non-abelian string involves L fundamental chiral multiplets with twisted masses M_i^F and L antifundamental multiplets with twisted masses M_i^{AF}, which are identified as:

$$M_i^F = m_i^f - \frac{3}{2}\epsilon, \qquad M_i^{AF} = m_i^{af} + \frac{1}{2}\epsilon. \tag{98}$$

The two-dimensional theory also contains an adjoint chiral multiplet with mass ϵ. The rank of the gauge group N (or equivalently the number of non-abelian strings) is given in terms of $\vec{n}$ vector by the relation:

$$N + L = \sum_{l=1}^{L} n_l. \tag{99}$$

The modular parameters of the four-dimensional and the two-dimensional theories are related as:

$$\tau_{2d} = \tau_{4d} + \frac{1}{2}(N+1). \tag{100}$$

The effective twisted worldsheet superpotential is given in terms of the four-dimensional superpotential:

$$\mathcal{W}_{4d}\left(a_i = m_i^f - \epsilon n_i, \epsilon\right) - \mathcal{W}_{4d}\left(a_i = m_i^f - \epsilon, \epsilon\right) = \mathcal{W}_{2d}\left(\{n_i\}\right). \tag{101}$$

The two-dimensional superpotential depends on the set of eigenvalues of the adjoint scalar in vector representation λ_i, $i = 1, \ldots, N$. The set of equations $\partial \mathcal{W}_{2d}/\partial \lambda = 0$ appears to be equivalent to the Bethe ansatz equations for the XXX spin chain:

$$\prod_{l=1}^{L}\left(\frac{\lambda_j - M_l^F}{\lambda_j - M_l^{AF}}\right) = \exp\left(2\pi i \tau_{4d}\right) \prod_{k \neq j}^{N}\left(\frac{\lambda_j - \lambda_k - \epsilon}{\lambda_j - \lambda_k + \epsilon}\right). \tag{102}$$

The Planck constant in the spin chain is identified with the ϵ deformation parameter. The complexified coupling parameter plays the role of twist in the spin chain. The renormalization of the spin chain amounts to decoupling of one fundamental and one anti-fundamental chiral multiplet. In the four-dimensional theory it corresponds to the decrease of the number of flavors $N_f \rightarrow N_f - 2$ simultaneously with reducing the rank of the gauge group $N_c \rightarrow N_c - 1$. Therefore the theory remains conformal. The renormalization of the coupling constant analogous to (73) derived from the relation (102) for $N = 1$ is:

$$\exp\left(2\pi i(\tau_L - \tau_{L-1})\right) = \frac{\lambda - M_L^F}{\lambda - M_L^{AF}}. \tag{103}$$

If we choose the masses to be equidistant with spacing δm, the change in the coupling constant during one step of RG flow is:

$$\exp\left(2\pi i(\tau_L - \tau_{L-1})\right) \propto \frac{\epsilon}{\delta m}. \tag{104}$$

Hence a number of nonperturbative scales emerges in a theory, analogously to the generation of the Efimov scaling in the Calogero model. These scales correspond to multiple gaps in the superconducting model:

$$\Delta_n \propto \Delta_0 \exp\left(-\frac{\pi n \delta m}{\epsilon}\right). \tag{105}$$

Note that the emergence of cyclic RG evolution is a feature caused by the Ω-deformation, since in a non-deformed theory a decoupling of the heavy flavor does not lead to any cyclic dynamic.

7.2. 3d gauge theories and theories on the brane worldvolumes

Let us briefly explain the main points concerning the geometrical engineering of the gauge theories on the D-branes suggesting the reader to consult the details in the review paper.[40] The Dp brane is the $(p+1)$-dimensional hypersurface in the ten-dimensional space-time which supports the $U(1)$ gauge field. This feature provides the possibility to built up the gauge theories with the desired properties. Let us summarize the key elements of the "building procedure".

- A stack of coinciding N D-branes supports $U(N)$ gauge theory with the maximal supersymmetry.

- Displacing some branes from the stack in the transverse direction corresponds to the Higgs mechanism in the $U(N)$ gauge theory and the distance between branes corresponds to the Higgs vev.
- To reduce the SUSY one imposes some boundary conditions at some coordinates using other types of branes or rotates some branes.
- All geometrical characteristics of the brane configurations have the meaning of parameters of the gauge theory like couplings or vevs of some operators in the gauge theory on their worldvolumes.
- If we move some brane through another one the brane of smaller dimension could be created. The Hanany-Witten move is the simplest example (see Fig. 2).
- Since generically we have branes of different dimensions in the configuration, for example, N $D2$ branes and M $D4$ branes we have simultaneously $U(N)$ 2+1 dimensional gauge theory and $U(M)$ dimensional 4+1 dimensional theory on the brane worldvolumes. These theories coexist simultaneously hence there is highly nontrivial interplay between two gauge theories.

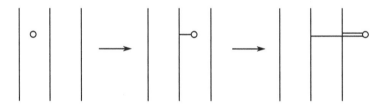

Fig. 2. Hanany-Witten transformation. Here vertical lines are $NS5$ branes, horizontal lines are $D3$ branes, and circles are $D5$ branes. When a $D5$ brane is moved through a sequence of $NS5$ branes the linking number between them is conserved hence additional $D3$ branes appear.

Let us explain now how these brane rules can be used to engineer the gauge theories which are related with the quantum spin chains. Our main example is a 3d $\mathcal{N} = 2$ quiver gauge theory.

The brane configuration relevant for this theory is built as follows. We have M parallel $NS5$ branes extended in (012456), N_i $D3$ branes extended in (0123) between i-th and $(i + 1)$-th $NS5$ branes, and M_i $D5$ branes extended in (012789) between i-th and $(i + 1)$-th $NS5$ branes (see Table 1). From this brane configuration we obtain the $\prod_i^M U(N_i)$ gauge group on the $D3$ branes worldvolume with M_i fundamentals for the i-th gauge group. The distances between the i-th and $(i + 1)$-th $NS5$ branes yield the complexified gauge coupling for $U(N_i)$ gauge group while the coordinates of the $D5$ branes in the (45) plane correspond to the masses of fundamentals. The positions of the $D3$ branes on (45) plane correspond to the coordinates on the Coulomb branch in the quiver theory. The additional Ω deformation reduces the theory with $\mathcal{N} = 4$ SUSY to the $\mathcal{N} = 2^*$ theory, i.e. an $\mathcal{N} = 2$ theory with massive adjoint. It is identified as 3d gauge theory when the distance between $NS5$ is assumed to be small enough. We assume that one coordinate is compact that is the theory lives on $\mathbb{R}^2 \times S^1$.

Table 1. Brane construction of the $3d$ quiver theory.

	0	1	2	3	4	5	6	7	8	9
D3	×	×	×	×						
NS5	×	×	×		×	×	×			
D5	×	×	×					×	×	×

The mapping of the gauge theory data into the integrability framework goes as follows. In the NS limit of the Ω-deformation the twisted superpotential in 3d gauge theory on the $D3$ branes gets mapped into the Yang-Yang function for the XXZ chain.[35] The minimization of the superpotential yields the equations describing the supersymmetric vacua and in the same time they are the Bethe ansatz equations for the XXZ spin chain, generally speaking the nested Bethe ansatz equations. That is $D3$ branes are identified with the Bethe roots which are distributed according to the ranks of the gauge groups at each of M steps of nesting $\prod_i^M U(N_i)$. The positions of the $D5$ branes in the (45) plane correspond to the inhomogeneities in the XXZ spin chain. The anisotropy of the XXZ chain is defined by the radius of the compact dimensions while the parameter of the Ω deformation plays the role of the Planck constant in the XXZ spin chain. At small radius the XXZ spin chain turns to the XXX spin chain. The twists in the spin chain correspond to the coordinates of the $NS5$ branes in the (78) plane, and the Fayet–Iliopoulos parameters in the three-dimensional theory.[30]

One step of the RG flow corresponds to elimination of one inhomogeneity in the spin chain resulting in renormalization of the twists. In the three-dimensional theory this means that the integration of one massive flavor leads to the renormalization of the FI parameters. In terms of the transformations of the brane configurations this process receives transparent geometrical interpretation:

- The RG step is the removing of one $D5$ brane which amounts to the renormalization of the position of $NS5$ branes or twists.
- The period of the RG cycle is fixed by the number of $NS5$ branes,[34] since it was identified with the Planck constant in the spin chain.
- At some scale the twists flow from $+\infty$ to $-\infty$.

8. Conclusion

Are there any general lessons which we could learn for the quantum field theory from the very existence of the cyclic RG flows? The most important point is that there is some fine structure at the UV scale which is reflected in the Efimov tower with the BKT scaling behavior. Moreover the cyclic flows imply the interplay between the UV and IR cutoffs in the theory which usually was attributed to the noncommutative theory. This mixing presumably could shed the additional light on the dimensional transmutation phenomena in the field theory and provide the examples for the simultaneous generation of the multiple scales.

The presence of two parameters in RG is quite common however probably some additional properties of these parameters are required. In particular in many (although not all) examples the period of the cycle is fixed by the "filling fraction" in some external field which could be magnetic field or parameter of Ω background. The latter has the meaning of the Planck constant in the auxiliary finite dimensional integrable model. This could suggest that the very issue can be formulated purely in terms of the quantum phase space since the Planck constant can be interpreted as the external field applied to the classical phase space.

Actually we could expect the relation of RG cycles with some refinement of the path integral in quantum mechanics. As an aside remark note that the attempt to get the rigorous mathematical formulation of the renormalization of the QFT leaded to the motivic generalization of the path integral. It corresponds to some fine structure at the regulator scale which has some similarities with the discussion above. The RG cycle in the quantum rational Calogero model implies the intimate relation with the knot theory since the knot invariants at the rational Calogero coupling are the characteristics of the Calogero spectrum (cf.[34]).

As we already mentioned, cyclic renormalization dynamic is connected with BKT-pairing of partons in two-dimensional model. One could wonder whether this connection is universal. One four-dimensional example of such pairing has to be mentioned. It is bion condensation in 3+1 dimensions. The RG analysis of the model involving the gas of bions and electrically charged W-bosons has been considered in[42] where the RG flows involves the fugacities for electric and magnetic components and the coupling constant. The coupled set of the RG equations has been solved explicitly in the self-dual case and the solution to the RG equations for the fugacities obtained in[42] is identical to the solution for the coupling in the RD model upon the analytic continuation. The period of the RG in the solution above is fixed by the RG invariant which has been identified with the product of the UV values of the electric and magnetic fugacities $y_e \times y_m$. The similarity between the RG behavior is not accidental since the mapping of the gauge theory and the perturbed XY model has been found in.[42]

We would like to emphasize that the investigation of various aspects of limit cycles in RG dynamics still remains on its early stage and there is a considerable number of open questions. The RG cycles can have numerous applications to different aspects of mathematical physics. In this case the RG dynamic is considered as an example of non-trivial dynamical system.

Acknowledgments

The work of A.G. and K.B. was supported in part by grants RFBR-12-02-00284 and PICS-12-02-91052. The work of K.B. was also supported by the "Dynasty" fellowship. A.G. thanks FTPI at University of Minnesota where the part of this work`has been done for the hospitality and support. We would like to thank N. Nekrasov and F. Popov for useful discussions and comments.

References

1. S. Glazek and K. Wilson, "Limit cycles in quantum theories," Phys. Rev. Lett. **89** (2002) 23401, arXiv:hep-th/0203088.
2. A. LeClair, J. M. Roman and G. Sierra, "Russian doll renormalization group and superconductivity," Phys. Rev. B **69**, 20505 (2004) arXiv:cond-mat/0211338.
3. E. Braaten, H. -W. Hammer, "Universality in few-body systems with large scattering length," Phys. Rept. **428**, 259-390 (2006), arXiv:cond-mat/0410417.
4. M. Bawin and S. A. Coon, "The Singular inverse square potential, limit cycles and selfadjoint extensions," Phys. Rev. A **67**, 042712 (2003), arXiv:quant-ph/0302199.
   E. Braaten and D. Phillips, "The Renormalization group limit cycle for the 1/r**2 potential," Phys. Rev. A **70**, 052111 (2004), arXiv:hep-th/0403168.
5. S. R. Beane, P. F. Bedaque, L. Childress, A. Kryjevski, J. McGuire and U. van Kolck, "Singular potentials and limit cycles," Phys. Rev. A **64**, 042103 (2001), arXiv:quant-ph/0010073.

6. A. Leclair, J. M. Roman and G. Sierra, "Russian doll renormalization group, Kosterlitz-Thouless flows, and the cyclic sine-Gordon model," Nucl. Phys. B **675**, 584 (2003), arXiv:hep-th/0301042.

7. D. B. Kaplan, J. -W. Lee, D. T. Son and M. A. Stephanov, "Conformality Lost," Phys. Rev. D **80**, 125005 (2009), arXiv:0905.4752 [hep-th].

8. A. Gorsky, "SQCD, Superconducting Gaps and Cyclic RG Flows," arXiv:1202.4306 [hep-th].

9. G. N. J. Ananos, H. E. Camblong, C. Gorrichategui, E. Hernadez and C. R. Ordonez, "Anomalous commutator algebra for conformal quantum mechanics," Phys. Rev. D **67**, 045018 (2003), arXiv:hep-th/0205191.

   H. E. Camblong and C. R. Ordonez, "Renormalization in conformal quantum mechanics," Phys. Lett. A **345**, 22 (2005), arXiv:hep-th/0305035.

   S. Moroz and R. Schmidt, "Nonrelativistic inverse square potential, scale anomaly, and complex extension," Annals Phys. **325**, 491 (2010), arXiv:0909.3477 [hep-th].

   G. N. J. Ananos, H. E. Camblong and C. R. Ordonez, "SO(2,1) conformal anomaly: Beyond contact interactions," Phys. Rev. D **68**, 025006 (2003), arXiv:hep-th/0302197.

10. T. L. Curtright, X. Jin and C. K. Zachos, "RG flows, cycles, and c-theorem folklore," Phys. Rev. Lett. **108**, 131601 (2012) [arXiv:1111.2649 [hep-th]].

11. V. Efimov, "Energy levels arising from resonant two-body forces in a three-body system," Phys. Lett. B **33**, 563 (1970).

    V. Efimov, "Energy levels of three resonantly interacting particles," Nucl. Phys. A **210**, 157 (1973).

12. H. -W. Hammer and L. Platter, "Efimov physics from a renormalization group perspective," Phil. Trans. Roy. Soc. Lond. A **369**, 2679 (2011), arXiv:1102.3789 [nucl-th].

13. E. Brezin, J. Zinn-Justin, "Renormalization group approach to matrix models," Phys. Lett. B **288**, 54-58 (1992), arXiv:hep-th/9206035.

14. A. Anfossi, A. Leclair and G. Sierra, "The elementary excitations of the exactly solvable Russian doll BCS model of superconductivity," Journal of Statistical Mechanics: 05011 (2005), arXiv:cond-mat/0503014 [cond-mat.supr-con].

15. K. Jensen, A. Karch, D. T. Son and E. G. Thompson, "Holographic Berezinskii-Kosterlitz-Thouless Transitions," Phys. Rev. Lett. **105**, 041601 (2010), arXiv:1002.3159 [hep-th].

16. V. A. Miransky, "Dynamics of Spontaneous Chiral Symmetry Breaking and Continuum Limit in Quantum Electrodynamics," Nuovo Cim. A **90**, 149 (1985).

    V. P. Gusynin, V. A. Miransky and I. A. Shovkovy, "Catalysis of dynamical flavor symmetry breaking by a magnetic field in (2+1)-dimensions," Phys. Rev. Lett. **73**, 3499 (1994) [Erratum-ibid. **76**, 1005 (1996)] [hep-ph/9405262].

17. D. Arean, I. Iatrakis, M. Jarvinen and E. Kiritsis, "The discontinuities of conformal transitions and mass spectra of V-QCD," JHEP **1311**, 068 (2013), arXiv:1309.2286 [hep-ph].

18. A. Gorsky and F. Popov, "Atomic collapse in graphene and cyclic RG flow", arxiv:1312.7399.

19. A. Shytov, M. Katsnelson and L. Levitov, "Vacuum Polarization and Screening of Supercritical Impurities in Graphene," Phys. Rev. Lett. **99**, 236801 (2007), arXiv:0705.4663 [cond-mat.mes-hall].

20. V. Pereira, V. Kotov and A. Castro Neto, "Supercritical Coulomb Impurities in Gapped Graphene,", Phys. Rev. B **78**, 085101 (2008), arXiv:0803.4195 [cond-mat.mes-hall].

21. M. Fogler, D. Novikov and B. Shklovskii, "Screening of a hypercritical charge in graphene," Phys. Rev. B **76**, 233402 (2007), arXiv:0707.1023 [cond-mat.mes-hall].

22. Y. Wang *et al.*, Nat. Phys **8**, 653 (2012).

23. Y. Wang *et al.*, Science **340**, 734 (2013).

24. A. Shytov, M. Katsnelson and L. Levitov, "Atomic Collapse and Quasi-Rydberg States in Graphene," Phys. Rev. Lett. **99**, 246802 (2007), arXiv:0708.0837 [cond-mat.mes-hall].

25. Y.Pomeranchuk and Y. Smorodinsky, J. Phys. USSR **9**, 97 (1945).
    Y. Zeldovich and V. Popov, Sov. Phys. Usp. **14**, 673 (1972).

26. R.Richardson, "A restricted class of exact eigenstates of the pairing-force Hamiltonian", Phys. Lett. **3**, (1963) 277.
    M. C. Cambiaggio, A. M. F. Rivas and M. Saraceno, "Integrability of the pairing hamiltonian," Nucl. Phys. A **624**, 157 (1997), arXiv:nucl-th/9708031.

27. M. C. Cambiaggio, A. M. F. Rivas and M. Saraceno, "Integrability of the pairing hamiltonian," Nucl. Phys. A **624**, 157 (1997), arXiv:nucl-th/9708031.

28. G. Sierra, "Conformal field theory and the exact solution of the BCS Hamiltonian," Nucl. Phys. B **572**, 517 (2000), arXiv:hep-th/9911078.
    M. Asorey, F. Falceto and G. Sierra, "Chern-Simons theory and BCS superconductivity," Nucl. Phys. B **622**, 593 (2002), arXiv:hep-th/0110266.

29. C. Dunning and J. Links, "Integrability of the Russian doll BCS model", Nucl. Phys. B **702** (2004) 481, arXiv:cond-mat/0406234 [cond-mat.stat-mech].

30. D. Gaiotto and P. Koroteev, *On Three Dimensional Quiver Gauge Theories and Integrability,* JHEP **1305**, 126 (2013) [arXiv:1304.0779 [hep-th]].

31. A. Gorsky, A. Zabrodin and A. Zotov, "Spectrum of Quantum Transfer Matrices via Classical Many-Body Systems," arXiv:1310.6958 [hep-th].

32. A. Veselov, *Calogero quantum problem, Knizhnik-Zamolodchikov equation, and Huygens principle,* Theor. Math. Phys. **98**, i.3 (1994) 368-376.

33. M. R. Adams, J. Harnad, J. Hurtubise, Lett. Math. Phys., Vol. 20, Num. 4, 299-308 (1990).
    E. Mukhin, V. Tarasov, A Varchenko, "Bispectral and (glN, glM) dualities, discrete versus differential", Advances in Mathematics **218**, 2008 216-265;

34. K. Bulycheva and A. Gorsky, "BPS states in the Omega-background and torus knots," arXiv:1310.7361 [hep-th].

35. N. Nekrasov and S. Shatashvili, "Supersymmetric vacua and Bethe ansatz," Nucl. Phys. B, Proc. Suppl. 192–193 2009:91–112 arXiv:0901.4744 [hep-th].
    N. A. Nekrasov and S. L. Shatashvili, "Quantization of Integrable Systems and Four Dimensional Gauge Theories," arXiv:0908.4052 [hep-th].

36. A. Gorsky, A. Mironov, "Integrable Many-Body Systems and Gauge Theories," arXiv:hep-th/0011197.

37. N. Nekrasov, "Seiberg-Witten Prepotential From Instanton Counting," Adv. Theor. Math. Phys. **7**, 831–864 (2004), arXiv:hep-th/0206161.

38. M. Shifman, A. Yung, "Supersymmetric Solitons and How They Help Us Understand Non-Abelian Gauge Theories," Rev. Mod. Phys. **79**, 1139 (2007), hep-th/0703267.

39. N. Dorey, T. Hollowood, S. Lee, "Quantization of Integrable Systems and a 2d/4d Duality," arXiv:1103.5726 [hep-th].
    N. Dorey, "The BPS spectra of two-dimensional supersymmetric gauge theories with twisted mass terms," JHEP **9811**, 005 (1998) [hep-th/9806056].

40. A. Giveon and D. Kutasov, "Brane dynamics and gauge theory," Rev. Mod. Phys. **71**, 983 (1999) [hep-th/9802067].

41. P. Bedaque, H. Hammer and U. van Kolck, "Renormalization of the three-body system with short-range interaction", Phys. Rev. Lett. **82** (1999) 463, arXiv:nucl-th/9809025.

42. E. Poppitz, M. Unsal, "Seiberg-Witten and 'Polyakov-like' magnetic bion confinements are continuously connected," JHEP **1107**, 082 (2011). [arXiv:1105.3969 [hep-th]].

COMPOSITE SYSTEMS IN MAGNETIC FIELD: FROM HADRONS TO HYDROGEN ATOM

B. O. KERBIKOV

Institute of Theoretical and Experimental Physics, B. Cheremushkinskaya, 25,
Moscow, 117218 Russia
E-mail: borisk@itep.ru
www.itep.ru

Moscow Institute of Physics and technology, Institutskii per., 9,
Moscow region, Dolgoprudny, 141701 Russia
www.mipt.ru

We briefly review the recent studies of the behavior of composite systems in magnetic field. The hydrogen atom is chosen to demonstrate the new results which may be experimentally tested. Possible applications to physics of anti-hydrogen are mentioned.

Keywords: Magnetic Field, Quarks, Hadrons, Hydrogen.

1. Introduction

We are witnessing an outburst of interest to the behavior of quantum systems in strong magnetic field (MF).[1] This is probably due to the fact that huge MF up to $eB \sim \Lambda_{QCD}^2 \sim 10^{19}G$ has become a physical reality. Such field is created (for a short time) in heavy ion collisions at RHIC and LHC.[2] The field about four orders of magnitude less is anticipated to operate in magnetars.[3] It is impossible in a brief presentation to cover the results of intensive studies performed by an impressive number of researches. I concentrate mainly on the work of ITEP group (M. A. Andreichikov B. O. Kerbikov, V. D. Orlovsky, and Yu. A. Simonov). And even more concise on the hydrogen atom in MF problem. The results concerning the quark systems in MF will be merely formulated.

The quantum mechanics of charged particle in magnetic field is presented in textbooks.[4] In a constant MF assumed to be along the z axis the transverse motion is quantized into Landau levels ($\hbar = c = 1$)

$$E_\perp = \left(n + \frac{1}{2} \right) \omega_H, \quad n = 0, 1, 2, ..., \tag{1}$$

where $\omega_H = \frac{|e|B}{m}$ is the cyclotron frequency. The quantity which in MF takes the role of the mechanical momentum, commutes with the Hamiltonian, and is therefore a constant of motion, is a pseudomomentum[4-8]

$$\hat{\mathbf{K}} = \mathbf{p} - e\mathbf{A} + e\mathbf{B} \times \mathbf{r} = -i\nabla - e\mathbf{A} + e\mathbf{B} \times \mathbf{r}. \tag{2}$$

In the London gauge $\mathbf{A}(\mathbf{r}) = \frac{1}{2}\mathbf{B} \times \mathbf{r}$ the pseudomomentum takes the form

$$\hat{\mathbf{K}} = \mathbf{p} + \frac{1}{2}\mathbf{B} \times \mathbf{r}. \tag{3}$$

Mathematically, the conservation of $\mathbf{K}$ reflects the invariance under the combined action of the spatial translation and the gauge transformation. Physically, $\mathbf{K}$ is conserved since it takes into account the Lorentz force acting on a particle in MF (motional electric field).

The importance of pseudomomentum becomes clear when we turn to a two-body, or many-body problems in MF.

2. The wave function factorization in MF

The total momentum of N mutually interacting particles with translation invariant interaction is a constant of motion and the center of mass motion can be separated in the Schrödinger equation. For the system with total electric charge $Q = 0$ embedded in MF factorization of the wave function can be performed making use of the pseudomomentum operator $\hat{K}$.[5-8] As a simple example consider two nonrelativistic particles with masses m_1 and m_2, charges $e_1 = e > 0, e_2 = -e$, and interparticle interaction $V(\mathbf{r}_1 - \mathbf{r}_2)$. The hydrogen atom is such a system. The Hamiltonian reads

$$\hat{H} = \frac{1}{2m_1}(\mathbf{p}_1 - e\mathbf{A}(\mathbf{r}_1))^2 + \frac{1}{2m_2}(\mathbf{P}_2 + e\mathbf{A}(\mathbf{r}_2))^2 + V(\mathbf{r}_1 - \mathbf{r}_2). \tag{4}$$

Choosing the gauge $\mathbf{A} = \frac{1}{2}\mathbf{B} \times \mathbf{r}$ and introducing

$$M = m_1 + m_2, \mu = m_1 m_2 (m_1 + m_2)^{-1}, s = (m_1 - m_2)(m_1 + m_2)^{-1},$$

$$\mathbf{r} = \mathbf{r}_1 - \mathbf{r}_2, \mathbf{R} = (m_1\mathbf{r}_1 + m_2\mathbf{r}_2)M^{-1}, \mathbf{P} = -i\partial/\partial\mathbf{R},$$

$\boldsymbol{\pi} = -i\partial/\partial\mathbf{r}$, we obtain

$$\hat{H} = \frac{1}{2M}(\mathbf{P} + \frac{e}{2}\mathbf{B} \times \mathbf{r})^2 + \frac{1}{2\mu}(\boldsymbol{\pi} + \frac{e}{2}\mathbf{B} \times \mathbf{R} + s\frac{e}{2}\mathbf{B} \times \mathbf{r})^2. \tag{5}$$

The two-body pseudomomentum operator is

$$\hat{\mathbf{K}} = \sum_{i=1}^{2}(\mathbf{p}_i + \frac{1}{2}e_i\mathbf{B} \times \mathbf{r}_i) = -i\frac{\partial}{\partial\mathbf{R}} - \frac{e}{2}\mathbf{B} \times \mathbf{r}. \tag{6}$$

Since $\hat{\mathbf{K}}$ commutes with $\hat{H}$, the full two-particle wave function $\Psi(\mathbf{R}, \mathbf{r})$ is the eigenfunction of $\hat{\mathbf{K}}$ with the eigenvalue $\mathbf{K}$.

$$\hat{\mathbf{K}}\Psi(\mathbf{R}, \mathbf{r}) = \mathbf{K}\Psi(\mathbf{R}, \mathbf{r}), \tag{7}$$

The wave function which satisfies (7) has the form

$$\psi(\mathbf{R}, \mathbf{r}) = \exp\{(\mathbf{K} + \frac{e}{2}\mathbf{B} \times \mathbf{r})\mathbf{R}\}\varphi(\mathbf{r}). \tag{8}$$

Substitution of the ansatz (8) into $\hat{H}$ leads to the equation

$$\left[\frac{\mathbf{K}^2}{2M} + \frac{e(\mathbf{K} \times \mathbf{B})\mathbf{r}}{M} + \frac{\pi^2}{2\mu} + \frac{es\mathbf{B}(\mathbf{r} \times \boldsymbol{\pi})}{2\mu} + \frac{e^2(\mathbf{B} \times \mathbf{r})^2}{8\mu} + V(\mathbf{r})\right]\varphi_K(\mathbf{r}) = E\varphi_K(\mathbf{r}). \tag{9}$$

The subscript K affixed to E reflects the fact that E has a residual dependence on $\mathbf{K}$ through the second term in (9).

For harmonic interaction $V(\mathbf{r}) = \frac{\sigma^2}{m}\mathbf{r}^2$ the problem has an analytical solution and the ground state energy corresponds to $\mathbf{K} = 0$. The simple calculation yields ($m_1 = m_2 = m$)

$$E = 2\Omega(n_x + n_y + 1) + 2\omega(n_2 + \frac{1}{2}) + \frac{1}{4m}\left[\frac{K_x^2 + K_y^2}{1 + \left(\frac{eB}{2\sigma}\right)^2} + K_z^2\right], \tag{10}$$

$$\Omega = \frac{\sigma}{m}\sqrt{1 + \left(\frac{eB}{2\sigma}\right)^2}, \omega = \frac{\sigma}{m}. \tag{11}$$

To complete this section, we present examples of the pseudomomentum for three- and four-body systems. Consider a model of the neutron as a system of two d-quarks with charges $-e/3$ and masses m_d, and one u-quark with a charge $2e/3$ and a mass m_u, $M = 2m_d + m_u$. This problem was formulated in[9] and is now under investigation in relativistic formalism. Following[9] we introduce the Jacobi coordinates.

$$\boldsymbol{\eta} = \frac{\mathbf{r}_1 - \mathbf{r}_2}{\sqrt{2}}, \boldsymbol{\xi} = \sqrt{\frac{m_u}{2M}}(\mathbf{r}_1 + \mathbf{r}_2 - 2\mathbf{r}_3), \mathbf{R} = \frac{1}{M}\sum_{i=1}^{3}m_i\mathbf{r}_i. \tag{12}$$

Then

$$\hat{\mathbf{K}} = \sum_{i=1}^{3}(\mathbf{p}_i + \frac{1}{2}e_i\mathbf{B} \times \mathbf{r}_i) = \mathbf{P} + \frac{e}{2}\sqrt{\frac{M}{2m_u}}\mathbf{B} \times \boldsymbol{\xi}. \tag{13}$$

As an example of a neutral four-body system consider hydrogen-antihydrogen $H - \bar{H}$.[10] Let $\mathbf{r}_1$ and $\mathbf{r}_2$ be the coordinates of p and $\bar{p}$, $\mathbf{r}_3$ and $\mathbf{r}_4$ be the coordinates of e^- and e^+. Then

$$\hat{\mathbf{K}} = \mathbf{P} + \frac{e}{2}\mathbf{B} \times \{(\mathbf{r}_1 - \mathbf{r}_2) + (\mathbf{r}_4 - \mathbf{r}_3)\} = \mathbf{P} + \frac{e}{2}\mathbf{B} \times \{(\mathbf{r}_1 - \mathbf{r}_3) + (\mathbf{r}_2 - \mathbf{r}_4)\}. \quad (14)$$

This obvious result corresponds to the two possible configurations of the system: a) $p\bar{p} + e^+ e^-$, b) $H - \bar{H}$. Transitions between these two configurations in MF as a Landau–Zener effect will be a subject of a forthcoming publication.

We have reminded the essential formalism needed to treat the composite system under MF. Now we turn to some physical problems.

3. The hierarchy of MF

The present interest to the effects induced by MF was triggered by the realization of the fact that MF generated in heavy ion collisions reaches the value $eB \sim 10^{19}G \sim \Lambda_{QCD}^2$. The highest MF which can be generated now in the laboratory is about $10^6 G$. From the physical point of view there are two characteristic values of MF strength. The Schwinger one is $B_{cr} = m_e^2/e = 4.414 \cdot 10^{13}G$. At $B = B_{cr}$ the distance between the lowest Landau level (LLL) of the electron and the next one is equal to m_e. This can be seen from (1), or from the relativistic dispersion relation

$$\omega_{n,\sigma}(p_z) = [p_z^2 + m^2 + qB(2n + 1 + \sigma)]^{1/2}. \quad (15)$$

Here MF is pointing along the z-axis, m is the particle mass, q is the absolute value of its electric charge, $\sigma = \pm 1$ depending on the spin projection. The LLL corresponds to $n = 0$, $\sigma = -1$. The second important benchmark is the atomic field $B_a = m_e^2 e^3 = 2.35 \cdot 10^9 G$. At $B = B_a$ the Bohr radius $a_B = (\alpha m)^{-1}$ becomes equal to the magnetic, or Landau, radius $l_H = (eB)^{-1/2}$, the oscillator energy $eB/2m_e$ becomes equal to Rydberg energy $Ry = m_e \alpha^2/2$. We use the system of units $\hbar = c = 1, \alpha = e^2 = 1/137$, dimensionless MF is defined as $H = B/B_a$. In this system of units $GeV^2 = 1.45 \cdot 10^{19}G$. The energy to change the electron spin from antiparallel to parallel to B is equal to $2H$ in units of Rydberg. In terms of H MF is classified[11] as low ($H < 10^{-3}$), intermediate, also called strong ($10^{-3} < H < 1$), and intense ($1 < H < \infty$). It seems natural to call MF $eB \sim \Lambda_{QCD}^2$ super-intense, and to say that in this region "QED meets QCD".[1]

4. Quarks in super-intense MF: a compendium of the results

A number of papers published on this subject in recent years is of the order of a hundred. Here we present in a very concise form the results of ITEP group (M.A.Andreichikov, B.O.Kerbikov, V.D.Orlovsky and Yu.A.Simonov).[12] Consider meson or baryon made of quarks embedded in strong MF. There are two parameters defining the transition to the regime when the mass and the geometrical shape of the hadron undergo important changes. The first one is the hadron size $r_h \simeq (0.6\text{--}0.8)$ fm. The strength of MF corresponding to it is defined by $l_h \simeq r_h$ which yields $B_h \simeq 10^{18}G$. Another related parameter is the string tension $\sigma \simeq 0.18$ GeV^2 responsible for the confinement. From the condition $eB/\sigma \simeq 1$ we obtain $B_\sigma \simeq 10^{19}G$. It is therefore clear that the problem of hadron properties in MF of the order of $(10^{18}\text{--}10^{19})G$ has to be formulated and solved at the quark level. The main questions is whether in super-strong MF the hadron mass, e.g., that of the ρ- meson, falls down to zero. For the quark system the question is whether MF induces the "fall to the center" phenomenon. It was shown by ITEP group that the answers to both questions are negative.

The relativistic few-body problem is hindered by well-known difficulties. Maybe the most efficient method to solve the problem is the Field-Correlator Method leading to the relativistic Hamiltonian.[12] To elucidate this formalism is beyond the scope of this presentation. The method includes the following steps:

a) Fock-Feynman-Schwinger proper time representation of the Green's function.

b) Derivation of the confinement and OGE (color Coulomb) interactions using minimal surface Wilson loop.

c) Introduction of the quark dynamical masses (einbein formalism).

d) Inclusion of the spin-dependent interactions σB and hyperfine.

e) Derivation of the relativistic Hamiltonian $\hat{H}$ as the end-result of a)-d).

f) Determination of the hadron mass and wave function

At step b) one obtains the confinement interaction in the form $\sigma|\mathbf{r}_i - \mathbf{r}_j|$ with σ being the string tension. In order to obtain analytical and physically transparent results we replaced the linear potential according to

$$V_{conf} = \sigma r \rightarrow \frac{\sigma}{2}\left(\frac{r^2}{\gamma} + \gamma\right), \qquad (16)$$

where γ is a variational parameter. Minimizing (18) with respect to it, one retrieves the original form of V_{conf}. As was shown by the numerical calculations, the accuracy of this procedure is $\lesssim 5\%$. With the account of MF and confinement, but without spin-dependent terms, the Hamiltonian has the form

$$\hat{H}_{q\bar{q}} = \frac{1}{\omega}\left(-\frac{\partial^2}{\partial \mathbf{r}^2} + \frac{e^2}{4}(\mathbf{B} \times \mathbf{r})^2\right) + \frac{\sigma}{2}\left(\frac{r^2}{\gamma} + \gamma\right). \tag{17}$$

This is a two-oscillator problem similar to (10)–(20). We are focusing on the ground state, hence the pseudomomentum can be taken equal to zero (see (20)), and it does not enter into (17). We note in passing that in the relativistic Hamiltonian approach we evade a subtle problem of the center-of-mass of the relativistic system. The mass eigenvalue $M(\omega)$ and the dynamical mass ω are determined from a set of equations.

$$\hat{H}\psi = M(\omega)\psi, \quad \frac{\partial M(\omega)}{\partial \omega} = 0. \tag{18}$$

The wave function which is a solution of (18) with the Hamiltonian (17) is

$$\Psi(\mathbf{r}) = \frac{1}{\sqrt{\pi^{3/2}a_\perp^2 a_z}}\exp\left(-\frac{r_\perp^2}{2a_\perp^2} - \frac{r_z^2}{2a_z^2}\right), \tag{19}$$

where at $eB \gg \sigma$ one has $a_\perp \simeq \sqrt{\frac{2}{eB}}$, $a_z \simeq \frac{1}{\sqrt{\sigma}}$. With MF increasing the $q\bar{q}$ system acquires the form of an elongated ellipsoid, see Fig. 1.

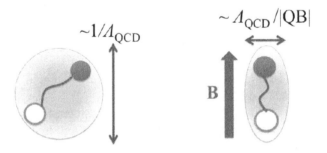

Fig. 1. A sketch of MF influence on the meson wave function (from Ref. 13).

A similar behavior was observed before for the hydrogen atom in strong MF.[14] The difference is that the longitudinal size of the $q\bar{q}$-meson is bounded by $a_z \sim 1/\sqrt{\sigma}$ in contrast to the hydrogen atom which in a strong MF takes the needle-like form with $a_z \sim (\ln H)^{-1}$.

The contribution of V_{OGE} (color Coulomb) was calculated as the average value of V_{OGE} over the wave function (19) with quark and gluon loop corrections taken into account. Hyperfine (hf) spin-spin interaction was treated in a similar way. Here a special care should be devoted to the δ-function. Taken literally, it would lead to a divergent $\psi^2(0) \sim eB$ factor (see the next section). Therefore the δ-function was smeared over the radius ~ 0.2 fm.

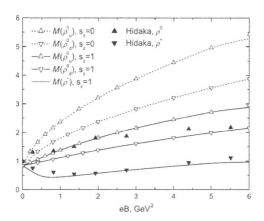

Fig. 2. The results of the calculations of the ρ- meson mass as a function of MF strength together with the lattice data.[15]

In Fig. 2 the results for the ρ-meson mass as a function of MF strength ar presented together with the lattice data.[15] We remind that MF violates both spin and isospin symmetries. In order to minimize the Zeeman energy the lowest state of $u\bar{u}$ (or $d\bar{d}$) in strong MF becomes spin polarized $|u \uparrow \bar{u} \downarrow\rangle$. In our somewhat oversimplified picture this state is a mixture of ρ^0 and π^0. The conclusion is that the mass of the quark-antiquark state does not reach zero no matter now strong MF is. The same result is true for the neutron made of three quarks.

Here we covered only few results of ITEP group on quarks in MF — see Ref. 12.

5. The new results on Zeeman levels in hydrogen

The spectrum of hydrogen atom (HA) in MF is a classical problem described textbooks.[4] The present wave of interest to superstrong MF inspired the reexamination of this problem.[16–19] Surprisingly enough, the new important results were obtained. It was shown that in superstrong MF radiative corrections screen the Coulomb potential thus leading to the freezing of the ground state energy at the value $E_0 = -1.7$ keV.[18,19]

Here we discuss the new correction to hyperfine (hf) splitting in HA.[14] In HA the dramatic changes of the ground state binding energy and the wave function geometry occurs starting from $H \simeq 1$. In this region magnetic confinement in the plane perpendicular to MF dominates the Coulomb binding to the proton. With MF strength growing, the binding energy rises.[16–19] The wave function squeezes and takes the needle-like form. The probability to find the electron near the proton increases. This means that the value of the wave function at the origin $|\psi(0)|^2$ depends on MF and in the limit $H \gg 1$ one has $|\psi(0)|^2 \sim H \ln H$.[14] This phenomenon may be called "Magnetic Focusing of Hyperfine Interaction in Hydrogen". In addition, the deviation of the HA ground state wave function from the spherical symmetry results in the appearance of the tensor forces. These two MF induced effects result in corrections to the standard picture (see Fig. 3) of the Zeeman splitting. The energies of the splitted levels are found by the diagonalization of the following Hamiltonian[14]

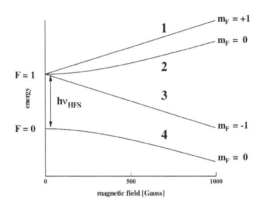

Fig. 3.   Hydrogen hf structure. Transition $|2\rangle \rightarrow |4\rangle$ at $B = 0$ corresponds to the $\lambda = 21$ cm (1420 MHz) line.

$$\hat{H}_{hf} = \frac{8\pi}{3} g\mu_B\mu_N |\psi_B(0)|^2 \boldsymbol{\sigma}_e\boldsymbol{\sigma}_p + \frac{8\pi}{3}\delta\psi_B(0)\sigma_{ez}\sigma_{pz} + \mu_B(\boldsymbol{\sigma}_e\mathbf{B}) - g\mu_N(\boldsymbol{\sigma}_p\mathbf{B}).$$
(20)

Here $g = 2.79$, $g\mu_n$ is the proton magnetic moment, μ_B is the Bohr magneton, index B affixed to ψ_B and $\delta\psi_B$ indicates the dependence on MF. At $B = 0$ one retrieves the standard expression with $|\psi_B(0)|^2 = m^3\alpha^3/\pi$ and $\delta\psi_B(0) = 0$. We do not discuss the corrections to the $\Delta E_{hf} = 1420 MHz$ line due to relativistic effects, QED, and nuclear structure. This subject is thoroughly elucidated in the literature.[20,21] For the frequency of the $\Delta F = 1, \Delta m_F = 0$ $|2 > \to |4 >$ transition diagonalization of the Hamiltonian (19) yields

$$\nu = E_2 - E_4 = \Delta E_{hf} \sqrt{\gamma^2 + \left(\frac{2\mu_B B}{\Delta E_{hf}}\right)^2 \left(1 + g\frac{m_e}{m_p}\right)^2}. \tag{21}$$

Here γ^2 is a new MF dependent parameter with the following asymptotic behavior.[14]

$$\gamma \to 1 + \left(1 - \frac{a_\perp^2}{a_z^2}\right), \quad H \ll \alpha^2 \frac{m_e}{m_p} \simeq 10^{-7}, \tag{22}$$

$$\gamma \to H \ln H, \quad H \gg 1, \tag{23}$$

$$\gamma \gg \frac{2\mu_B B}{\Delta E_{nf}}, \quad \ln H \gg 10^7. \tag{24}$$

In the standard picture without magnetic focusing $\gamma \equiv 1$.

The question is whether magnetic focusing in HA can be experimentally detected in the laboratory conditions. A very preliminary positive answer relies on extremely accurate experiments in search of Zeeman frequency variation using the hydrogen maser.[22] It typically operates with constant MF of the order of ~ 1 mG. In this regime the frequencies of the $F = 1, \Delta m_F = \pm 1$ Zeeman transitions were measured with a precision of ~ 1 mHz.[22] This subject deserves a detailed discussion to be presented in another publication.

This presentation is based on the work of the ITEP team: M. A. Andreichikov, B. K., V. D. Orlovsky and Yu. A. Simonov.

Acknowledgments

The author gratefully acknowledges the encouraging discussions with M. I. Vysotsky, S. I. Godunov, V. S. Popov, B. M. Karnakov, A. E. Shabad and A. Yu. Voronin.

References

1. D. E. Kharzeev, K. Landsteiner, A. Schmitt, and H.-U. Yee, Lect. Notes Phys. **871**, 1 (2013).
2. D. E. Kharzeev, L. D. McLerran and H. J. Warringa, Nucl. Phys. A **803**, 227 (2008); V. Skokov, A. Illarionov and V. Toneev, Int. J. Mod. Phys. A **24**, 5925 (2009).

3. A. Y. Potekhin, Phys. Usp. **53**, 1235 (2010); A. K. Harding and Dong Lai, Rept. Prog. Phys. **69**, 2631 (2006).

4. L. D. Landau and E. M. Lifshitz, Quantum mechanics. Course of Theoretical Physics, vol. 3, Pergamon Press, Oxford (1978).

5. W. E. Lamb, Phys. Rev. **85**, 259 (1952); L. P. Gor'kov and I. E. Dzyaloshinskii, Soviet Physics JETP **26**, 449 (1968); J. E. Avron, I. W. Herbst, and B. Simon, Ann. Phys. (NY) **114**, 431 (1978); H. Grotsch and R. A. Hegstrom, Phys. Rev. A **4**, 59 (1971).

6. H. Herold, H. Ruder, and G. Wunner, J. of Phys. B **14**, 751 (1981).

7. Dong Lai, Rev. Mod. Phys. **73**, 629 (2001).

8. J. Alford and M. Strickland, Phys. Rev. D **88**, 105017 (2013).

9. Yu. A. Simonov, Phys. Lett. B **719**, 464 (2012).

10. A. Yu. Voronin and P. Froelich, Phys. Rev. A **77**, 022505 (2008).

11. M. D. Jones, G. Ortiz, and D. M. Ceperley, Phys. Rev. A **54**, 219 (1996).

12. M. A. Andreichikov, B. O. Kerbikov, V. D. Orlovsky, and Yu. A. Simonov, Phys. Rev. D **87**, 094029 (2013); M. A. Andreichikov, V. D. Orlovsky, and Yu. A. Simonov, Phys. Rev. Lett. **110**, 162002 (2013); V. D. Orlovsky and Yu. A. Simonov, JHEP 1309, 136 (2013), arXiv:1306.2232 [hep-ph]; Yu. A. Simonov, Phys. Rev. D **88**, 025028 (2013), arXiv:1303.4952 [hep-ph]; M. A. Andreichikov, B. O. Kerbikov, Yu. A. Simonov, arXiv:1304.2516 [hep-ph]; Yu. A. Simonov, arXiv:1308.5553 [hep-ph]; Yu. A. Simonov, Phys. Rev. D **88**, 053004 (2013).

13. Toru Kojo and Nan Su, The quark mass gap in a magnetic field, arXiv:1211.7318 [hep-ph].

14. M. A. Andreichikov, B. O. Kerbikov, and Yu. A. Simonov, Magnetic field focussing of hyperfine interaction in hydrogen, arXiv:1304.2516 [hep-ph].

15. Y. Hidaka and A. Yamamoto, Phys. Rev. D **87**, 094502 (2013).

16. B. M. Karnakov, V. S. Popov, J. Exp. Theor. Phys. **114**, 1 (2012); Physics uspekhi, accepted for publication.

17. A. E. Shabad and V. V. Usov, Phys. Rev. Lett. **98**, 180403 (2007); Phys. Rev. D **77**, 025001 (2008).

18. B. Machet and M. I. Vysotsky, Phys. Rev. D **83**, 025022 (2011).

19. S. I. Godunov, B. Machet, M. I. Vysotsky, Phys. Rev. D **85**, 044058 (2012).

20. S. G. Karshenboim, Phys. Rept. **422**, 1 (2005).

21. M. I. Eides, H. Grotch, and V. A. Shelyuto, Phys. Rept. **342**, 63 (2001).

22. D. F. Phillips *et al.*, Phys. Rev. D **63**, 111101 (2001); M. A. Humphrey *et al.*, Phys. Rev. A **68**, 063807 (2003).

GRAVITATIONAL FOUR-FERMION INTERACTION IN THE EARLY UNIVERSE

I. KHRIPLOVICH and A. RUDENKO

Budker Institute of Nuclear Physics,
Novosibirsk State University,
630090 Novosibirsk, Russia
E-mail: khriplovich@inp.nsk.su
a.s.rudenko@inp.nsk.su

If torsion exists, it generates gravitational four-fermion interaction (GFFI), essential on the Planck scale. We analyze the influence of this interaction on the Friedmann-Lemaitre-Robertson-Walker cosmology. Explicit analytical solution is derived for the problem where both the energy-momentum tensor generated by GFFI and the common ultrarelativistic energy-momentum tensor are included. We demonstrate that gravitational four-fermion interaction does not result in Big Bounce.

Keywords: Gravitational Four-Fermion Interaction, Big Bounce.

1. Introduction

According to the common belief, present expansion of the Universe is the result of Big Bang. The idea is popular that this expansion had been preceded by the compression with subsequent Big Bounce. We analyze here the assumption that the Big Bounce is due to the gravitational four-fermion interaction.

The observation that, in the presence of (non-propagating) torsion, the interaction of fermions with gravity results in the four-fermion interaction of axial currents, goes back at least to.[1,2]

The most general form of the gravitational four-fermion interaction is as follows:

$$
S_{ff} = \frac{3\pi G \gtrsim^2}{2(\gtrsim^2 + 1)} \int d^4 x \sqrt{-g}\, \eta_{IJ} \times
$$
$$
\times \left[\left(1 - \beta^2 + \frac{2\beta}{\gtrsim}\right) A^I A^J - 2\alpha \left(\beta - \frac{1}{\gtrsim}\right) V^I A^J - \alpha^2 V^I V^J \right];
$$

(1)

here and below G is the Newton gravitational constant; g is the determinant of the metric tensor; A^I and V^I are the total axial and vector neutral currents, respectively:

$$A^I = \sum_a A_a^I = \sum_a \bar{\psi}_a \gtrsim^5 \gtrsim^I \psi_a \,, \qquad V^I = \sum_a V_a^I = \sum_a \bar{\psi}_a \gtrsim^I \psi_a \,, \quad (2)$$

the sums over a in (2) extend over all sorts of elementary fermions with spin $1/2$. α, β, $\gtrsim$ are numerical parameters of the problem. The values of α, β are unknown. As to the so-called Barbero-Immirzi parameter $\gtrsim$, we assume the value $\gtrsim = 0.274$. In fact, the exact numerical values of these parameters are inessential for our problem.

The AA contribution to expression (1) corresponds (up to a factor) to the action derived long ago in.[1,2] Then, this contribution was obtained in the limit $\beta \to 0$, $\gtrsim \to \infty$ in.[4] The present form of the AA interaction, given in (1), was derived in.[5,6] The VV and VA terms in (1) were derived in[7] and,[6,7] respectively.

Simple dimensional arguments demonstrate that interaction (1), being proportional to the Newton constant G and to the particle number density squared n^2, could get essential and comparable to the common interactions only at very high densities, i.e. on the Planck scale.

Quite extensive list of references on the papers, where the gravitational four-fermion interaction is discussed in connection with cosmology, can be found in.[6,8,9]

2. Energy-momentum tensor

The energy-momentum tensor (EMT) $T_{\mu\nu}^{ff}$ generated by action (1) is:

$$
\begin{aligned}
T_{\mu\nu}^{ff} &= -\frac{3\pi G \gtrsim^2}{2(\gtrsim^2 + 1)} \, g_{\mu\nu} \eta_{IJ} \times \\
&\quad \times \left[\left(1 - \beta^2 + \frac{2\beta}{\gtrsim} \right) A^I A^J - 2\alpha \left(\beta - \frac{1}{\gtrsim} \right) V^I A^J - \alpha^2 V^I V^J \right].
\end{aligned}
\tag{3}
$$

The nonvanishing components of this expression, written in the locally inertial frame, are energy density $T_{00}^{ff} = \rho_{ff}$ and pressure $T_{11}^{ff} = T_{22}^{ff} = T_{33}^{ff} = p_{ff}$ (they are marked here and below by ff to indicate their origin from the four-fermion interaction; for the correspondence between ρ, p and EMT components see,[10] §35).

Let us analyze the expressions for ρ_{ff} and p_{ff} in our case of the interaction of two ultrarelativistic fermions (labeled a and b) in their locally inertial center-of-mass system. We follow here the line of reasoning of.[11]

The axial and vector currents of fermion a are, respectively,

$$A_a^I = \frac{1}{4E^2}\, \phi_a^\dagger \{E\,\boldsymbol{\sigma}_a\,(\partial' + \partial),\ (E^2 - (\partial'\partial))\,\boldsymbol{\sigma}_a + \partial'(\boldsymbol{\sigma}_a\partial)$$

$$+ \partial\,(\boldsymbol{\sigma}_a\partial') - i\,[\partial' \times \partial]\}\,\phi_a = \frac{1}{4}\,\phi_a^\dagger\,\{\boldsymbol{\sigma}_a\,(\mathbf{n}' + \mathbf{n}), \tag{4}$$

$$(1 - (\mathbf{n}'\mathbf{n}))\,\boldsymbol{\sigma}_a + \mathbf{n}'(\boldsymbol{\sigma}_a\mathbf{n}) + \mathbf{n}\,(\boldsymbol{\sigma}_a\mathbf{n}') - i\,[\mathbf{n}' \times \mathbf{n}]\}\,\phi_a;$$

$$V_a^I = \frac{1}{4E^2}\,\phi_a^\dagger\,\{E^2 + (\partial'\partial) + i\,\boldsymbol{\sigma}_a\,[\partial' \times \partial],\ E\,(\partial' + \partial - i\,\boldsymbol{\sigma}_a \times (\partial' - \partial))\}\,\phi_a$$

$$= \frac{1}{4}\,\phi_a^\dagger\,\{1 + (\mathbf{n}'\mathbf{n}) + i\,\boldsymbol{\sigma}_a\,[\mathbf{n}' \times \mathbf{n}],\ \mathbf{n}' + \mathbf{n} - i\,\boldsymbol{\sigma}_a \times (\mathbf{n}' - \mathbf{n})\}\,\phi_a;$$

$$\tag{5}$$

here E is the energy of fermion a, ϕ_a is two-component spinor, $\mathbf{n}$ and $\mathbf{n}'$ are the unit vectors of its initial and final momenta ∂ and ∂', respectively; under the discussed extreme conditions all fermion masses can be neglected. In the center-of-mass system, the axial and vector currents of fermion b are obtained from these expressions by changing the signs: $\mathbf{n} \to -\mathbf{n}$, $\mathbf{n}' \to -\mathbf{n}'$. Then, after averaging over the directions of $\mathbf{n}$ and $\mathbf{n}'$, we arrive at the following semiclassical expressions for the nonvanishing components of the energy-momentum tensor, i.e. for the energy density ρ_{ff} and pressure p_{ff}:

$$\rho_{ff} = T_{00} = -\frac{\pi}{48}\,G\,\frac{\gtrless^2}{\gtrless^2 + 1}\,\sum_{a,b} n_a\,n_b\,\left[\left(1 - \beta^2 + \frac{2\beta}{\gtrless}\right)(3 - 11 <\boldsymbol{\sigma}_a\boldsymbol{\sigma}_b>)\right.$$

$$\left. -\alpha^2(60 - 28 <\boldsymbol{\sigma}_a\boldsymbol{\sigma}_b>)\right]$$

$$= -\frac{\pi}{48}\,G\,\frac{\gtrless^2}{\gtrless^2 + 1}\,n^2\,\left[\left(1 - \beta^2 + \frac{2\beta}{\gtrless}\right)(3 - 11\,\zeta) - \alpha^2(60 - 28\,\zeta)\right];$$

$$\tag{6}$$

$$p_{ff} = T_{11} = T_{22} = T_{33} = \frac{\pi}{48}\,G\,\frac{\gtrless^2}{\gtrless^2 + 1}\,\sum_{a,b} n_a\,n_b\,\times$$

$$\times \left[\left(1 - \beta^2 + \frac{2\beta}{\gtrless}\right)(3 - 11 <\boldsymbol{\sigma}_a\boldsymbol{\sigma}_b>) - \alpha^2(60 - 28 <\boldsymbol{\sigma}_a\boldsymbol{\sigma}_b>)\right]$$

$$= \frac{\pi}{48}\,G\,\frac{\gtrless^2}{\gtrless^2 + 1}\,n^2\,\left[\left(1 - \beta^2 + \frac{2\beta}{\gtrless}\right)(3 - 11\,\zeta) - \alpha^2(60 - 28\,\zeta)\right];$$

$$\tag{7}$$

here and below n_a and n_b are the number densities of the corresponding sorts of fermions and antifermions, $n = \sum_a n_a$ is the total density of fermions and antifermions, the summation $\sum_{a,b}$ extends over all sorts of fermions and antifermions; $\zeta = <\boldsymbol{\sigma}_a\boldsymbol{\sigma}_b>$ is the average value of the product of corresponding $\boldsymbol{\sigma}$-matrices, presumably universal for any $a \neq b$. Since the number of sorts of fermions and antifermions is large, one can neglect here for numerical reasons the contributions of exchange and annihilation contributions, as well as the fact that if $\boldsymbol{\sigma}_a$ and $\boldsymbol{\sigma}_b$ refer to the same particle, $<\boldsymbol{\sigma}_a\boldsymbol{\sigma}_b> = 3$. It is only natural that after the performed averaging over all momenta orientations, the P-odd contributions of VA to ρ_{ff} and p_{ff} vanish.

Thus, the equation of state (EOS) is here

$$\rho_{ff} = -p_{ff} = -\frac{\pi}{48} \frac{G \gtrsim^2 n^2}{\gtrsim^2 +1} \left[\left(1 - \beta^2 + \frac{2\beta}{\gtrsim}\right)(3 - 11\,\zeta) - \alpha^2(60 - 28\,\zeta)\right] \tag{8}$$

The four-fermion energy density (8) can be conveniently rewritten as

$$\rho_{ff} = \varepsilon\, G\, n^2,$$

$$\text{with } \varepsilon = -\frac{\pi}{48} \frac{\gtrsim^2}{\gtrsim^2 +1} \left[\left(1 - \beta^2 + \frac{2\beta}{\gtrsim}\right)(3 - 11\,\zeta) - \alpha^2(60 - 28\,\zeta)\right]. \tag{9}$$

Parameter $\zeta = <\boldsymbol{\sigma}_a\boldsymbol{\sigma}_b>$ for $a \neq b$, just by its physical meaning, in principle can vary in the interval from 0 (which corresponds to the complete thermal incoherence or to the antiferromagnetic ordering) to 1 (which corresponds to the complete ferromagnetic ordering). Correspondingly, ε varies from

$$\varepsilon = -\frac{\pi}{16} \frac{\gtrsim^2}{\gtrsim^2 +1} \left(1 - \beta^2 + \frac{2\beta}{\gtrsim} - 20\alpha^2\right) \text{ for } \zeta = 0 \tag{10}$$

to

$$\varepsilon = \frac{\pi}{6} \frac{\gtrsim^2}{\gtrsim^2 +1} \left(1 - \beta^2 + \frac{2\beta}{\gtrsim} + 4\alpha^2\right) \text{ for } \zeta = 1. \tag{11}$$

The absolute numerical value of the parameter ε is inessential for the analysis below. Its sign, however, is crucial for the physical implications, and depends on α, β and ζ. As to ζ, most probably, at the discussed extreme conditions of high densities and high temperatures, this correlation function is negligibly small.

We go over now to the contributions of common matter to the energy density and pressure. For the extreme densities, where gravitational four-fermion interaction gets essential, this matter is certainly ultrarelativistic, and its contribution to the energy density can be written, for simple dimensional reasons, as

$$\rho = \nu\, n^{4/3}, \tag{12}$$

where ν is a numerical factor. One power of $n^{1/3}$ is here an estimate for the energy per particle. Another factor n in this expression is the total density of ultrarelativistic particles and antiparticles, fermions and bosons, contributing to (12). Since bosons also contribute to the total energy density, this factor should exceed the fermion density n entering the above four-fermion expressions. This difference, however, is absorbed in (12) by the factor ν. As it was the case with ρ_{ff}, it is natural to assume that ρ as well is independent of the spin correlations.

Let us consider now the energy-momentum tensor of the common ultrarelativistic matter in our problem. Since the problem is isotropic, the mixed components of the energy-momentum tensor should vanish:

$$T_{0m} = T_{m0} = 0 \quad (m = 1, 2, 3).$$

Then, the space components of the energy-momentum tensor can be diagonalized, and due to the same isotropy, we arrive at

$$T_{11} = T_{22} = T_{33}.$$

At last, the trace of the energy-momentum tensor of this ultrarelativistic matter should vanish, $T^{\mu}_{\mu} = 0$. Thus, the discussed energy-momentum tensor can be written as

$$T^{\mu}_{\nu} = \rho \, \mathrm{diag}(1, -1/3, -1/3, -1/3), \quad \text{or} \quad \mathrm{T}_{\mu\nu} = \rho \, \mathrm{diag}(1, 1/3, 1/3, 1/3); \quad (13)$$

here and below ρ is the energy density of the common ultrarelativistic matter, and its pressure is $p = \rho/3$.

With $\rho_{ff} \sim Gn^2$, close to the Planck scale gravitational four-fermion interaction is quite comparable to $\rho \sim n^{4/3}$, so that on this scale both contributions should be included. Unfortunately, in our previous papers on the subject the contribution of the common ultrarelativistic matter was not taken into account.

3. FLRW equations

We will assume that, even on the scale close to Planck one, the Universe is homogeneous and isotropic, and thus can be described by the Friedmann-Lemaitre-Robertson-Walker (FLRW) metric

$$ds^2 = dt^2 - a(t)^2 \, [dr^2 + f(r)(d\theta^2 + \sin^2\theta \, d\phi^2)]; \quad (14)$$

here $f(r)$ depends on the topology of the Universe as a whole:

$$f(r) = r^2, \quad \sin^2 r, \quad \sinh^2 r$$

for the spatial flat, closed, and open Universe, respectively.

Now the total energy density and total pressure are, respectively,

$$\rho_{tot} = \rho_{ff} + \rho, \qquad p_{tot} = -\rho_{ff} + \frac{1}{3}\rho.$$

The fact that ρ_{ff} and ρ enter the expression for the total pressure with opposite signs can be traced back to the difference between algebraic structures of the tensors T^{ff} and T. The first of them is proportional in the mixed components to $\delta^\mu_\nu = \text{diag}(1,1,1,1)$, and the second one is proportional to $\text{diag}(1, -1/3, -1/3, -1/3)$.

Thus, the Einstein equations for the FLRW metric (14) are

$$\left(\frac{\dot{a}}{a}\right)^2 + \frac{k}{a^2} = \frac{8\pi G}{3}\rho_{tot} = \frac{8\pi G}{3}\left(\rho_{ff} + \rho\right), \tag{15}$$

$$\frac{\ddot{a}}{a} = -\frac{4\pi G}{3}\left(\rho_{tot} + 3p_{tot}\right) = \frac{8\pi G}{3}\left(\rho_{ff} - \rho\right); \tag{16}$$

parameter k in equation (15) equals 0, 1, and -1 for the spatial flat, closed, and open Universe, respectively. These equations result in the covariant conservation law for the total energy-momentum tensor:

$$\dot{\rho}_{tot} + 3\frac{\dot{a}}{a}\left(\rho_{tot} + p_{tot}\right) = \dot{\rho}_{ff} + \dot{\rho} + 4\frac{\dot{a}}{a}\rho = 0. \tag{17}$$

Let us note here that in the absence of the four-fermion interaction, i.e. for $\rho_{ff} = 0$, this equation reduces to the well-known one for the common ultrarelativistic matter: $\dot{\rho} + 4\left(\dot{a}/a\right)\rho = 0$.

On the other hand, without the common matter, i.e. for $\rho = 0$, equation (17) degenerates into $\dot{\rho}_{ff} = 0$. This is quite natural since energy-momentum tensor (3), generated by the four-fermion interaction, can be conserved by itself only with $\rho_{ff} = \text{const.}$[12]

In fact, observational data strongly favor the idea that our Universe is spatial flat, i.e. that $k = 0$. Then equation (15) simplifies to

$$\left(\frac{\dot{a}}{a}\right)^2 = \frac{8\pi G}{3}\left(\rho_{ff} + \rho\right). \tag{18}$$

Obviously, if the gravitational four-fermion interaction exists, our equations (15)–(18) are as firmly established as the common FLRW equations in the absence of the gravitational four-fermion interaction.

4. Solutions and conclusions

Let us go over now to the solution of FLRW equations. With substitution

$$a(t) = a_0 \exp(f(t)), \tag{19}$$

equations (16) and (18) transform to

$$\frac{8\pi G}{3}(\rho_{ff} + \rho) = \dot{f}^2, \tag{20}$$

$$\frac{8\pi G}{3}\rho = -\frac{1}{2}\ddot{f}. \tag{21}$$

Now, differentiating equation (20) over t and combining the result with (21), we arrive at the following solution:

$$f = -\frac{3}{4\nu}\varepsilon G n^{2/3} - \frac{1}{3}\ln n, \tag{22}$$

the numerical factor ν was introduced in (12). A comment on the ratio $\varepsilon G/\nu$ in this expression is pertinent. It can be easily demonstrated that in the absence of the four-fermion interaction, it is just relation $f = -\frac{1}{3}\ln n$ which results in the law $a(t) = \sqrt{t}$. Then, it is only natural that the relative weight of the four-fermion interaction enters formula (27) via the ratio $\varepsilon G/\nu$.

Thus we obtain

$$a(t) = a_0 \exp(f(t)) \sim n^{-1/3} \exp\{-\frac{3}{4\nu}\varepsilon G n^{2/3}\}. \tag{23}$$

Let us introduce the dimensionless ratio $\xi(t)$ of the four-fermion energy density ρ_{ff} and the energy density ρ of ultrarelativistic matter:

$$\xi(t) = \frac{\rho_{ff}}{\rho} = \frac{\varepsilon G}{\nu} n^{2/3}. \tag{24}$$

Then

$$a(t) \sim \frac{1}{\sqrt{\xi(t)}} e^{-\frac{3}{4}\xi(t)}. \tag{25}$$

Combining equations (20) and (27), we arrive at

$$\dot{\xi} = \mp\frac{4}{3}\sqrt{\frac{8\pi G}{3}} \frac{\nu^{3/2}}{\varepsilon G} \xi^2 \frac{\sqrt{\xi+1}}{\xi+2/3}, \tag{26}$$

which results in relations between ξ and t:

$$\ln\left(\frac{\sqrt{\xi(t)}}{1+\sqrt{1+\xi(t)}}\right) - \frac{1}{2}\frac{\sqrt{1+\xi(t)}}{\xi(t)} = \mp\sqrt{\frac{8\pi G}{3}}\frac{\nu^{3/2}}{\varepsilon G} t + \text{const} \tag{27}$$

for $\varepsilon > 0$,

130

and

$$-\ln\left(\frac{\sqrt{|\xi(t)|}}{1+\sqrt{1-|\xi(t)|}}\right) - \frac{1}{2}\frac{\sqrt{1-|\xi(t)|}}{|\xi(t)|} = \mp\sqrt{\frac{8\pi G}{3}}\frac{\nu^{3/2}}{|\varepsilon|\,G}\,t + \text{const} \tag{28}$$

for $\varepsilon < 0$.

The constants on right-hand sides of (27) and (28) are fixed by initial conditions. As to the signs in formulas (26), (27) and (28), $-$ and $+$ therein refer to expansion and compression, respectively.

The physical implications of formula (23) for positive and negative values of parameter ε are quite different.

For positive ε, both factors in (23), $n^{-1/3}$ and $\exp\{-3/(4\,\nu)\,\varepsilon\,G\,n^{2/3}\}$, and of course their product $a(t)$, shrink to zero together with increasing density n. To analyze the compression, we rewrite equations (16), (18) as follows:

$$\dot{a} = -\sqrt{\frac{8\pi G}{3}}\,a\,\sqrt{\rho_{ff} + \rho}, \tag{29}$$

$$\ddot{a} = \frac{8\pi G}{3}\,a\,(\rho_{ff} - \rho). \tag{30}$$

At the initial moment, when $\rho_{ff} \ll \rho$, both $\dot{a}$ and $\ddot{a}$ are negative, therefore the Universe shrinks with acceleration. Then at $\rho_{ff} = \rho$ acceleration $\ddot{a}$ changes sign, while $\dot{a}$ remains negative, therefore the compression of the Universe decelerates. According to relations (23) and (27), it takes finite time for a to shrink to zero. Due to the exponential factor in (23), $\dot{a}$ and $\ddot{a}$ also vanish at the same moment (the curve $\varepsilon > 0$ in Fig. 1a). Therefore, repulsive gravitational four-fermion interaction does not stop the collapse, but only reduces its rate. The asymptotic behavior of $a(t)$ is

$$a(t) \sim (t_1 - t)\exp\left(-\frac{9\,\varepsilon^2 G}{128\pi\nu^3}\frac{1}{(t_1 - t)^2}\right); \tag{31}$$

here t_1 is the moment of the collapse for $\varepsilon > 0$.

For negative ε, the situation is different. Here the right-hand side of (25)

$$a(t) \sim \frac{1}{\sqrt{|\xi(t)|}}\,e^{\frac{3}{4}|\xi(t)|}$$

reaches its minimum value at $|\xi_m| = 2/3$, i.e., $a(t)$ cannot decrease further. It follows from (18), however, that the compression rate $\dot{a}$ at this point does not vanish and remains finite (the curve $\varepsilon < 0$ in Fig. 1a). In a sense, the situation here resembles that in the standard cosmology with ultrarelativistic particles: therein $a(t) \sim \sqrt{t_0 - t} \to 0$ for $t \to t_0$ (t_0 is the moment of the

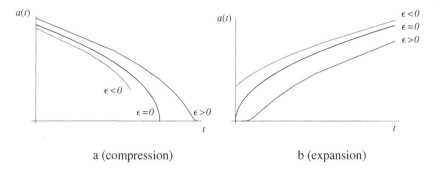

a (compression)   b (expansion)

Fig. 1.   Time-dependence of scale factor

collapse in this case), though at this point $\dot{a}$ does not vanish, but tends to infinity (the curve $\varepsilon = 0$ in Fig. 1a). In the standard cosmology one does not expect that this compression to the origin is followed by expansion. Therefore, in the present case, with $\varepsilon < 0$, it looks natural to assume as well that the compression will not change to expansion.

Thus, contrary to possible naïve expectations,[11] the gravitational four-fermion interaction does not result in Big Bounce.

We note in conclusion that it is difficult (if possible at all) to imagine a realistic possibility to detect any effect of the gravitational four-fermion interaction.

Acknowledgments

We are grateful to V.V. Sokolov for a useful discussion.

The investigation was supported in part by the Foundation for Basic Research through Grant No. 11-02-00792-a, by the Ministry of Education and Science of the Russian Federation, and by the Grant of the Government of Russian Federation, No. 11.G34.31.0047.

References

1. T.W.B. Kibble, Lorentz invariance and the gravitational field, *J. Math. Phys.* **2** (1961) 212.
2. V.I. Rodichev, Twisted space and nonlinear field equations, *J. Exp. Theor. Phys.* **40** (1961) 1469.
3. I.B. Khriplovich and R.V. Korkin, How is the maximum entropy of a quantized surface related to its area?, *J. Exp. Theor. Phys.* **95** (2002) 1 [arXiv:gr-qc/0112074].
4. G.D. Kerlick, Cosmology and particle pair production via gravitational spin-spin interaction in the Einstein-Cartan-Sciama-Kibble theory of gravity, *Phys. Rev. D* **12** (1975) 3004.

5. A. Perez and C. Rovelli, Physical effects of the Immirzi parameter, *Phys. Rev. D* **73** (2006) 044013 [arXiv:gr-qc/0505081].

6. J. Magueijo, T.G. Zlosnik and T.W.B. Kibble, Cosmology with a spin, *Phys. Rev. D* **87** (2013) 063504 [arXiv:1212.0585[astro-ph.CO]].

7. L. Freidel, D. Minic and T. Takeuchi, Quantum gravity, torsion, parity violation and all that, *Phys. Rev. D* **72** (2005) 104002 [arXiv:hep-th/0507253].

8. I.L. Shapiro, Physical aspects of the space-time torsion, *Phys. Rep.* **357** (2002) 113 [arXiv:hep-th/0103093].

9. G. de Berredo-Peixoto, L. Freidel, I.L. Shapiro and C.A. de Souza, Dirac fields, torsion and Barbero-Immirzi parameter in Cosmology, *JCAP* **06** (2012) 017 [arXiv:1201.5423[gr-qc]].

10. L.D. Landau and E.M. Lifshitz, *The Classical Theory of Fields*, Butterworth - Heinemann, (1975).

11. I.B. Khriplovich, Big bounce and inflation from gravitational four-fermion interaction, in V.G. Gurzadyan, A. Klumper and A.G. Sedrakyan, eds., *Low Dimensional Physics and Gauge Principles: Matinyan Festschrift*, World Scientific, (2012), pg. 183 [arXiv:1203.5875[gr-qc]].

12. I.B. Khriplovich and A.S. Rudenko, Cosmology constrains gravitational four-fermion interaction, *JCAP* **11** (2012) 040 [arXiv:1210.7306[astro-ph]].

NON-PERTURBATIVE SCHWINGER–DYSON EQUATIONS: FROM BPS/CFT CORRESPONDENCE TO THE NOVEL SYMMETRIES OF QUANTUM FIELD THEORY

N. NEKRASOV[*]

Simons Center for Geometry and Physics
Stony Brook University, Stony Brook NY 11794-3636, USA
E-mail: nikitastring@gmail.com

We report on recent advances in understanding the non-local symmetries of quantum field theory, notably gauge theory. The symmetries relate topologically distinct sectors of the field space. We study these symmetries in some detail in the context of the BPS/CFT correspondence.

Keywords: Gauge theories, Instantons, Non-local symmetries, Quantum integrability, Quantum Groups.

1. Introduction

In Euclidean quantum field theory the correlation functions are defined by the suitably regularized and renormalized path integral:

$$\langle \mathcal{O}_1(x_1) \ldots \mathcal{O}_n(x_n) \rangle = \frac{1}{Z} \int_\Gamma D\Phi \, \mathcal{O}_1(x_1) \ldots \mathcal{O}_n(x_n) \, e^{-S[\Phi]} \qquad (1)$$

The classical theory is governed by the Euler-Lagrange equations, which are derived from the variational principle:

$$\delta S[\Phi_{cl}] = 0 \qquad (2)$$

These equations are modified in the quantum theory: consider an infinitesimal transformation

$$\Phi \longrightarrow \Phi + \delta\Phi \qquad (3)$$

---

[*]on leave of absence from: IHES, Bures-sur-Yvette, France
ITEP and IITP, Moscow, Russia

Assuming (3) preserves the measure $D\Phi$ in (1) (no anomaly), then

$$\langle \mathcal{O}_1(x_1)\ldots\mathcal{O}_n(x_n)\delta S[\Phi]\rangle =$$

$$\sum_{i=1}^{n}\langle \mathcal{O}_1(x_1)\ldots\mathcal{O}_{i-1}(x_{i-1})\delta\mathcal{O}(x_n)\mathcal{O}_{i+1}(x_{i+1})\ldots\mathcal{O}_n(x_n)\rangle \quad (4)$$

2. Non-perturbative Dyson-Schwinger identities

The discovery of H. Nakajima[1] that the cohomology of the moduli spaces of instantons carries representations of the infinite-dimensional algebras naturally found as symmetries of two dimensional conformal theories suggests the existence of a novel kind of symmetry in quantum field theory, acting via some sort of permutation of the integration regions in the path integral. The transformations of cohomology classes do not, typically, come from the symmetries of the underlying space. Indeed, the infinitesimal symmetries, e.g. generated by some vector field $v \in Vect(X)$ act trivially on the de Rham cohomology $H^*(X)$ of X, as closed differential forms change by the exact forms:

$$\delta\omega = Lie_v\omega = d(\iota_v\omega) \implies [\delta\omega] = 0 \in H^*(X) \quad (5)$$

The symmetries of the cohomology spaces come, therefore, from the *large* transformations $f : X \to X$ or, more generally, the correspondences $L \subset X \times X$:

$$\phi_L = t_*\left(\delta_L \wedge s^*\right) : H^*(X) \to H^*(X)$$
$$\delta_L = \text{Poincare dual to } L,$$
$$s, t : \text{ the projections}$$
$$X \times X$$
$$s \swarrow \qquad \searrow t \quad (6)$$
$$X \qquad X$$

and we assume the compactness and smoothness. There are modifications and generalizations, which we cannot consider due to the lack of space.

The physical realization of the symmetries generated by (6) is yet to be understood. It was proposed in[2] that the domain walls in quantum field theory separating different phases of one theory or even connecting, e.g. in a supersymmetric fashion, two different quantum field theories, can be used to generate the generalized symmetries of the sort we discussed earlier. More precisely, one exchanges the spatial and the temporal directions, producing the S-brane[3] version of the domain wall.

This paper will be dealing with another type of *large* symmetries, having to do with the transformations (3) changing the topological sector, i.e. mapping one connected component of the space of fields to another. We shall be concerned with gauge theories, i.e. the Yang-Mills theory on some space-time X,

$$Z = \int DA \exp - \frac{1}{4g^2} \int_X \mathrm{tr} F_A \wedge \star F_A \qquad (7)$$

with the gauge group G, and its supersymmetric generalizations. The connected components of the space of gauge fields are labelled by the topology types of the principal G-bundles. Now, there is no way, a priori, to deform a connection A_0 on a principal bundle P_0 to a connection A_1 on a principal bundle P_1, which is not isomorphic to P_0. However, imagine that we modify P_0 in a small neighborhood of a point $x \in X$ so that it becomes isomorphic to P_1. It means that outside a small disk D_x there is a gauge transformation, which makes A_0 deformable to A_1. We shall loosely call such a modification *adding a point-like instanton at x*.

The specific realization of such modification is possible in the context of $\mathcal{N} = 2$ supersymmetric gauge theories subject to Ω-deformation. For such a theory one can compute the supersymmetric partition function

$$\mathcal{Z}(a; \mu; \tau; \varepsilon_1, \varepsilon_2) =$$
$$\mathcal{Z}^{\mathrm{tree}}(a; \mu; \tau; \varepsilon_1, \varepsilon_2) \mathcal{Z}^{1-\mathrm{loop}}(a; \mu; \varepsilon_1, \varepsilon_2) \mathcal{Z}^{\mathrm{inst}}(a; \mu; \mathfrak{q}; \varepsilon_1, \varepsilon_2) \qquad (8)$$

where a belongs to the complexified Cartan subalgebra of the gauge group of the theory, μ stands for the set of complex masses of the matter multiplets, τ is the set of complexified gauge couplings,

$$\mathfrak{q} = \exp 2\pi \tau$$

stands for the instanton factors, and $\varepsilon_1, \varepsilon_2$ are the complex numbers, the parameters of the Ω-deformation of the theory.[23] The function (8) contains non-trivial information about the theory. For example, the asymptotics at $\varepsilon_1, \varepsilon_2 \to 0$ produces the prepotential[5,6] of the low-energy effective action of the theory:

$$\mathcal{Z}(a; \mu; \tau; \varepsilon_1, \varepsilon_2) \sim \exp \frac{1}{\varepsilon_1 \varepsilon_2} \mathcal{F}(a; \mu; \tau) + \text{less singular in } \varepsilon_1, \varepsilon_2. \qquad (9)$$

the low-energy effective action being given by the superspace integral

$$S^{\mathrm{eff}} = \int_{\mathbb{R}^{4|4}} d^4 x d^4 \vartheta \, \mathcal{F}(a + \vartheta \psi + \vartheta \vartheta F^- + \ldots; \mu; \tau) \qquad (10)$$

The prepotential $\mathcal{F}(a; \mu; \tau)$ as a function of a determines the special geometry of the moduli space $\mathcal{M}^{\text{vector}}$ of Coulomb vacua:

$$d \left(\frac{a}{\frac{\partial \mathcal{F}(a;\mu;\tau)}{\partial a}} \right) = \text{periods of } \varpi^{\mathbb{C}} \tag{11}$$

along the 1-cycles on the abelian variety A_b, the fiber $p^{-1}(b)$ of the Lagrangian projection

$$p : \mathcal{P} \longrightarrow \mathcal{M}^{\text{vector}} \tag{12}$$

of a complex symplectic manifold $(\mathcal{P}, \varpi^{\mathbb{C}})$, the moduli space of vacua of the same gauge theory, compactified on a circle.[7] The asymptotics of (8) at $\varepsilon_2 \to 0$ with $\varepsilon_1 = \hbar$ fixed gives the effective twisted superpotential

$$\mathcal{Z}(a; \mu; \tau; \hbar, \varepsilon_2) \sim \exp \frac{1}{\varepsilon_2} \mathcal{W}(a; \mu; \tau; \hbar) + \text{less singular in } \varepsilon_2 \tag{13}$$

of a two dimensional effective theory. This function plays an important role in quantization of the symplectic manifold $\mathcal{P}$.[2,8–11] The full partition function (8) is expected to be a remarkable special function, generalizing all the known special functions given by the periods, matrix integrals, matrix elements of group, Kac-Moody, and quantum group representations etc.[12,14,23,24] This feature is the subject of the BPS/CFT correspondence, and its particular implementations, the AGT conjecture,[15,16] and the Bethe/gauge correspondence.[2,8] We shall not discuss all the aspects of these relations in this short note. The interested reader may consult the references in, e.g.[11]

3. Supersymmetric quiver gauge theories

The class of theories we wish to consider here is described in detail in.[11,17] The theory is characterized by a quiver diagram. The quiver is an oriented graph γ, with the set Vert_γ of vertices and the set $\text{Edge}_\gamma \subset \text{Vert}_\gamma \times \text{Vert}_\gamma$ of oriented edges. We have two maps $s, t : \text{Edge}_\gamma \longrightarrow \text{Vert}_\gamma$, which send each edge e to its source $s(e)$ and the target $t(e)$, respectively. The gauge theory is characterized by the vectors $\mathbf{n}, \mathbf{m}$:

$$\mathbf{n} = (n_{\mathbf{i}})_{\mathbf{i} \in \text{Vert}_\gamma} \in \mathbb{Z}_{>0}^{\text{Vert}_\gamma}, \qquad \mathbf{m} = (m_{\mathbf{i}})_{\mathbf{i} \in \text{Vert}_\gamma} \in \mathbb{Z}_{\geq 0}^{\text{Vert}_\gamma} \tag{14}$$

to which we associate the vector spaces $N_{\mathbf{i}} = \mathbb{C}^{n_{\mathbf{i}}}, M_{\mathbf{i}} = \mathbb{C}^{m_{\mathbf{i}}}$. The gauge group $G_{\mathbf{g}}$ of the theory is the product

$$G_{\mathbf{g}} = \times_{\mathbf{i} \in \text{Vert}_\gamma} U(N_{\mathbf{i}}) \tag{15}$$

The theory has the global symmetry which is usually called the flavor symmetry. The flavor symmetry group G_f is a quotient:

$$G_f = \left(\times_{i \in \mathrm{Vert}_\gamma} U(M_i) \times U(1)^{\mathrm{Edge}_\gamma} \right) / U(1)^{\mathrm{Vert}_\gamma} \tag{16}$$

The field content of the theory is the set of $\mathcal{N} = 2$ vector multiplets Φ_i, $i \in \mathrm{Vert}_\gamma$, transforming in the adjoint representation of G_g, the set Q_i, $i \in \mathrm{Vert}_\gamma$ of hypermultiplets transforming in the fundamental representation $\mathbb{C}^{n_i}$ of G_g, and the (anti-)fundamental representation $\mathbb{C}^{m_i}$ of G_f, and the set Q_e, $e \in \mathrm{Edge}_\gamma$ of hypermultiplets transforming in the bifundamental representation $\left(\overline{\mathbb{C}^{n_{s(e)}}}, \mathbb{C}^{n_{t(e)}} \right)$ of G_g. The Lagrangian of the theory is parametrized by the complexified gauge couplings $\tau = (\tau_i)_{i \in \mathrm{Vert}_\gamma}$, the masses $\mu = (\mu_e)_{e \in \mathrm{Edge}_\gamma} \oplus (\mu_i)_{i \in \mathrm{Vert}_\gamma}$, where $\mu_e \in \mathbb{C}$, $\mu_i \in \mathbb{C}^{m_i} \subset \mathrm{End}(\mathbb{C}^{m_i})$. The vacua of the theory are parametrized by the Coulomb moduli $a = (a_i)_{i \in \mathrm{Vert}_\gamma}$, $a_i = \mathrm{diag}(\mathfrak{a}_{i,1}, \ldots, \mathfrak{a}_{i,n_i}) \in \mathrm{End}(\mathbb{C}^{n_i})$.

3.1. *Perturbative consistency and asymptotic freedom*

The description of the tree level and the perturbative contributions to the partition function (the latter is given exactly by one loop computation) can be found in.[11]

The theory defined by the quiver data is perturbatively asymptotically free if the one-loop beta function of all gauge couplings is not positive. For this to be possible we must restrict the gauge group to be the product of special unitary groups

$$G_g \to \prod_{i \in \mathrm{Vert}_\gamma} SU(n_i) \tag{17}$$

since the abelian factors will never be asymptotically free. For the $SU(n_i)$ gauge coupling the beta function is easy to compute:

$$\beta_i = \frac{d}{d\mu} \tau_i = -2n_i + m_i + \sum_{e \in t^{-1}(i)} n_{s(e)} + \sum_{e \in s^{-1}(i)} n_{t(e)} \tag{18}$$

The requirement $\beta_i \leq 0$ for all $i \in \mathrm{Vert}_\gamma$ implies[17–19] that γ is a Dynkin graph of finite or affine type of a simply-laced finite dimensional or affine Lie algebra $\mathfrak{g}_\gamma$. In the latter case all $\mathbf{m} = 0$.

4. Integration over instanton moduli spaces

For the purposes of this paper we only need the instanton partition function $\mathcal{Z}^{\mathrm{inst}}$. Mathematically it can be described as follows. Let $\mathbf{k} = (k_i)_{i \in \mathrm{Vert}_\gamma} \in$

$\mathbb{Z}_+^{\mathrm{Vert}_\gamma}$ be the vector of instanton charges for the gauge group $G_{\mathbf{g}}$. Let $\mathcal{M}_{\mathbf{k}}$ be the moduli space of framed quiver torsion free sheaves on $\mathbb{CP}^2$, that is:

$$
\begin{aligned}
\mathcal{E} &= (\mathcal{E}_{\mathbf{i}})_{\mathbf{i} \in \mathrm{Vert}_\gamma} \\
\mathcal{E}_{\mathbf{i}} &\text{ is a torsion free sheaf on} \quad \mathbb{CP}^2 \\
\mathcal{E}_{\mathbf{i}}|_{\mathbb{CP}^1_\infty} &\xrightarrow{\sim} N_i \\
\mathrm{ch}_2(\mathcal{E}_{\mathbf{i}}) &= k_{\mathbf{i}}
\end{aligned}
\tag{19}
$$

Let $\mathcal{E}_i$ be the universal i'th sheaf over $\mathcal{M}_{\mathbf{k}} \times \mathbb{CP}^2$. Define the *obstruction sheaf* $\mathcal{F}$ over $\mathcal{M}_{\mathbf{k}}$ by:

$$
\mathcal{F}_{\mathbf{k}} = R\pi_* \bigoplus_{e \in \mathrm{Edge}_\gamma} \mathrm{Hom}(\mathcal{E}_{s(e)}, \mathcal{E}_{t(e)}) \oplus \bigoplus_{i \in \mathrm{Vert}_\gamma} \mathrm{Hom}(\mathcal{E}_i, M_i)
$$

$$
\pi : \mathcal{M}_{\mathbf{k}} \times \mathbb{CP}^2 \longrightarrow \mathcal{M}_{\mathbf{k}} \tag{20}
$$

All the sheaves above are $H = (G_{\mathbf{g}} \times G_{\mathbf{f}})_{\mathbb{C}} \times \mathbb{C}^\times \times \mathbb{C}^\times$-equivariant. The complexification of $G_{\mathbf{g}}$ acts on the isomorphisms $\mathcal{E}_i|_{\mathbb{CP}^1_\infty} \xrightarrow{\sim} N_i$, the complexification of $G_{\mathbf{f}}$ acts on the fibers of (20) in the natural way. The torus $\mathbb{C}^\times \times \mathbb{C}^\times$ acts by the symmetries of $\mathbb{CP}^2$, with the fixed point $0 \in \mathbb{C}^2 = \mathbb{CP}^2 \backslash \mathbb{CP}^1_\infty$.

The instanton factor in the partition function can be shown to reduce to the generating function of the equivariant integrals

$$
\mathcal{Z}^{\mathrm{inst}}(\mathbf{a}; \boldsymbol{\mu}; \mathbf{q}; \varepsilon_1, \varepsilon_2) = \sum_{\mathbf{k}} \mathbf{q}^{\mathbf{k}} \int_{\mathcal{M}_{\mathbf{k}}} \mathrm{Euler}(\mathcal{F}_{\mathbf{k}}) \tag{21}
$$

Each term of the q-expansion, with

$$
\mathbf{q}^{\mathbf{k}} = \prod_{\mathbf{i} \in \mathrm{Vert}_\gamma} q_{\mathbf{i}}^{k_{\mathbf{i}}}
$$

are rational functions on $\mathrm{Lie}(H)$. For the asymptotically free theories these are rational functions of negative homogeneity degree. For the asymptotically conformal theories these are degree zero rational functions.

The fixed points $\mathcal{M}_{\mathbf{k}}^H$ of the H-action on $\mathcal{M}_{\mathbf{k}}$ are the sheaves which split as direct sums of monomial ideals:

$$
\mathcal{E} \in \mathcal{M}_{\mathbf{k}}^H \Leftrightarrow \mathcal{E}_{\mathbf{i}} = \oplus_{\alpha=1}^{n_{\mathbf{i}}} \mathcal{I}_{i,\alpha},
$$

$$
\mathcal{I}_{i,\alpha} = I_{\lambda_{i,\alpha}} \tag{22}
$$

In (22) we denote by I_λ a monomial ideal in $\mathbb{C}[z_1, z_2]$, i.e. the ideal, spanned by the monomials. The contribution of such an ideal to the instanton charge is its codimension

$$
|\lambda| = \dim\left(\mathbb{C}[z_1, z_2]/I_\lambda\right) < \infty \tag{23}
$$

The correspondence between the monomial ideals and Young diagrams, or partitions goes as follows:

The monomial $z_1^{i-1} z_2^{j-1}$, with $i, j \geq 1$ belongs to the ideal I_λ iff $j > \lambda_i$

Here $\lambda = (\lambda_1 \geq \lambda_2 \geq \ldots \lambda_{\ell_\lambda} > 0)$ is the partition of the non-negative integer

$$|\lambda| = \sum_{i=1}^{\ell_\lambda} \lambda_i \qquad (24)$$

also called the *size of the partition* λ. The number ℓ_λ of non-zero entries λ_i is called the *length of the partition* λ.

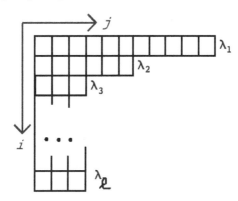

Fig. 1. *Young diagram with the (i,j)-coordinates.*

Thus, the set of fixed points $\mathcal{M}_k^H$ is in one-to-one correspondence with the set of *quiver γ-colored partitions*:

$$\mathcal{E}_\lambda \leftrightarrow \boldsymbol{\lambda} =$$

$$\left\{ \lambda_{\mathbf{i},\alpha} \,\middle|\, \mathbf{i} \in \mathrm{Vert}_\gamma, \alpha = 1, \ldots, n_{\mathbf{i}}, \ \lambda_{\mathbf{i},\alpha} \text{ is a partition}, \ \sum_{\alpha=1}^{n_{\mathbf{i}}} |\lambda_{\mathbf{i},\alpha}| = k_{\mathbf{i}} \right\} \qquad (25)$$

The fixed point formula expresses the gauge theory path integral as the sum over the set of quiver-colored partitions.

Now that the path integration is reduced to a finite sum, the non-perturbative field redefinitions involving adding a point-like instanton can be discussed rigorously.

4.1. Characters, tangent spaces

The contribution of a given fixed point to the partition function can be conveniently expressed using the characters of various vector spaces involved in the local analysis of the path integral measure. Let us use the notation: for a virtual representation R of H, we denote by $I[R]$ the rational function on $\mathrm{Lie}(H)$ defined as follows:

$$I[R](\xi) = \frac{\prod_{w \in W(R^-)} w(\xi)}{\prod_{w \in W(R^+)} w(\xi)}, \qquad \xi \in \mathrm{Lie}(H)$$

$$R = R^+ \ominus R^-,$$

$$\mathrm{Tr}_{R^\pm} e^\xi = \sum_{w \in W(R^\pm)} e^{w(\xi)} \qquad (26)$$

where $R^\pm$ are the vector spaces, the actual representations of H, and $W(R^\pm)$ are the sets of corresponding weights (the linear functions on $\mathrm{Lie}(H)$). Note that in (26) the weights of R^+ go in the denominator.

The instanton partition function can be then written as:

$$\mathcal{Z}^{\mathrm{inst}}(\boldsymbol{a}; \boldsymbol{\mu}; \mathsf{q}; \varepsilon_1, \varepsilon_2) =$$

$$\sum_{\boldsymbol{\lambda} = (\lambda_{\mathbf{i},\alpha})} \prod_{\mathbf{i} \in \mathrm{Vert}_\gamma} \prod_{\mathbf{i} \in \mathrm{Vert}_\gamma} \left(\prod_{\alpha=1}^{n_{\mathbf{i}}} \mathsf{q}_{\mathbf{i}}^{|\lambda_{\mathbf{i},\alpha}|} \right) I\left[\mathcal{T}_{\boldsymbol{\lambda}}\right] \qquad (27)$$

where

$$\mathcal{T}_{\boldsymbol{\lambda}} = \sum_{\mathbf{i} \in \mathrm{Vert}_\gamma} \left[N_{\mathbf{i}} K_{\mathbf{i}}^* + N_{\mathbf{i}}^* K_{\mathbf{i}} q_1 q_2 - P K_{\mathbf{i}} K_{\mathbf{i}}^* \right]$$

$$- \sum_{\mathbf{i} \in \mathrm{Vert}_\gamma} M_{\mathbf{i}}^* K_{\mathbf{i}}$$

$$- \sum_{e \in \mathrm{Edge}_\gamma} e^{m_e} \left[N_{t(e)} K_{s(e)}^* + N_{s(e)}^* K_{t(e)} q_1 q_2 - P K_{t(e)} K_{s(e)}^* \right] \qquad (28)$$

where we denote the characters of the vector spaces by the same letters as the vector spaces themselves,

$$q_1 = e^{\varepsilon_1}, q_2 = e^{\varepsilon_2}, \qquad P = (1 - q_1)(1 - q_2) \qquad (29)$$

and:

$$N_{\mathbf{i}} = \sum_{\alpha=1}^{n_{\mathbf{i}}} e^{a_{\mathbf{i},\alpha}}, \qquad M_{\mathbf{i}} = \sum_{f=1}^{m_{\mathbf{i}}} e^{\mu_{\mathbf{i},f}}$$

$$K_{\mathbf{i}} = \sum_{\alpha=1}^{n_{\mathbf{i}}} \left(e^{a_{\mathbf{i},\alpha}} \sum_{\square \in \lambda_{\mathbf{i},\alpha}} e^{c_\square} \right) \qquad (30)$$

and we denote by $c_\square = \varepsilon_1(i-1) + \varepsilon_2(j-1)$ the *content* of the box $\square = (i,j)$, $i,j \geq 1$, $j \leq \lambda_i$ in the Young diagram of the partition λ.

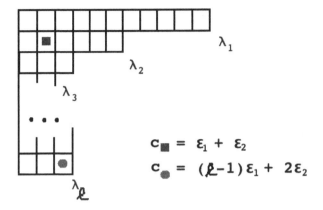

Fig. 2. *Some examples of the contents.*

In (30) we use the convention:

$$R^* = \sum_{w \in W(R)} e^{-w}, \quad \text{for} \quad R = \sum_{w \in W(R)} e^{w} \tag{31}$$

4.2. Observables

The important observables in the gauge theory are the characteristic polynomials of the adjoint Higgs fields:

$$\mathcal{Y}_i(x) = x^{n_i} \exp - \sum_{l=1}^{\infty} \frac{x^{-l}}{l} \operatorname{Tr}(\Phi_i|_0)^l \tag{32}$$

Here we denote by $\Phi_i|_0$ the lowest component of the vector multiplet $\mathbf{\Phi}_i$ corresponding to the node $i \in \mathrm{Vert}_\gamma$, evaluated at the specific point $0 \in \mathbb{C}^2$ in the Euclidean space-time. This is the fixed point of the rotational symmetry $SO(4)$ of which the maximal torus $U(1) \times U(1)$ is generated by the rotations in the two orthogonal two-planes.

In the $\mathcal{N} = 2$ theory the gauge-invariant polynomials of the scalar components of the vector multiplets, i.e.

$$\mathcal{O}_l(\mathbf{x}) = \operatorname{Tr}\Phi_i^l(\mathbf{x}) \ ,$$

for $\mathbf{x} \in \mathbb{R}^4$ are invariant under some supersymmetry transformations, which are nilpotent on the physical states. Moreover, the $\mathbf{x}$-variation of such operators is in itself a supersymmetry variation. Therefore, in the cohomology of such a supercharge, the observable $\mathcal{O}_l(\mathbf{x})$ is $\mathbf{x}$-independent. The supersymmetry of the Ω-deformed $\mathcal{N} = 2$ gauge theory is such that the operator $\mathcal{O}_l(\mathbf{x})$ is invariant only at $\mathbf{x} = 0$, i.e. at the fixed point of the rotations.

Classically, i.e. for the ordinary matrix-valued function $\Phi_{\mathbf{i}}(\mathbf{x})$ the exponential (34) evaluates to the characteristic polynomial of this matrix:

$$\mathcal{Y}_{\mathbf{i}}(x)^{\text{tree}} = \text{Det}_{n_{\mathbf{i}}}(x - \Phi_{\mathbf{i}}|_0) \tag{33}$$

Mathematically $\mathcal{Y}_{\mathbf{i}}(x)$ is defined using the virtual Chern polynomials of the universal sheaves, localized at the point $0 \in \mathbb{C}^2$:

$$\mathcal{Y}_{\mathbf{i}}(x) = c_x(R\pi_* \left[E_{\mathbf{i}} \to E_{\mathbf{i}} \otimes T \to E_{\mathbf{i}} \otimes \wedge^2 T \right]) \tag{34}$$

Here we used the Koszul resolution of the skyscrape sheaf $\mathcal{S}_0$ supported at $0 \in \mathbb{C}^2$:

$$0 \to \wedge^2 \mathcal{T}^* \to \mathcal{T}^* \to \mathcal{O} \to \mathcal{S}_0 \tag{35}$$

where the second and the third maps are the contraction with the Euler vector field $z_1 \partial_{z_1} + z_2 \partial_{z_2}$.

For our calculations, we need the fixed point expression, i.e. the value $\mathcal{Y}_{\mathbf{i}}^{\lambda}(x)$ of the observable $\mathcal{Y}_{\mathbf{i}}(x)$ on the special instanton configuration $\mathcal{E}_\lambda$:

$$\mathcal{Y}_{\mathbf{i}}^{\lambda}(x) = \prod_{\alpha=1}^{n_{\mathbf{i}}} \left(\frac{\prod_{\blacksquare \in \partial_+ \lambda_{\mathbf{i},\alpha}} (x - \mathfrak{a}_{\mathbf{i},\alpha} - c_\blacksquare)}{\prod_{\blacksquare \in \partial_- \lambda_{\mathbf{i},\alpha}} (x - \mathfrak{a}_{\mathbf{i},\alpha} - \varepsilon_1 - \varepsilon_2 - c_\blacksquare)} \right) \tag{36}$$

where for a monomial ideal I_λ, corresponding to the partition λ the *outer boundary* $\partial_+ \lambda$ and the *inner boundary* $\partial_- \lambda$ are the monomials corresponding to the generators, and the relations (divided by the factor $z_1 z_2$) of the ideal. Explicitly, given the character K_λ of the quotient $\mathbb{C}[z_1, z_2]/I_\lambda$

$$\chi_\lambda = \sum_{(i,j) \in \lambda} q_1^{i-1} q_2^{j-1} \tag{37}$$

the contents of the inner and the outer boundaries can be read off the character of the *tautological sheaf*:

$$S_\lambda = 1 - P\chi_\lambda = \sum_{\square \in \partial_+ \lambda} e^{c_\square} - q_1 q_2 \sum_{\blacksquare \in \partial_- \lambda} e^{c_\blacksquare} \tag{38}$$

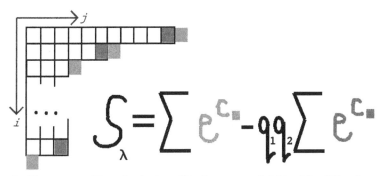

Fig. 3. *Generators* ■ *and relations* ■ *of a monomial ideal* I_λ. *The character of the tautological sheaf* $S_\lambda = 1 - P\chi_\lambda$.

The $\mathcal{Y}$-observable is the essentially the character of the localized tautological complex $\mathcal{S}_i$, which is the cohomology (along $\mathbb{CP}^2$) of the complex

$$\mathcal{S}_i = \left(\mathcal{E}_i \to \mathcal{E}_i \otimes \mathcal{T} \to \mathcal{E}_i \otimes \wedge^2 \mathcal{T} \right) [-1] \qquad (39)$$

which is the dual Koszul complex tensored with the universal sheaf $\mathcal{E}_i$. The relevant character S_i is easy to calculate:

$$S_i = N_i - PK_i = \sum_{\alpha=1}^{n_i} e^{a_{i,\alpha}} S_{\lambda_{i,\alpha}} \qquad (40)$$

The previous formulae can be succinctly written as:

$$\mathcal{Y}_i^\lambda(x) = I[e^x S_i^*] \qquad (41)$$

The relevance of S_i for the analysis of the non-perturbative Schwinger-Dyson equations is the following. The large field redefinitions we shall employ involve adding a point-like instanton at the i'th gauge factor, or, conversely, removing a point-like instanton of the i'th type. This transformation maps one allowed γ-colored partition λ of instanton charge $\mathbf{k} = (k_j)_{j \in \mathrm{Vert}_\gamma}$ to another one $\tilde{\lambda}$, which corresponds to the instanton charge $\tilde{\mathbf{k}} = (k_j \pm \delta_{i,j})_{j \in \mathrm{Vert}_\gamma}$. Inspecting the picture Fig. 3 one easily gets convinced that the modifications of the indicated type consist of adding a box ■ $\in \partial_+ \lambda_{i,\alpha}$ for some $\alpha = 1, \ldots, n_i$, or removing a box ■ $\in \partial_- \lambda_{i,\beta}$ for some $\beta = 1, \ldots, n_i$. In other words, the allowed modifications of λ at the i'th node correspond to the zeroes and poles of $\mathcal{Y}_i^\lambda(x)$.

The measures $I[\mathcal{T}_\lambda]$ and $I[\mathcal{T}_{\tilde{\lambda}}]$ are related to each in a simple manner. Indeed, the character $\mathcal{T}_\lambda$ is quadratic in K_i, more precisely, it is sesquilinear. The variation $\mathcal{T}_{\tilde{\lambda}} - \mathcal{T}_\lambda$ is, therefore, linear in K_i and K_i^*. In fact, it is linear in $S_{\mathbf{j}}$'s and $S_{\mathbf{j}}^*$'s. The ratio of the measures can be, therefore, expressed as a product of the values and residues of various $\mathcal{Y}_{\mathbf{j}}^\lambda$ functions.

Remarkably, these relations can be summarized in the following proposition:

For any γ-graded vector space $\boldsymbol{W} = \oplus_{\mathbf{i}\in\mathrm{Vert}_\gamma} W_{\mathbf{i}}$, with the corresponding dimension vector $\mathbf{w} \in \mathbb{Z}_{\geq 0}^{\mathrm{Vert}_\gamma}$, $W_{\mathbf{i}} = \mathbb{C}^{w_{\mathbf{i}}}$, and a choice of ℓ-weights $\boldsymbol{\nu} = (\nu_{\mathbf{i}})_{\mathbf{i}\in\mathrm{Vert}_\gamma}$, $\nu_{\mathbf{i}} = \mathrm{diag}\,(\nu_{\mathbf{i},1}, \ldots, \nu_{\mathbf{i},w_{\mathbf{i}}}) \in \mathrm{End}(W_{\mathbf{i}})$, there is a Laurent polynomial (Laurent power series for affine γ) in $\mathcal{Y}_{\mathbf{j}}(x + \xi_{\mathbf{j},\kappa})$'s with possibly shifted arguments,

$$\mathcal{X}_{\mathbf{w},\boldsymbol{\nu}}[\mathcal{Y}_{\mathbf{j}}(x + \ldots)] \tag{42}$$

such that its expectation value in the γ-quiver gauge theory is a polynomial in x:

$$\langle \mathcal{X}_{\mathbf{w},\boldsymbol{\nu}}[\mathcal{Y}(x + \ldots)] \rangle =$$

$$\frac{1}{\mathcal{Z}^{\mathrm{inst}}} \sum_\lambda \mathcal{X}_{\mathbf{w},\boldsymbol{\nu}}[\mathcal{Y}^\lambda(x + \ldots)]\, \mathfrak{q}^{\mathbf{k}}\, I[\mathcal{T}_\lambda] =$$

$$T_{\mathbf{w},\boldsymbol{\nu}}(x), \text{ polynomial in } x$$

$$\deg T_{\mathbf{w},\boldsymbol{\nu}}(x) = \mathbf{w} \cdot \mathbf{n} = \sum_{\mathbf{i}\in\mathrm{Vert}_\gamma} w_{\mathbf{i}} n_{\mathbf{i}} \tag{43}$$

We shall call $\mathcal{X}_{\mathbf{w},\boldsymbol{\nu}}$ the $(\varepsilon_1, \varepsilon_2)$-character of $Y(\mathfrak{g}_\gamma)$. In the limit $\varepsilon_2 \to 0$ it will reduce to the Yangian version of the q-characters of finite-dimensional representations of the Yangian $Y(\mathfrak{g}_\gamma)$, constructed for finite γ in.[20] In[21,22] the q-characters for the quantum affine algebras $U_q(\mathfrak{g}_\gamma)$ for finite γ's and in[23] for affine γ's are constructed. These correspond to the K-theoretic version of our story, which would produce the (q_1, q_2)-characters. The physical applications of these (q_1, q_2)-characters are the five dimensional supersymmetric gauge theories compactified on a circle.[24] We shall not discuss them in this note in detail, as the generalizations are straightforward.

4.3. Examples

Let us start with a couple of examples for the theories with a single factor gauge group, i.e. either the A_1 theory or the $\widehat{A}_0$ theory. In the A_1 case the

$(\varepsilon_1, \varepsilon_2)$-character of $Y(\mathrm{sl}_2)$ is given by:

$$\mathcal{X}_{w,(\nu_1,\ldots,\nu_w)} =$$

$$\sum_{I\subset\{1,\ldots,w\}} \mathsf{q}^{\#I} \prod_{i\in I, j\in\bar{I}} \frac{(\nu_i - \nu_j + \varepsilon_1)(\nu_i - \nu_j + \varepsilon_2)}{(\nu_i - \nu_j)(\nu_i - \nu_j + \varepsilon_1 + \varepsilon_2)} \times$$

$$\times \prod_{i\in I} \frac{P(x + \nu_i)}{\mathcal{Y}(x + \nu_i)} \times \prod_{j\in\bar{I}} \mathcal{Y}(x + \varepsilon_1 + \varepsilon_2 + \nu_j)$$

$$\bar{I} = \{1,\ldots,w\}\backslash I \quad (44)$$

For the $\widehat{A}_0$ theory (also known as the $\mathcal{N} = 2^*$ theory) we shall give the expression for $\mathcal{X}_{1,0}$:

$$\mathcal{X}_{1,0} =$$

$$\sum_\lambda \mathsf{q}^{|\lambda|} \prod_{\square\in\lambda} \frac{(-\varepsilon_2 + a_\square\mu + (l_\square + 1)(\varepsilon_1 + \varepsilon_2 + \mu))}{(a_\square\mu + (l_\square + 1)(\varepsilon_1 + \varepsilon_2 + \mu))} \times$$

$$\times \frac{(-\varepsilon_2 - l_\square(\varepsilon_1 + \varepsilon_2 + \mu) - (a_\square + 1)\mu)}{(-l_\square(\varepsilon_1 + \varepsilon_2 + \mu) - (a_\square + 1)\mu)} \times$$

$$\times \frac{\prod_{\square=(i,j)\in\partial_+\lambda} \mathcal{Y}(x + (\varepsilon_1 + \varepsilon_2 + \mu)(i-1) - \mu(j-1))}{\prod_{\square=(i,j)\in\partial_-\lambda} \mathcal{Y}(x + \varepsilon_1 + \varepsilon_2 + (\varepsilon_1 + \varepsilon_2 + \mu)(i-1) - \mu(j-1))} \quad (45)$$

Here $a_\square$, $l_\square$ are the arm-length and the leg-length of the box $\square \in \lambda$.

4.4. *Quiver variety*

In order to write the general formula we shall need an auxiliary object, the quiver variety.

Given two dimension vectors $\mathbf{v} = (v_\mathbf{i})$, $\mathbf{w} = (w_\mathbf{i})$, $\mathbf{v} \in \mathbb{Z}_{>0}^{\mathrm{Vert}_\gamma}$, $\mathbf{w} \in \mathbb{Z}_{\geq 0}^{\mathrm{Vert}_\gamma}$ and a choice of *stability parameter* $\zeta \in \mathbb{R}^{\mathrm{Vert}_\gamma}$, H. Nakajima defines the quiver variety

$$\mathfrak{M}_\zeta(\mathbf{v}, \mathbf{w}) = \mu_{\mathbb{C}}^{-1}(0) \cap \mu_{\mathbb{R}}^{-1}(\zeta)/G_\mathbf{v} \quad (46)$$

where $G_\mathbf{v} = \times_{\mathbf{i}\in\mathrm{Vert}_\gamma} U(V_\mathbf{i})$, and $\mu_{\mathbb{C}}, \mu_{\mathbb{R}}$ are the quadratic maps $\mathcal{H}_\gamma \to \mathrm{Lie}G_\mathbf{v}^* \otimes \mathbb{C}, \mathbb{R}$, respectively, with $\mathcal{H}_\gamma$ the vector space

$$\mathcal{H}_\gamma = T^* \left(\bigoplus_{\mathbf{i}\in\mathrm{Vert}_\gamma} \mathrm{Hom}(V_\mathbf{i}, W_\mathbf{i}) \oplus \bigoplus_{e\in\mathrm{Edge}_\gamma} \mathrm{Hom}(V_{s(e)}, V_{t(e)}) \right) \quad (47)$$

of linear operators (matrices) $(I_\mathbf{i}, J_\mathbf{i}, B_e, \tilde{B}_e)$:

$$I_\mathbf{i} : W_\mathbf{i} \to V_\mathbf{i}, \quad J_\mathbf{i} : V_\mathbf{i} \to W_\mathbf{i}$$
$$B_e : V_{s(e)} \to V_{t(e)}, \quad \tilde{B}_e : V_{t(e)} \to V_{s(e)} \quad (48)$$

Explicitly: $\mu_{\mathbb{R},\mathbb{C}} = (\mu_\mathbf{i}^\mathbb{R}, \mu_\mathbf{i}^\mathbb{C})_{\mathbf{i}\in\text{Vert}_\gamma}$, with

$$\mu_\mathbf{i}^\mathbb{C} = I_\mathbf{i} J_\mathbf{i} + \sum_{e\in s^{-1}(\mathbf{i})} \tilde{B}_e B_e - \sum_{e\in t^{-1}(\mathbf{i})} B_e \tilde{B}_e$$

$$\mu_\mathbf{i}^\mathbb{R} = I_\mathbf{i} I_\mathbf{i}^\dagger - J_\mathbf{i}^\dagger J_\mathbf{i} + \sum_{e\in s^{-1}(\mathbf{i})} B_e^\dagger B_e - \tilde{B}_e \tilde{B}_e^\dagger + \tag{49}$$

$$+ \sum_{e\in t^{-1}(\mathbf{i})} \tilde{B}_e^\dagger \tilde{B}_e - B_e B_e^\dagger$$

The definition (46) translates to the set of equations:

$$\mu_\mathbf{i}^\mathbb{R} = \zeta_\mathbf{i} \mathbf{1}, \qquad \mu_\mathbf{i}^\mathbb{C} = 0, \qquad \mathbf{i} \in \text{Vert}_\gamma \tag{50}$$

with the identification of solutions related by the $G_\mathbf{v}$ transformations:

$$(B_e, \tilde{B}_e, I_\mathbf{i}, J_\mathbf{i}) \mapsto (h_{t(e)} B_e h_{s(e)}^{-1}, h_{s(e)} \tilde{B}_e h_{t(e)}^{-1}, h_\mathbf{i} I_\mathbf{i}, J_\mathbf{i} h_\mathbf{i}^{-1}), \qquad h_\mathbf{i} \in U(V_\mathbf{i}) \tag{51}$$

4.4.1. *Symmetries of* $\mathfrak{M}_\zeta(\mathbf{v}, \mathbf{w})$

We have the following group $G_\mathbf{w} \times (\mathbb{C}^\times)^{1+b_1}$ of symmetries acting on $\mathfrak{M}_\zeta(\mathbf{v}, \mathbf{w})$. Here

$$G_\mathbf{w} = \prod_{\mathbf{i}\in\text{Vert}_\gamma} U(W_\mathbf{i})$$

acts on the I, J maps:

$$(g_\mathbf{i})_{\mathbf{i}\in\text{Vert}_\gamma} : (I_\mathbf{i}, J_\mathbf{i})_{\mathbf{i}\in\text{Vert}_\gamma} \mapsto (I_\mathbf{i} g_\mathbf{i}, g_\mathbf{i}^{-1} J_\mathbf{i})_{\mathbf{i}\in\text{Vert}_\gamma} \tag{52}$$

The $\mathbb{C}^\times$-factor acts by scaling $J_\mathbf{i}, \tilde{B}_e$ while keeping $I_\mathbf{i}, B_e$ intact. The group $(\mathbb{C}^\times)^{b_1} \approx (\mathbb{C}^\times)^{\text{Edge}_\gamma}/(\mathbb{C}^\times)^{\text{Vert}_\gamma}$ acts on the $(B, \tilde{B})$ maps:

$$(t_e)_{e\in\text{Edge}_\gamma} : (B_e, \tilde{B}_e)_{e\in\text{Edge}_\gamma} \mapsto (t_e B_e, t_e^{-1} \tilde{B}_e)_{e\in\text{Edge}_\gamma} \tag{53}$$

4.4.2. *The canonical complexes and bundles*

The vector spaces $V_\mathbf{i}$ form the vector bundles over $\mathfrak{M}_\zeta(\mathbf{v}, \mathbf{w})$. There are also the canonical complexes:

$$\mathcal{C}_\mathbf{i} = \left[V_\mathbf{i} \to W_\mathbf{i} \bigoplus \oplus_{e\in t^{-1}(\mathbf{i})} V_{s(e)} \bigoplus \oplus_{e\in s^{-1}(\mathbf{i})} V_{t(e)} \to V_\mathbf{i} \right] \tag{54}$$

where the first and the second maps are given by:

$$d_2 = J_\mathbf{i} \bigoplus \oplus_{e\in t^{-1}(\mathbf{i})} \tilde{B}_e \bigoplus \oplus_{e\in s^{-1}(\mathbf{i})} B_e ,$$

$$d_1 = I_\mathbf{i} \bigoplus \oplus_{e\in t^{-1}(\mathbf{i})} B_e \bigoplus \oplus_{e\in s^{-1}(\mathbf{i})} -\tilde{B}_e \tag{55}$$

4.4.3. The bi-observables

Define the character (in fact, it is the character of an equivariant complex over $\mathfrak{M}_{\mathbf{k}} \times \mathfrak{M}_\zeta(\mathbf{v}, \mathbf{w})$):

$$\mathcal{F} = \sum_{\mathbf{i} \in \mathrm{Vert}_\gamma} S_\mathbf{i}^* C_\mathbf{i} + M_\mathbf{i}^* V_\mathbf{i} \tag{56}$$

We identify the equivariant parameters of the $G_{\mathbf{w}}$ group with $\boldsymbol{\nu}$, the equivariant parameters for the $\mathbb{C}^\times$ factor with $\varepsilon_1 + \varepsilon_2$, the equivariant parameters for the $(\mathbb{C}^\times)^{\mathrm{Edge}_\gamma}$-group with $m_e + \varepsilon_1 + \varepsilon_2$.

4.4.4. The formula

Finally, the formula for $\mathcal{X}_{\mathbf{w}, \boldsymbol{\nu}}$ reads:

$$\mathcal{X}_{\mathbf{w}, \boldsymbol{\nu}} = \sum_{\mathbf{v}} \mathsf{q}^{\mathbf{v}} \int_{\mathfrak{M}_\zeta(\mathbf{v}, \mathbf{w})} e_{-\varepsilon_2}(T\mathfrak{M}_\zeta(\mathbf{v}, \mathbf{w})) I[\mathcal{F}] \tag{57}$$

Here $I[\mathcal{F}]$ is understood as the $G_{\mathbf{w}} \times (\mathbb{C}^\times)^{1+b_1}$-equivariant cohomology class of $\mathfrak{M}_\zeta(\mathbf{v}, \mathbf{w})$: represent $C_\mathbf{i}$ as the virtual bundle $C_\mathbf{i}^+ - C_\mathbf{i}^-$ over $\mathfrak{M}_\zeta(\mathbf{v}, \mathbf{w})$, where $C_\mathbf{i}^+$, $C_\mathbf{i}^-$ are the actual bundles, with the formal Chern roots $\xi_{\mathbf{i}, \kappa_\pm}^\pm$. Then

$$I[\mathcal{F}] = \prod_{\mathbf{i} \in \mathrm{Vert}_\gamma} \prod_\kappa P_\mathbf{i}(x + v_{\mathbf{i}, \kappa}) \frac{\prod_{\kappa_+} \mathcal{Y}_\mathbf{i}(x + \xi_{\mathbf{i}, \kappa_+}^+)}{\prod_{\kappa_-} \mathcal{Y}_\mathbf{i}(x + \xi_{\mathbf{i}, \kappa_-}^-)} \tag{58}$$

where $v_{\mathbf{i}, \kappa}$ are the formal Chern roots of $V_\mathbf{i}$.

For the examples we considered so far the quiver varieties $\mathfrak{M}_\zeta(\mathbf{v}, \mathbf{w})$ are the cotangent bundles to the Grassmanian variety $T^* Gr(v, w)$ and the Hilbert scheme $Hilb^{[v]}(\mathbb{C}^2)$ of points on $\mathbb{C}^2$, respectively.

5. Applications and implications

One obvious application of our formalism is the solution of the low-energy theories. Indeed, in the limit $\varepsilon_2 \to 0$ the integrals (57) simplify (the Chern polynomial of the tangent bundle drops out). In addition, the sum over all γ-colored partitions (43) is dominated by a single *limit shape*[11] which maps (57) to a system of difference equations for the $\mathcal{Y}_\mathbf{i}$-functions. These equations were studied in.[11] In this case it suffices to study the equations for the dimension vectors $\mathbf{w}$ corresponding to the dominant weights of $\mathfrak{g}_\gamma$.

In the quantum case where $\varepsilon_1, \varepsilon_2$ are finite we need all $\mathbf{w}$. The equations (57) can be viewed as a system of Hirota difference equations, which should fix $\mathcal{Z}$ uniquely. This direction is currently investigated. Note that for the dominant $\mathbf{w}$ for the A-type quivers the similar equations were found in[25] by somewhat different methods.

The K-theoretic version, the (q_1, q_2)-characters are defined in a similar fashion, except that one should use the $\chi_{q_2^{-1}}$-genus instead of the Chern polynomial and push forwards in equivariant K-theory instead of the equivariant integrals. It would be interesting to compare our (q_1, q_2)-characters to the t-deformed q-characters, cf.[26]

Of course, the most interesting question is to extend our formalism of non-perturbative Dyson-Schwinger equations beyond the BPS limit, even beyond the realm of supersymmetric theories.

Another exciting problem is to find the string theory analogue of our $(\varepsilon_1, \varepsilon_2)$-characters and the stringy version of the large field redefinitions.

Acknowledgements

I have greatly benefited from discussions with H. Nakajima and V. Pestun. Research was supported in part by RFBR grants 12-02-00594, 12-01-00525, by Agence Nationale de Recherche via the grant ANR 12 BS05 003 02, by Simons Foundation, and by Stony Brook Foundation.

References

1. H. Nakajima, *Internat. Math. Res. Notices*, 61 (1994).
2. N. Nekrasov and S. Shatashvili, *AIP Conf. Proc.* **1134**, 154 (2009).
3. M. Gutperle and A. Strominger, *JHEP* **0204**, 018 (2002).
4. N. A. Nekrasov, *Adv. Theor. Math. Phys.* **7**, 831 (2004), To Arkady Vainshtein on his 60th anniversary.
5. N. Seiberg and E. Witten, *Nucl. Phys. B* **426**, 19 (1994).
6. N. Seiberg and E. Witten, *Nucl. Phys. B* **431**, 484 (1994).
7. N. Seiberg and E. Witten (1996).
8. N. A. Nekrasov and S. L. Shatashvili (2009).
9. N. Nekrasov and E. Witten, *JHEP* **09**, 092 (2010).
10. N. Nekrasov, A. Rosly and S. Shatashvili, *Nucl. Phys. Proc. Suppl.* **216**, 69 (2011).
11. N. Nekrasov, V. Pestun and S. Shatashvili (2013).
12. A. S. Losev, A. Marshakov and N. A. Nekrasov (2003).
13. N. Nekrasov and A. Okounkov (2003).
14. N. Nekrasov, lecture at the string theory group seminar, University of Amsterdam (Feb. 3, 2004).
15. L. F. Alday, D. Gaiotto and Y. Tachikawa, *Lett. Math. Phys.* **91**, 167 (2010).
16. N. Wyllard, *JHEP* **0911**, 002 (2009).
17. N. Nekrasov and V. Pestun (2012).
18. S. Katz, P. Mayr and C. Vafa, *Adv. Theor. Math. Phys.* **1**, 53 (1998).

19. A. E. Lawrence, N. Nekrasov and C. Vafa, *Nucl.Phys. B* **533**, 199 (1998).
20. H. Knight, *J. Algebra* **174**, 187 (1995).
21. E. Frenkel and N. Reshetikhin, ArXiv Mathematics e-prints (1998).
22. E. Frenkel and N. Reshetikhin, *Recent developments in quantum affine algebras and related topics (Raleigh, NC, 1998)* **248**, 163 (1999).
23. D. Hernandez, *Selecta Math. (N.S.)* **14**, 701 (2009).
24. N. Nekrasov, *Nucl. Phys. B* **531**, 323 (1998).
25. S. Kanno, Y. Matsuo and H. Zhang, *JHEP* **1308**, 028 (2013).
26. H. Nakajima, 196 (2001).

UNDERSTANDING CHIRAL MAGNETIC EFFECT

V. I. SHEVCHENKO*

NRC Kurchatov Institute
ac. Kurchatova sq., 1, Russia 123182 Moscow
* E-mail: Vladimir.Shevchenko@cern.ch

Anisotropy of electric currents fluctuations in magnetic field is discussed using quantum measurements theory language. Possible interpretation of parity-odd phenomena like Chiral Magnetic Effect in this framework is given. It is argued that quantum-to-classical transition caused by decoherence plays crucial role. This talk on Pomeranchuk-100 conference is based on the papers.[1–3]

Keywords: Chiral Magnetic Effect, Heavy Ion Collisions, Decoherence.

1. Introduction

Vacuum of any quantum field theory (and the Standard Model in particular) can often be viewed as a sort of medium. Needless to say that this medium is in many respects different from the conventional media we study in condensed matter physics — in particular, it has no rest frame and it looks identically for all uniformly moving observers. In spite of that, two main approaches to study properties of this (and actually of any) media stay with us since Galileo: one can either send some test particles and look how they move and interact (with each other and with the medium) or one should put some external conditions on the medium (e.g., to heat it) and study what happens. Of particular interest is a question about the fate of discrete symmetries under this or that choice of external conditions.

It is well known that all three most important discrete symmetries: charge conjugation **C**, time reverse **T** and parity **P** are broken both at micro- and at macro-level in our world. First, matter dominates over anti-matter in the Universe. Second, there are several arrows of time: entropic (order → chaos), cosmological (Big Bang → Armageddon), psychological (dreams versus memories), electroweak (CKM complex phase) etc. Third, left and right chiral particles act differently both in the Standard Model Lagrangian (lefts are doublets while rights are singlets) and in biological systems (DNA spiral etc). Let us take a closer look at the parity invariance. The fundamental result due to C.Vafa and E.Witten[4,5] says that vacuum expectation value of any local **P**-odd observable has to vanish[a] in vector-like theories such as QCD, e.g.

$$\langle \bar{\psi}\gamma^5\psi \rangle = 0 \ ; \ \langle G_{\mu\nu}\tilde{G}^{\mu\nu} \rangle = 0 \tag{1}$$

In principle, one would not expect violation of (1) at finite temperature/baryon density. The situation with **P**-parity is different from **C**- and **T**-symmetries in this respect (for example, **C**-invariance is intact at finite temperature, but gets broken at finite density and there is no Furry theorem at nonzero chemical potential) and it is of special interest to study possible **P**-odd effects in strong interaction context under nontrivial conditions. Despite Vafa-Witten theorem puts serious constraints on possible **P**-parity violating phenomena in the domain of strong interactions physics, such effects have been under discussion for a long time. One can mention T.D.Lee's idea of **P**-odd bubbles[15] and A.B.Migdal's hypothesis of pion condensate in nuclei.[7] Closely related effects of $\rho - \pi$ mixing at finite temperature[8] and sphaleron dynamics in QCD[9] were discussed.

In recent years the interest to this topic has been raised again after a series of papers by D.E.Kharzeev and coauthors[10–13] and numerous subsequent publications. The two mentioned approaches to study any vacuum nicely work together in heavy ion collision experiments with respect to the QCD vacuum. Test particles used in these experiments — heavy ions — are able to create, in the first instants after the collision, nontrivial multiparticle state, which itself plays a role of external conditions. Besides temperature and density there is also extremely strong magnetic field, of the order of $(10^3–10^4)$ MeV^2 in about 0.2 Fm/c after the moment of collision.[14] The main qualitative result can be formulated as follows: if by whatever dynamical mechanism there is an excess of quarks of definite chirality inside a fireball, electric current flows along the magnetic field, whose main effect is charge asymmetry of final particles distribution between upper and lower (with respect to the interaction plane) hemispheres. On quantitative level, for free massless spinors with charge e, chemical potentials μ_L, μ_R for left-handed and right-handed ones, respectively, in constant and spatially uniform magnetic field **B** electric current is given by the following expression, known as chiral magnetic effect (CME):

$$\mathbf{J} = \frac{e^2}{2\pi^2}\mu_5\mathbf{B} \ ; \ \mu_5 = \frac{\mu_R - \mu_L}{2} \tag{2}$$

---

[a]It does not mean of course that correlators of the corresponding operators vanish.

In electroweak context the expression (2) was obtained for the first time in[15] and later reappeared in various contexts (see e.g.[16]). It is worth stressing that (2) is a robust theoretical result and as such can be reproduced in many different ways. What is special and important about the result (2) is that proportionality coefficient there is universal and fixed by the famous triangle anomaly.[1,18]

The current (2) has many interesting properties. The most important one is that the corresponding transport coefficient is **P**-odd, but **T**-even. According to general arguments, this means that the current is non-dissipative and the corresponding entropy production rate equals to zero. This has close resemblance with superconductivity and superconducting current, despite the latter is by itself a temperature-dependent phenomenon, while the chiral current (2) is (at least naively) temperature-independent.

In terms of correlators CME can be seen as a consequence of correlation between the vector current and divergence of the axial current. Let us define

$$\Pi^5_{\mu\nu}(k) = i \int d^4x\, e^{ik(x-y)} \left\langle \mathrm{T}\{j_\mu(x) j^5_\nu(y)\}\right\rangle \tag{3}$$

This correlator vanishes in the vacuum: $\Pi^5_{\mu\nu}(k) = 0$, but it does not if external abelian field is applied:

$$k^\nu \Pi^5_{\mu\nu}(k) = \frac{\tilde{F}_{\mu\rho} k^\rho}{2\pi^2}[1 + ...] \tag{4}$$

where the coefficient is fixed by the anomaly and does not depend on the physics behind fluctuations of the currents.

In experiments, both real and numerical,[19] one addresses (2) indirectly, via measuring some **P**-even charge dependent correlations, which comes out, roughly speaking, as **P**-odd effect like (2) squared. The charge-dependent asymmetries have been studied on heavy ion collision facilities in STAR and PHENIX experiments at RHIC and in ALICE experiment at LHC.[20–23] The effect can be described as follows. For noncentral collision one can fix the reaction plane by two vectors: beam momentum and impact parameter. By convention angular momentum of the beams (and the corresponding magnetic field) is orthogonal to the reaction plane, while the azimuthal angle $\phi \in [0, 2\pi)$ is defined in the plane, orthogonal to particles momenta. With this notation, in any particular event one studies charged particles distribution in ϕ using the following conventional parametrization

$$\frac{dN_\pm}{d\phi} \propto 1 + 2v_{1,\pm}\cos\phi + 2v_{2,\pm}\cos 2\phi + 2a_\pm \sin\phi + ... \tag{5}$$

The coefficients $v_{1,\pm}$ and $v_{2,\pm}$ account for the so called direct and elliptic flow. They are believed to be universal for positively and negatively charged particles with good accuracy. The coefficients a_+ and a_- describe asymmetric charge flow along the axis, normal to the reaction plane. This **P**-parity forbidden correlation between a polar vector (electric current) and an axial one (angular momentum) is sometimes interpreted in the literature as a signature of **P**-parity violation in a given event with $a_\pm \neq 0$. On the other hand, the random nature of the process dictates $\langle a_+\rangle_e = \langle a_-\rangle_e = 0$ (there the averaging over events is taken). The averages over events and over azimuthal angle of sine and cosine functions with the weight (5) provides information about correlation coefficients. The most recent results from ALICE experiment can be found in.[24]

2. CME and decoherence

There are a few important questions related to CME, which are left open by the expression (2). Needless to say that the list is subjective.

- How to proceed in a reliable way from nice qualitative picture of CME to quantitative predictions for charge particle correlations measured in experiments?
- What is quantum physics behind μ_5?
- How to disentangle the genuine nonabelian physics from dynamics of free massless fermions in magnetic field?
- How is quantum anomaly in microscopic current encoded in dynamics for macroscopic, effective currents (anomalous hydrodynamics)?

The first thing worth noting in this context is that one should carefully distinguish symmetry breaking caused by dynamics from event-by-event fluctuations. A simple quantum-mechanical analogy can be useful to illustrate the point. In one-dimensional bound state problem with **P**-parity even potential $V(x) = V(-x)$ one has $\langle x\rangle = \int x dx |\psi_0(x,t)|^2 = 0$ where $\psi_0(x,t)$ is the ground state **P**-parity even wave function. On the other hand, performing particle position measurements on ensemble of N identical systems all in the ground state one gets sequence of positive and negative numbers $x_1, x_2, .., x_N$ (with some uncertainties determined by the measuring device properties). Quantum mechanics does not predict the result of a single measurement, but guarantees $\langle x\rangle = \lim_{N\to\infty} \frac{1}{N}\sum_{i=1}^{N} x_i = 0$. For each measurement with the outcome $x_i \neq 0$ one could say that **P**-invariance is broken in this particular experiment, "event-by-event". In this simple case "breaking" is clearly of statistical origin and has nothing to do with dynamics —

i.e. with the properties of the potential $V(x)$. It is common in quantum mechanics not to use such terminology and compute instead nonzero P-parity even observables, e.g. $\langle x^2 \rangle = \int x^2 dx |\psi_0(x,t)|^2 \neq 0$, characterizing the pattern of quantum fluctuations.

Exactly in the same sense quantum average of P-odd quantity like μ_5 has to vanish due to Vafa-Witten theorem and one would have from (2) $\langle \mathbf{J} \rangle = 0$. In a particular "event", however, the current could be nonzero due to both quantum and classical (i.e. thermal and combinatorial) fluctuations. We are interested in a very special P-odd "part" of this fluctuation pattern. Thus we are forced to focus our attention to the process of current measurement.

According to common lore a measurement is a story about interaction between two quantum systems where the first ("an object") has a few degrees of freedom while the second ("a device") has a lot. As is well known the role of the device is often played just by the medium an object is immersed into. Interactions with the medium lead to decoherence and transition from quantum to classical fluctuations in the process of continuous measurement. It is important in this respect not to mix quantum and classical fluctuations. In quantum case all histories (field configurations in QFT context) coexist together and simultaneously. For classical fluctuations one has some concrete random value of fluctuating observable at any given moment of time (and at a point in space). There exist various theoretical frameworks to describe quantum measurements in relativistic setting. We consider two in the following: point-like detectors (Unruh-DeWitt) and filter functions (quantum corridors).

2.1. Point-like detectors

The standard Unruh–DeWitt detector coupled to vector current is described by the Hamiltonian:

$$H = \int\limits_{-\infty}^{\infty} f(\tau) d\tau \, m(\tau) \bar{\psi}(x(\tau)) \Gamma \psi(x(\tau)) \tag{6}$$

Here $x(\tau)$ parameterizes the detector's world-line, τ is the proper time along it, $m(\tau)$ — internal quantum variable (monopole momentum) of the detector whose evolution in τ is described by the standard two-level Hamiltonian with the levels E_0 and E_1, $E_1 - E_0 = \omega > 0$. The Lorentz structure of the coupling is fixed by the matrix Γ, in what follows we concentrate on the vector case, $\Gamma = n_\mu \gamma^\mu$. The dimensionless "window function" $f(\tau)$ encodes the fact that any realistic measurement takes finite time (which we denote λ in this paper), so the window function has the important property $f(\tau \lesssim \lambda) \approx 1$, $f(\tau \gg \lambda) \to 0$. By convention the measurement window is located around the point $\tau = 0$. The interested reader is refereed to[25] for detailed discussion of the finite measurement time effects.

An amplitude for the detector to "click" is given by

$$\mathcal{A} = i \int\limits_{-\infty}^{\infty} f(\tau)d\tau \, \langle 1|m(\tau)|0\rangle \cdot \langle \Omega|j(x(\tau))|\Omega_0\rangle \tag{7}$$

where $j(x(\tau)) = \bar{\psi}(x(\tau))\Gamma\psi(x(\tau))$ and $|\Omega_0\rangle$ stays for initial state of the field sub-system, while $|\Omega\rangle$ represents final (after the measurement) state. The probability is proportional to $|\mathcal{A}|^2$, and the corresponding response function reads:

$$\mathcal{F}(\omega) = \int\limits_{-\infty}^{\infty} f(\tau)d\tau \int\limits_{-\infty}^{\infty} f(\tau')d\tau' \, e^{-i\omega(\tau-\tau')} \cdot G^+(\tau - \tau') \tag{8}$$

where

$$G^+(\tau - \tau') = \langle \Omega_0|j(x(\tau))j(x(\tau'))|\Omega_0\rangle \tag{9}$$

For infinite measurement time, i.e. in the limit $f(\tau) \equiv 1$ one is interested in detector's excitation rate in unit time. It is determined by the power spectrum of the corresponding Wightman function:

$$\dot{\mathcal{F}}(\omega) = \int\limits_{-\infty}^{\infty} ds \, e^{-i\omega s} \, G^+(s) \tag{10}$$

In general case one has to work with the original expression (8).

First, we consider infinite measurement time case with $f(\tau) \equiv 1$. Usually in quantum measurements theory one compares response functions of a given detector in a state of inertial movement versus some non-inertial one. We are interested in another kind of asymmetry, namely between the detector oriented to measure current along the magnetic field direction and perpendicular to it. This choice is fixed by the vector $n_\mu = (0, \mathbf{n})$. With respect to its spatial movement the detector is supposed to be always at rest, so we can take $x(\tau) = (\tau, 0, 0, 0)$. Therefore it is convenient to switch to the coordinate space and as in (10) we denote $s = \tau - \tau'$. It is convenient to use the exact fermion propagator in external magnetic field given by:[26]

$$S(s) = \frac{is\gamma^0}{32\pi^2} \int\limits_0^{\infty} \frac{du}{u^3} \left(\frac{qBu}{\tan(qBu)} + \gamma^1\gamma^2 qBu \right) e^{-i\frac{s^2}{4u}} \tag{11}$$

where q stays for quark electric charge and we take $m = 0$.

The response function asymmetry given by $\delta\dot{\mathcal{F}}(\omega) = \dot{\mathcal{F}}_{33}(\omega) - (\dot{\mathcal{F}}_{11}(\omega) + \dot{\mathcal{F}}_{22}(\omega))/2$ is quadratic in B for all values of the magnetic field.[b] Explicitly, one gets:

$$G_{33}^+(s) = -\left[\frac{s}{16\pi^2}\int_0^\infty \frac{du}{u^3}\frac{qBu}{\tan(qBu)}e^{-i\frac{s^2}{4u}}\right]^2 - \frac{(qB)^2}{16\pi^4 s^2} \tag{12}$$

$$G_{11}^+(s) = -\left[\frac{s}{16\pi^2}\int_0^\infty \frac{du}{u^3}\frac{qBu}{\tan(qBu)}e^{-i\frac{s^2}{4u}}\right]^2 + \frac{(qB)^2}{16\pi^4 s^2} \tag{13}$$

and $G_{11}^+(s) = G_{22}^+(s)$. These results are exact for free fermions in external magnetic field in the massless limit.

To compute the response function one needs to take into account $s \to s - i\epsilon$ prescription corresponding to definition of the Wightman function (9) and switch on the temperature introducing sum over periodic shifts in imaginary time with $\beta = 1/kT$ and Fermi-Dirac statistics factor $(-1)^k$ for fermions:

$$\delta\dot{\mathcal{F}}(\omega) = -\frac{(qB)^2}{8\pi^4}\int_{-\infty}^{+\infty} ds\, e^{-i\omega s}\left[\sum_{k=-\infty}^{+\infty}\frac{(-1)^k}{(s - i\epsilon + ik\beta)}\right]^2 \tag{14}$$

where we denoted $\delta\dot{\mathcal{F}}(\omega) = \dot{\mathcal{F}}_{33}(\omega) - \dot{\mathcal{F}}_{11}(\omega)$.

Taking into account that $\sum_{k=-\infty}^{\infty}\frac{(-1)^k}{x+ik} = \frac{\pi}{\sinh \pi x}$ and doing the integral with the help of residues one gets

$$\delta\dot{\mathcal{F}}(\omega) = \frac{(qB)^2}{4\pi^3}\frac{\omega}{e^{\beta\omega} - 1} \tag{15}$$

Expression (15) is worth commenting. It is positive, which corresponds to the fact that the detector measuring the current along magnetic field clicks more often than measuring perpendicular currents. It is also worth noticing the change of statistics from Fermi-Dirac to Bose-Einstein — what is relevant is the statistic of operators whose fluctuations are being measured by the detector (Bose-currents in our case) and not the statistics of primary fluctuating fields.

The fact that current fluctuations are suppressed in perpendicular direction is obvious from general physics: the charged particle moving in the

---

[b]Notice that $\dot{\mathcal{F}}_{11}(\omega) = \dot{\mathcal{F}}_{22}(\omega)$ for our choice of the field along the third axis.

orthogonal plane is deflected by the magnetic field (or, using quantum mechanical language, confined to Landau levels). What is less obvious is that fluctuations along the field are enhanced (exactly by the same amount), since classically (i.e. neglecting spin effects) magnetic field has no influence on a charge moving in parallel direction. This enhancement is caused by spin interaction with the magnetic field.

At large magnetic fields

$$\frac{\dot{\mathcal{F}}_{33}(\omega) - \dot{\mathcal{F}}_{11}(\omega)}{\dot{\mathcal{F}}_{33}(\omega) + \dot{\mathcal{F}}_{11}(\omega)} \to 1 \tag{16}$$

This pattern is easy to understand in term of the corresponding energy-momentum tensor. Indeed, at zero magnetic field one has the standard thermal pressure for massless fermions

$$T_{11} = T_{22} = T_{33} = \frac{7\pi^2 T^4}{180} \tag{17}$$

which is isotropic. At large magnetic field, however, all the pressure is along the magnetic field and there is no pressure in orthogonal directions:

$$T_{11} = T_{22} \to 0 \ ; \ T_{33} = \frac{qBT^2}{12} \tag{18}$$

It is interesting to note that for strong but slowly varying magnetic field the plasma as a whole is to experience buoyancy force in the direction of the field gradient:

$$f_3 = - \int_V d^3 x \frac{\partial T_{33}}{\partial z} \tag{19}$$

This effect can be called magnetic Archimedes force. Since in real scattering events the fields are indeed highly inhomogeneous, this effect could be important for such phenomena as charge dependence of the elliptic flow etc.

Now, let us study effects of finite measurement time. Again, we are to compute the main quantity of our interest: $\delta\mathcal{F}(\omega)$, the difference of the response functions corresponding to the currents parallel to the magnetic field to transverse ones. Roughly speaking, this quantity tells us which detector clicks more often (or, in other words, which one has more clicks for a given time interval): the detector measuring the currents along the field or the one oriented to measure the orthogonal currents. This difference is an integral of the following integrand:

$$\delta G^+(\tau - \tau') = \langle \Omega_0 | j_3(x(\tau))j_3(x(\tau')) - j_1(x(\tau))j_1(x(\tau')) | \Omega_0 \rangle =$$

$$= J^2 + \frac{(eB)^2}{8\pi^4} \frac{1}{(\tau - \tau')^2} \tag{20}$$

where J is given by (2). The first term corresponds to the usual CME current and due to its stationary nature it cannot be detected by stationary detector. The same is true for the second term, since

$$\int_{-\infty}^{\infty} ds \, e^{-i\omega s} \frac{1}{(s - i\xi)^2} = 0 \tag{21}$$

for $\omega > 0$. It this respect the situation at finite density is different from that at finite temperature. The thermal state has excitations of any desired energy (corresponding to poles along imaginary time axes in integral (10)), which can excite the detector. The finite density state is a state of the lowest energy which has to be excited "by hands" (i.e. by nontrivial $f(\tau)$) to get observable results.

Inserting (20) into (8) we get the following answer:

$$\delta \mathcal{F}(\omega) = \frac{(eB)^2}{4\pi^4} \left[\mu_5^2 I_0 - \frac{1}{2} I_2 \right] \tag{22}$$

where

$$I_n = \int f(\tau)d\tau \int f(\tau')d\tau' e^{-i\omega(\tau - \tau')} \cdot \frac{1}{(\tau - \tau' - i\xi)^n} \tag{23}$$

Once again, the result (22) is exact in B, i.e. the asymmetry $\delta \mathcal{F}(\omega)$ gets no contributions from higher powers of magnetic field.

The integral (23) depends on typical measurement time λ encoded in $f(\tau)$ and also on dimensionless variables $\omega\lambda$ and ξ/λ. The infinitesimal parameter ξ from Wightman's prescription has physical meaning of inverse maximal particle energy which can be measured by the detector and which is necessary a finite quantity for any realistic detector. From another point of view, it can be seen as related to the finite size of the detector. The quantum measurement theory with finite measurement windows has physical sense and can be applicable only for $\lambda \gg \xi$. However to keep $\xi \neq 0$ is important to get formally correct $\lambda \to 0$ limit, i.e. when the detector is not switched on at all and should, correspondingly, return zero count.

The concrete form of (23) depends on a chosen detector switching time profile. In the problem in question it is natural to identify it with the time profile of the magnetic field itself. The latter was computed by many authors, for example, it was suggested in[27] to use the following Ansatz:

$$B(\tau) = \frac{B_0}{(1 + (\tau/\lambda)^2)^{3/2}} \tag{24}$$

where B_0 and λ are functions of impact parameter and rapidity, whose typical scale for RHIC setup is given by $B_0 \sim 10^5 \, \text{MeV}^2$, $\lambda \sim 0.1 \, \text{Fm/c}$. It is of course not quite correct simply replace $B \to B(\tau)$ in (20) since the exact Green's function is written for time-independent B. But since our approach anyway depends on a detector's model, we consider it is a reasonable approximation to use (24) as a profile for the window function $f(\tau)$. The result for (23) with $n = 0$ follows trivially:

$$I_0 = 4\omega^2\lambda^4 \, K_1^2(\kappa) \tag{25}$$

where $\kappa = \omega\lambda$. The maximum of this function in κ (i.e. optimal measurement time) is reached at $\lambda \approx 1.33/\omega$. It is easy to see that both at $\lambda \to 0$ (no measurement at all) and $\lambda \to \infty$ (stationary measurement) I_0 vanishes.

We now can rewrite (22) as

$$\delta\mathcal{F}(\omega) = \frac{(eB)^2}{4\pi^4}I_0\left[\mu_5^2 + \frac{1}{\lambda^2}g(\kappa)\right] \tag{26}$$

with the dimensionless function $g(\kappa)$ given by

$$g(\kappa) = \frac{\kappa^2}{2}\int_1^\infty dy\,y^2(y-1)\left(\frac{K_1(y\kappa)}{K_1(\kappa)}\right)^2 \tag{27}$$

We have put $\xi = 0$, assuming finiteness of κ.

The expression in square brackets in (26) can be called effective axial chemical potential. The physical meaning of it is quite transparent: by energy-momentum uncertainty relation the finite observation time λ makes quarks Fermi energies uncertain, Dirac sea becomes wavy, and these fluctuations provide the vector current excess along the magnetic field even if "bare" axial chemical potential is absent. While the concrete form of the functions I_0, $g(\kappa)$ depends, of course, on the chosen window function profile, the qualitative form of the result (26) is robust.

In quantitative terms, $g(\kappa \sim 1) \approx 0.15 \div 0.25$ and the model (26) with (27) corresponds to effective axial chemical potential of order of 1 GeV, even if "bare" $\mu_5 = 0$.

2.2. Filter functions

The formalism of quantum corridors introduced in seminal paper[28] allows to take a look at the problem of interest from another prospective. Imagine one is monitoring some **P**-odd observable for a given quantum system. Then it can lead to nonzero result for measurement of correlated **P**-odd quantity.

The simplest way to see it is to use a language of decoherence functionals[28,29] and path integral formalism. Generally, for some filter function $\alpha[\Phi]$ the amplitude is given by

$$\Psi[\alpha] = \int \mathcal{D}\Phi \, \alpha[\Phi] \, e^{iS[\Phi]} \qquad (28)$$

To illustrate the point again on quantum-mechanical example, consider three-dimensional system given by Lagrangian $L = \dot{q}^2 - V(q)$, where $q = (x, y, z)$ and the potential $V(q)$ is invariant under **P**-parity transformation: $V(q) = V(-q)$, but not invariant under separate reflections $x \to -x$ or $y \to -y$ or $z \to -z$. Suppose that an external observer is monitoring the y-coordinate continuously in time. To describe this situation in path integral representation one has to introduce quantum corridor:

$$\int \mathcal{D}y(t) \to \int \mathcal{D}y(t) \, \exp\left(-\zeta \int\limits_0^T (y(t) - \bar{y}(t))^2 dt\right) \qquad (29)$$

where the corridor width is given by $\Delta y \propto (\zeta T)^{-1/2}$. Consequently, all amplitudes and correlators computed with (1.11) become dependent on the function $\bar{y}(t)$ (which has the meaning of continuous observation result) and quantum corridor width Δy. For example, for x-coordinate one would have

$$\langle x(T) \rangle = X[\bar{y}(t), \zeta] \qquad (30)$$

where the functional $X[\bar{y}(t), \zeta]$ depends, generally speaking, on the function $\bar{y}(t)$ in all past moments of time and vanishes at the point $\zeta = 0$, corresponding to no measurement: $X[\bar{y}(t), 0] = 0$. Its exact form of course depends on the potential $V(q)$ and is of no importance for us at the moment. What is crucial is the fact that monitoring **P**-odd quantity (coordinate y in our example) can result in nonzero quantum average for some other **P**-odd quantity (coordinate x), despite the interaction is still strictly **P**-even. One can say that **P**-parity is broken by measuring apparatus.

Coming back to our discussion of CME in the previous sections it is clear that crucial missing ingredient is of course the fact that in strong interaction domain the singlet axial vector current is not conserved because of triangle nonabelian anomaly:

$$\partial^\nu j_\nu^5(x) = -\eta(x) = -\frac{g^2 N_f}{16\pi^2} G_{\alpha\beta}^a(x)\tilde{G}^{a\alpha\beta}(x) \qquad (31)$$

We are interested to find common distribution for the vector current and some **P**-odd quantity, which we have chosen in this section to be the field

$\eta(x)$ from (31). The corresponding amplitude reads:

$$\Psi[\lambda, \zeta] = \int \mathcal{D}\bar{\psi}\mathcal{D}\psi\mathcal{D}A_\mu \, e^{iS_{QCD}+i\int dx\lambda(x)n_\mu j^\mu(x)+i\int dx\zeta(x)\eta(x)} \qquad (32)$$

The vector current is given by the standard expression $j_\mu = \bar{\psi}Q\gamma_\mu\psi$, where Q is quark electric charges diagonal matrix in flavor space. The closed-time-path functional is given by

$$e^{iW[\lambda,\zeta;\lambda',\zeta']} = \Psi[\lambda, \zeta]\Psi^*[\lambda', \zeta'] \qquad (33)$$

and the mean current is

$$\langle n_\mu j^\mu(x)\rangle[\lambda, \zeta] = -i\,\frac{\delta}{\delta\lambda(x)}e^{iW[\lambda,\zeta;\lambda',\zeta']}\bigg|_{\substack{\zeta=\zeta' \\ \lambda=\lambda'}} \qquad (34)$$

It is a functional of **P**-even field $\lambda(x)$ and **P**-odd field $\zeta(x)$ in the same sense as $\langle x(T)\rangle$ from (30) is a functional of $\bar{y}(t)$.

It is easy to compute $\Psi[\lambda, \zeta]$ in Gaussian approximation. It reads:

$$\Psi_{Gauss}[\lambda, \zeta] = e^{\frac{i}{2}\int dp(\lambda(p),\zeta(p))D(p)(\lambda(-p),\zeta(-p))^T} \qquad (35)$$

where

$$D(p) = \begin{pmatrix} \Pi(p) & \Delta(p) \\ \Delta(p) & \Pi^5(p) \end{pmatrix} \qquad (36)$$

with the components

$$\Pi(p) = i\int dx\, e^{ipx}\langle T\{j_\mu(x)j_\nu(0)\}\rangle\, n^\mu n^\nu \qquad (37)$$

$$\Pi^5(p) = i\int dx\, e^{ipx}\,\langle T\{\eta(x)\eta(0)\}\rangle \qquad (38)$$

$$\Delta(p) = \frac{e^2}{2\pi^2}n^\mu p^\alpha \tilde{F}_{\alpha\mu}\cdot N_c\text{Tr}Q^2 \qquad (39)$$

The non-diagonal terms of the matrix $D(p)$ arise due to correlation of fluctuations of the quantities of opposite **P**-parity in external abelian field.

To make the above picture suitable for concrete computations let us take a model profile for the ζ-field, corresponding to the measurement taking place inside 3-dimensional "decoherence volume" $V \sim R^3$ for the time period τ starting from the moment $t = 0$. We chose Gaussian Ansatz

$$\zeta(t, \mathbf{x}) = \zeta \cdot f(t, \mathbf{x}) = \zeta \cdot \exp\left(-\mathbf{x}^2/2R^2\right)\exp\left(-t^2/2\tau^2\right) \qquad (40)$$

It leads to the following expression for the current parallel to magnetic field:

$$\langle j_3(t, \mathbf{x})\rangle[0, \zeta] = -\zeta B \cdot \frac{N_c\text{Tr}Q^2}{2\pi^2}\cdot\frac{t\cdot f(t, \mathbf{x})}{\tau^2}\cdot e^{-\int dp\,\zeta(p)\Im\{\Pi^5(p)\}\zeta(-p)} \qquad (41)$$

where we switched off the **P**-even filter ($\lambda = 0$), but has kept the **P**-odd one. The above expression is a generalization of (2). The current is linear both in ζ and in B and vanishes being integrated over ζ in symmetric limits. Notice that due to the form-factor the current flows only inside the volume (where the measurement has been done).

Expression (41) remarkably demonstrates two different faces of decoherence. On one hand, the mere existence of this quantum current is due to classical nature of the field ζ. On the other hand, the last exponent in the right hand side of (41) is responsible for current damping due to decoherence. This is analogous to measuring electric current by applying external electromagnetic field. If the field is weak and/or momentum of its dominant Fourier modes k is small, linear response theory works well and one gets conductivity as proportionality coefficient between applied field and induced current via Green-Kubo formula. If however the quanta of the field have $k^2 > 4m^2$, they decay into real charged particles with the mass m instead of being absorbed by charged current carriers and the linear response picture is not valid anymore. Quantitatively this process is controlled by the discussed factor with the imaginary part of the corresponding polarization operator in the exponent.

3. P-odd effects in central collisions

We have shown in the previous sections that finite time effects as well as thermal fluctuations can lead to charge fluctuation asymmetry very similar to the one we would expect for CME. Since the effect is related to abelian anomaly, one might say that CME *by definition is nothing but these asymmetries*. This would not be terminologically correct, however. In the original framework CME has been seen as a manifestation of **P**-odd effects *of nonabelian nature*. In some sense, the role of magnetic field in noncentral collisions is just to make **P**-odd chirality imbalance manifest by electric charge asymmetry of final particles, while the imbalance itself (encoded, e.g. in nonzero μ_5) is supposed to be generated by some nonperturbative mechanism of QCD origin. In the former cases, on the contrary, free fermions have been considered. Roughly speaking, we demonstrated that one does not need to have nonzero μ_5 to get CME-like charge asymmetry. This however does not at all mean, that there is no effective nonzero μ_5 in real heavy-ion collision events and all the corresponding **P**-odd physics like CME (understood in its original way) related to that. In the light of the above it is worth to think (to rethink, see e.g.[31]) about possible **P**-odd signatures, *not related* to the magnetic field.

The following analogy could be useful. In order to measure the standard electric conductivity of some medium one can proceed along the following two ways. The first recipe is to apply external electric field $\mathcal{E}(\omega)$, measure induced electric current $J(\omega)$ and get the (complex) resistivity from the Ohm's law $\mathcal{E}(\omega) = Z(\omega) \cdot J(\omega)$. The second way is according to Nyquist — measure fluctuations (thermal and quantum) of electric current in the medium *without* any external field and extract conductivity from

$$\left(J^2\right)_\omega = \hbar\omega \frac{R(\omega)}{|Z(\omega)|^2} \cot \frac{\hbar\omega}{2kT} \tag{42}$$

where $R(\omega) = \Re\{Z(\omega)\}$. The relation between the above two approaches is of course the essence of fluctuation-dissipation theorem.[32] Despite the chiral conductivity is **T**-even and the corresponding current is non-dissipative, general linear response theory arguments are applicable here as well. Indeed, one has for the induced current in linear response approximation:[27]

$$\langle j_\mu(x)\rangle = \int d^4x' \, \Pi^{(R)}_{\mu\nu}(x,x') A^\nu(x') \tag{43}$$

where the retarded polarization operator is given by

$$\Pi^{(R)}_{\mu\nu}(x,x') = i\theta(x_0 - x'_0)\langle[j_\mu(x), j_\nu(x')]\rangle \tag{44}$$

Assuming there is a distinguished rest frame, parameterized by unit vector u_μ, the Fourier transform of (44) can be decomposed as follows (we are using tilde sign to distinguish Fourier components in what follows):

$$\tilde\Pi^{(R)}_{\mu\nu}(q,u) = \sum_{i=1}^{3} \Psi^{(i)}_{\mu\nu} \, \tilde\Pi_i(q^2, qu) \tag{45}$$

with the following tensor structures:

$$\Psi^{(1)}_{\mu\nu} = g_{\mu\nu}q^2 - q_\mu q_\nu$$
$$\Psi^{(2)}_{\mu\nu} = \left((qu)q_\mu - q^2 u_\mu\right)\left((qu)q_\nu - q^2 u_\nu\right)$$
$$\Psi^{(3)}_{\mu\nu} = i\epsilon_{\mu\nu\alpha\beta}q^\alpha u^\beta$$

The decomposition (45) respects current conservation $q^\mu\tilde\Pi^{(R)}_{\mu\nu}(q,u) = 0$ and Bose-symmetry $\tilde\Pi^{(R)}_{\mu\nu}(q,u) = \tilde\Pi^{(R)}_{\nu\mu}(-q,u)$. The electric chiral current corresponds to the third term of this decomposition:

$$\langle j_\mu(x)\rangle_\chi = \int \frac{d^4q}{(2\pi)^4} e^{-iqx} \, \tilde\Pi^{(3)}(q^2, qu)u^\beta \tilde F_{\mu\beta}(q) \tag{46}$$

where

$$\tilde F_{\mu\beta}(q) = \frac{1}{2}\epsilon_{\mu\beta\nu\alpha} \int d^4x \, e^{iqx} \, F^{\nu\alpha}(x) \tag{47}$$

For free fermions with chemical potentials μ_L, μ_R the chiral conductivity is exactly computed in this way in[27] and in static limit coincides with (1), as it should be:

$$\lim_{q \to 0} \tilde{\Pi}^{(3)}(q^2, qu) = \frac{e\mu_5}{2\pi^2} \tag{48}$$

Without loss of generality one can take $u_\mu = (1, 0, 0, 0)$. To extract a part proportional to the invariant function $\Pi^{(3)}$ it is convenient to rewrite the definition (44) for particular spatial components of the commutator $i = 1$, $k = 2$ and to choose $(x - x') = (t, 0, 0, z)$ with $t > 0$. The result reads:

$$\langle [j_1(t, z/2), j_2(0, -z/2)] \rangle = \int \frac{d^4q}{(2\pi)^4} \, e^{-i(q_0 t - q_3 z)} q_3 \, \tilde{\Pi}^{(3)}(q^2, qu) \tag{49}$$

Of course, any $\mathbb{O}(3)$-rotated choice of indices instead of (49) is equally legitimate.

The formula (49) expresses correlation of current fluctuations of some particular form in terms of the transport coefficient of interest. One can check that both sides of this expression change sign for $z \to -z$ and vanish at $z = 0$. The skew-symmetric form of the l.h.s of (49) corresponds to non-dissipative nature of these correlations. It is also convenient to rewrite (49) in coordinate form:

$$\Pi^{(3)}(t, t) - \Pi^{(3)}(t, 0) = i \int\limits_0^t dz \langle [j_1(t, z), j_2(0, 0)] \rangle \tag{50}$$

To get possible intuitive physical picture behind (49), imagine outgoing radial flow of massless fermions, all of the same chirality, emitted from the origin and absorbed ("detected") by a rigid spherical shell. For uniform flow total angular momentum of the shell after interaction is equal to zero by symmetry. Suppose now that the shell is cut along its equator.[c] Then after interaction both hemispheres start to rotate with angular momenta equal in magnitude and opposite in sign. It is obvious that for chirally balanced matter (with $\mu_5 = 0$) there is no such effect. Observation of the hemispheres rotation in such experiment would be a direct sign of **P**-odd physics behind. All that is in some correspondence with such phenomena as Chiral Magnetic Spiral[33] and Chiral Magnetic Wave[34] but there is no magnetic field in the discussed example.

---

[c]In terms of (49) that equator plane has $z = 0$. Notice that due to spherical symmetry of the problem the choice of equator is arbitrary.

It would be tempting to try *to measure* the r.h.s. of (50) and hence be able *to compute* $\Pi^{(3)}$ without any reference to non-centrality and magnetic field (i.e. to proceed in Nyquistic way, not in Ohmic one). However one could not expect simple correspondence between expressions like (50) containing quantum commutators and final particles distributions which are essentially classical.[d]

As is well known the final particle distribution in any heavy-ion collision event is conventionally parameterized as $dN(y, p_\perp, \phi)$ where y is rapidity $y = (\log(E + p_\parallel) - \log(E - p_\parallel))/2$, ϕ — azimuthal angle, and $p_\perp$, $p_\parallel$ — transverse and longitudinal momenta, respectively. **P**-inverse of this fluctuation map can be obtained by the replacement $y \to -y$, $\phi \to \pi + \phi$ and in general it does not, of course, coincide with the original distribution dN even for spherical nuclei experiencing exact central collision because of quantum, thermal and combinatorial fluctuations, which present at all stages starting from initial state of colliding nuclei till the freeze-out phase. One can introduce positive definite distance between the original and **P**-inverted distributions and study it as a function of collision's parameters (energy, type of nuclei etc). The question to answer is whether also *additional* source of **P**-odd asymmetry, related to internal strong interaction dynamics, is present, *besides* all mentioned above statistical factors.

4. Conclusion

Isaak Pomeranchuk often said to his pupils: "Study vacuum, this boiling operator liquid!". In 1951 he published an article[35] entitled "On the theory of multiple particle production in a single collision". This paper was in line with seminal work of E.Fermi, L.D.Landau and others on hydrodynamic theory of multiple production of particle in high-energy collisions. Much later it was experimentally discovered, to a great surprise for many, that the medium created in heavy ion collisions indeed had many properties of liquid with extremely low viscosity. We have seen Renaissance of these old ideas about importance of classical patterns for understanding the dynamics of such extremely quantum object as quark-gluon liquid. The central role here is played by the decoherence. As we tried to argue, this quantum-to-classical transition could also be crucial for interpreting such a delicate phenomenon as Chiral Magnetic Effect.

---

[d]Like, for example, measuring commutator of x- and y-components of angular momentum $\langle [L_x, L_y] \rangle$ has to do with measuring L_z, but not L_x and L_y.

166

References

1. V. D. Orlovsky and V. I. Shevchenko, Phys. Rev. D **82** 094032 (2010)
2. V. Shevchenko, Nucl. Phys. B **870** 1 (2013)
3. V. Shevchenko, Annals of Physics **339** 371 (2013)
4. C. Vafa and E. Witten, Nucl. Phys. B **234** 173 (1984)
5. C. Vafa and E. Witten, Phys. Rev. Lett. **53** 535 (1984)
6. T. D. Lee and G. C. Wick, Phys. Rev. **148** 1385 (1966)
7. A. B. Migdal, Rev. Mod. Phys. **50** 107 (1978)
8. M. Dey, V. L. Eletsky and B. L. Ioffe, Phys. Lett. B **252** 620 (1990)
9. L. D. McLerran, E. Mottola and M. E. Shaposhnikov, Phys. Rev. D **43** 2027 (1991)
10. D. Kharzeev, R. D. Pisarski and M. H. G. Tytgat, Phys. Rev. Lett. **81** 512 (1998)
11. D. Kharzeev and R. D. Pisarski, Phys. Rev. D **61** 111901 (2000)
12. D. Kharzeev, A. Krasnitz and R. Venugopalan, Phys. Lett. B **545** 298 (2002)
13. D. Kharzeev, Phys. Lett. B **633** 260 (2006)
14. V. Skokov, A. Y. .Illarionov and V. Toneev, Int. J. Mod. Phys. A **24** 5925 (2009)
15. A. Vilenkin, Phys. Rev. D **22** 3080 (1980)
16. M. Giovannini and M. E. Shaposhnikov, Phys. Rev. D **57** 2186 (1998)
17. S. L. Adler, Phys. Rev. **177** 2426 (1969)
18. J. S. Bell and R. Jackiw, Nuovo Cim. A **60** 47 (1969)
19. P. V. Buividovich *et al.*, Phys. Rev. D **80** 054503 (2009)
20. S. A. Voloshin, Phys. Rev. C **62** 044901 (2000)
21. S. A. Voloshin, Phys. Rev. C **70** 057901 (2004)
22. S. A. Voloshin [STAR Collaboration], Nucl. Phys. A **830** 377C (2009)
23. G. Wang [STAR Collaboration], Nucl. Phys. A **830** 19C (2009)
24. B. Abelev *et al.* [ALICE Collaboration], Phys. Rev. Lett. **110** 012301 (2013)
25. L. Sriramkumar and T. Padmanabhan, Class. Quant. Grav. **13** 2061 (1996)
26. J. S. Schwinger, Phys. Rev. **82** 664 (1951)
27. D. E. Kharzeev and H. J. Warringa, Phys. Rev. D **80** 034028 (2009)
28. R. P. Feynman and F. L. Vernon, Jr., Annals Phys. **24** 118 (1963)
29. E. Calzetta and B. L. Hu, arXiv:hep-th/9501040.
30. D. E. Kharzeev and H. -U. Yee, arXiv:1207.0477 [cond-mat.mes-hall].
31. S. A. Voloshin, Phys. Rev. C **62** (2000) 044901
32. H. B. Callen and T. A. Welton, Phys. Rev. **83** 34 (1951)
33. G. Basar, G. V. Dunne and D. E. Kharzeev, Phys. Rev. Lett. **104** 232301 (2010)
34. D. E. Kharzeev and H. -U. Yee, Phys. Rev. D **83** 085007 (2011)
35. I. Y. .Pomeranchuk, Dokl. Akad. Nauk Ser. Fiz. **78** 889 (1951)

A NEW METHOD FOR SOLVING THE Z > 137 PROBLEM AND FOR DETERMINATION OF ENERGY LEVELS OF HYDROGEN-LIKE ATOMS

V.P. NEZNAMOV[*] and I.I. SAFRONOV

*Russian Federal Nuclear Center – All-Russian Research Institute of Experimental Physics, 37, Mira eve., Sarov, Nizhny Novgorod region, 607188, Russia
Email: neznamov@vniief.ru

The "catastrophe" in solving the Dirac equation for an electron in the field of a point electric charge, which emerges for the charge numbers Z > 137, is removed in this work by new method of accounting of finite dimensions of nuclei. For this purpose, in numerical solutions of equations for Dirac radial wave functions, we introduce a boundary condition at the nucleus boundary such that the components of the electron current density is zero. As a result, for all nuclei of the periodic table the calculated energy levels practically coincide with the energy levels in standard solutions of the Dirac equation in the external field of the Coulomb potential of a point charge. Further, for $Z > 105$, the calculated energy level functions $E(Z)$ are monotone and smooth. The lower energy level reaches the energy $E = -mc^2$ (the electron "drop" on a nuclei) at $Z_c = 178$. The proposed method of accounting of the finite size of nuclei can be easily used in numerical calculations of energy levels of many-electron atoms.

1. Introduction

A century ago, in 1913, Niels Bohr developed the postulates of a new quantum theory. As early as in three years, based on the theory of Bohr's orbits, A.Sommerfeld [1] developed a formula for the fine structure of energy levels of hydrogen-like atoms,

$$E = \frac{mc^2}{\left(1 + \dfrac{\alpha_{em}^2 Z^2}{\left(n - |\kappa| + \sqrt{\kappa^2 - \alpha_{em}^2 Z^2}\right)^2}\right)^{1/2}}. \tag{1}$$

Following the development of the Dirac theory in 1928, Dirac [2], Darwin [3] and Gordon [4] obtained expression (1) as a result of exact solution of the Dirac equation in the Coulomb field of a point charge $(-Ze)$.

In (1), m is the mass of electron, c is the speed of light, $\alpha_{em} = e^2/\hbar c$ is the electromagnetic constant of the fine structure, Z is the atomic number, $n = 1, 2...$ is the main quantum number, κ is the quantum number of the Dirac equation:

$$\kappa = \pm 1 \pm 2... = \begin{cases} -(l+1), & j = l + \dfrac{1}{2} \\ l, & j = l - \dfrac{1}{2} \end{cases}. \tag{2}$$

In (2), j, l are the quantum numbers of the total and orbital momentum of the electron.

Formula (1) is became a complex number if

$$Z > \frac{|\kappa|}{\alpha_{em}} \simeq 137|\kappa|. \tag{3}$$

From the practical viewpoint of the existence of real nuclei in the periodic table, of interest in (3) are the electron states with $|\kappa| = 1$, i.e. the $1S_{1/2}$ and $2P_{1/2}$- states. For these states, the complexity of energy levels in (1) is often called the "$Z > 137$ catastrophe".

It was established fairly quickly that the "catastrophe" results from the ignorance of the finite size of the nuclei.

In 1945, Pomeranchuk and Smorodinsky [5] considered an atomic system with potential

$$U = \begin{cases} -\dfrac{Ze^2}{r_N} & \text{for } r \le r_N \\ -\dfrac{Ze^2}{r} & \text{for } r > r_N, \end{cases} \tag{4}$$

where r_N is the nucleus radius.

As a result, they estimated Z_c, at which the lower energy level of the $1S_{1/2}$- state reaches the limiting value of $E = -mc^2$.

$$Z_c = 175 \text{ at } r_N = 0,8 \cdot 10^{-12} \text{ cm.} \tag{5}$$

This lead to an important conclusion that in the range of $Z_c \ge Z > 137$ there must exist a real function of $E(Z)$, and the "catastrophe" in (1) indeed occurs as a result of the ignorance of the finite size of nuclei.

In 1959, Zeldovich [6] demonstrated that variations in the Coulomb potential near the origin of coordinates produce minor effects on the energy spectrum of hydrogen-like atoms.

An overview of subsequent papers devoted to the structure of hydrogen-like atoms at $Z\alpha > 1$ is presented in [7], [9], [10].

In [7], to analyze the structure of energy levels, in addition to the potential (4) Zeldovich and Popov used a potential corresponding to the potential of a uniformly charged sphere.

$$U_1 = \begin{cases} \dfrac{Ze^2}{r_N}\left(-\dfrac{3}{2}+\dfrac{1}{2}\left(\dfrac{r}{r_N}\right)^2\right) & \text{for } r \leq r_N \\ -\dfrac{Ze^2}{r} & \text{for } r > r_N \end{cases} \tag{6}$$

The authors [8] numerically calculated the energy levels of the first nine states $\left(1S_{\frac{1}{2}}, 2S_{\frac{1}{2}}......3D_{\frac{5}{2}}\right)$ as a function of Z for the potential (6). The value of Z_c determined in [8] is $Z_c = 169$ for $r_N = 9.5 \cdot 10^{-12}$ cm. This value is close to the values of $Z_c = 170 \div 175$ obtained by other researchers (see [5], [7]).

Now it is known more than 30 electrostatic potentials, which take into account the finite distribution of electric charge in the atom nuclei, have been offered by the different authors. These potentials are used in the various machine codes for determination of electronic structure of atoms and molecules. A review of developed potentials and their use in numerical calculations of Dirac and Schrödinger equations are in [9] (see also [10]). In the reviews [9], [10] there are also the wide range of literature of analytical and numerical determination of electronic structure of atoms and molecules.

The solutions of the Dirac and Schrödinger equations with using of the finite potentials of the atomic nucleus are determined by the standard method. Firstly the wave function of electrons are calculated within a nucleus in a filed of electrostatic potential of interest. Then, the values of these functions at the boundary of the nucleus are equated with the similar values of wave function of electrons in the Coulomb field. The boundary condition for radial wave at functions $r \to \infty$ and $r = 0$ determine the energy spectrum of the atomic and molecular systems.

According to [5]–[9] the use of different introduced electrostatic potentials of atomic nuclei lead to relatively little change (tenths of a percent) both absolute values of the energy levels, and differences of the energy levels. These changes grow when Z increases.

In the present paper, the problem of determination of the energy spectrum of hydrogen-like atoms, including nucleus with $Z > 137$, is solved by a new approach to numerical calculations of the Dirac equation in the Coulomb field by introducing a boundary condition for wave functions at the boundary of the nuclei of interest.

The boundary condition at the nucleus boundary is taken by analogy with the analysis of the possibility of existence of stationary bound states in the Schwarzschild gravitational field [7]. It involves zeroing of the φ-component of Dirac current density at the boundary of the nucleus of interest, which resolves itself in zeroing of one of two radial wave functions at the nucleus boundary in the Coulomb field. In this case the calculations are simplified and its are made from $r \to \infty$ up to the boundary of the nucleus r_N.

This paper has the following structure. For completeness of presentation, Section 2 contains the Dirac equation in the Coulomb field, outlines the procedure of separation of variables, and gives a system of equations for radial wave functions.

Section 3 explores the behavior of the components of the vector of current density of Dirac particles and introduces the boundary condition for wave functions at the boundary of the nucleus.

Section 4 reviews the results of numerical calculations of energy spectra of hydrogen-like atoms with various Z.

The Conclusion summarizes the results of this study.

2. Dirac Equation in the Coulomb Field of the Charge $(-Ze)$

Below we will use the system of units $\hbar = c = 1$, the signature

$$g^{\alpha\beta} = diag[1, -1, -1, -1], \tag{7}$$

β, α^k, $k = 1,2,3$ are 4×4 Dirac matrices in the Dirac-Pauli representation, and σ^k are 2×2 Pauli matrices.

We consider the stationary case, when the wave function can be written as $\psi(\mathbf{r},t) = \psi(\mathbf{r})e^{-iEt}$.

The Dirac equation in the Coulomb field of the point charge $(-Ze)$ in spherical coordinates (r,θ,φ) can be expressed as:

$$E\psi(\mathbf{r}) = \left[\beta m - i\alpha^1\left(\frac{\partial}{\partial r} + \frac{1}{r}\right) - i\alpha^2\frac{1}{r}\left(\frac{\partial}{\partial\theta} + \frac{1}{2}\mathrm{ctg}\,\theta\right) - \right.$$
$$\left. -i\alpha^3\frac{1}{r\sin\theta}\frac{\partial}{\partial\varphi} - \frac{Ze^2}{r}\right]\psi(\mathbf{r}). \tag{8}$$

Eq. (8) allows for the separation of variables, if the bispinor $\psi(\mathbf{r}) = \psi(r,\theta,\varphi)$ is given by

$$\psi(r,\theta,\varphi) = \begin{pmatrix} F(r)\xi(\theta) \\ -iG(r)\sigma^3\xi(\theta) \end{pmatrix}e^{im_\varphi\varphi} \tag{9}$$

and the following equation is used (see, e.g., [12]):

$$\left[-\sigma^2\left(\frac{\partial}{\partial\theta} + \frac{1}{2}\mathrm{ctg}\,\theta\right) + i\sigma^1 m_\varphi\frac{1}{\sin\theta}\right]\xi(\theta) = i\kappa\xi(\theta). \tag{10}$$

In (9), (10), $\xi(\theta)$ are spherical harmonics for spin ½, m_φ is the magnetic quantum number, κ is the quantum number (2). $\xi(\theta)$ can be represented as in [13].

$$\xi(\theta) = \begin{pmatrix} _{-\frac{1}{2}}Y_{jm_\varphi}(\theta) \\ _{\frac{1}{2}}Y_{jm_\varphi}(\theta) \end{pmatrix} = (-1)^{m_\varphi+\frac{1}{2}}\sqrt{\frac{1}{4\pi}\frac{(j-m_\varphi)!}{(j+m_\varphi)!}}\begin{pmatrix} \cos\dfrac{\theta}{2} & \sin\dfrac{\theta}{2} \\ -\sin\dfrac{\theta}{2} & \cos\dfrac{\theta}{2} \end{pmatrix}\times$$
$$\times\begin{pmatrix} \left(\kappa - m_\varphi + \dfrac{1}{2}\right)P_l^{m_\varphi-\frac{1}{2}}(\theta) \\ P_l^{m_\varphi+\frac{1}{2}}(\theta) \end{pmatrix}. \tag{11}$$

In (11), $P_l^{m_\varphi\pm\frac{1}{2}}(\theta)$ are Legendre polynomials.

The separation of variables gives a system of equations for real radial functions $F(r),G(r)$. We write these equations in dimensionless variables $\varepsilon = E/m$, $\rho = r/l_c$, where $l_c = \hbar/mc$ is the Compton wavelength of the electron.

$$\frac{dF}{d\rho} + \frac{1+\kappa}{\rho}F - \left(\varepsilon + 1 + \frac{\alpha_{em}Z}{\rho}\right)G = 0$$
$$\frac{dG}{d\rho} + \frac{1-\kappa}{\rho}G + \left(\varepsilon - 1 + \frac{\alpha_{em}Z}{\rho}\right)F = 0. \tag{12}$$

If we introduce the phase from the definition

$$\operatorname{tg}\Phi = \frac{F(\rho)}{G(\rho)}, \tag{13}$$

then the energy spectrum ε_n can also be defined from the equation for the phase $\Phi = \operatorname{arctg}\left(F(\rho)/G(\rho)\right) + k\pi,\ k = 0, \pm 1, \pm 2,\ldots$ in the form proposed by Vronsky [11]

$$\frac{d\Phi}{d\rho} = \varepsilon + \frac{\alpha_{em}Z}{\rho} + \cos 2\Phi - \frac{\kappa}{\rho}\sin 2\Phi. \tag{14}$$

For the finite motion of the electron, asymptotics of solutions to Eqs. (12) for $\rho \to \infty$ is given by

$$F(\rho) = C_1 e^{-\rho\sqrt{1-\varepsilon^2}}$$

$$G(\rho) = -\sqrt{\frac{1-\varepsilon}{1+\varepsilon}}F(\rho). \tag{15}$$

The phase Φ for $\rho \to \infty$ equals

$$\Phi = -\operatorname{arctg}\sqrt{(1+\varepsilon)/(1-\varepsilon)}. \tag{16}$$

3. Electron Current Density, Boundary Condition for the Wave Functions

In the course of separation of variables when deriving Eqs. (10), (12) from Eq. (8), we performed an equivalent substitution of the Dirac matrices

$$\alpha^1 \to \alpha^3;\ \alpha^2 \to \alpha^1;\ \alpha^3 \to \alpha^2. \tag{17}$$

Then, considering (9), (11), components of the Dirac current density equal

$$j^r = \psi^+\alpha^3\psi = -iF(\rho)G(\rho)\left[\xi^+(\theta)\left(\sigma^3\sigma^3 - \sigma^3\sigma^3\right)\xi(\theta)\right] = 0, \tag{18}$$

$$j^\theta = \psi^+\alpha^1\psi = -2F(\rho)G(\rho)\left[\xi^+(\theta)\sigma^2\xi(\theta)\right] = 0, \tag{19}$$

$$j^\varphi = \psi^+\alpha^2\psi = 2F(\rho)G(\rho)\left[\xi^+(\theta)\sigma^1\xi(\theta)\right] \neq 0. \tag{20}$$

The equalities (18)–(20) coincide with previously obtained results in [14].

Our boundary condition involves zeroing of the current component j^φ at the nucleus boundary ρ_N, which resolves itself into zeroing of one of the two wave functions $F(\rho_N),\ G(\rho_N)$:

$$F(\rho_N)G(\rho_N) = 0. \tag{21}$$

The boundary condition (21) is similar to the condition near the "event horizon" introduced in the numerical calculations of the solution to the Dirac equation in the Schwarzschild field [11].

As a result, for the values of the gravitational coupling constant $\alpha \ll 1$, calculations [11] yield energy levels close to the energy levels in the hydrogen atom.

4. Results of Numerical Calculations of the Energy Spectrum of Hydrogen-Like Atoms with Effective Accounting of the Finite Size of Nuclei

In the calculations, the size of nuclei were determined from the relationships

$$r_N = \left(0.836 A^{\frac{1}{3}} + 0,57\right) \cdot 10^{-13} \text{ cm,} \quad A > 9 \text{ [16]}$$

$$r_N = 1.3 \cdot 10^{-13} \cdot A^{\frac{1}{3}} \text{ cm,} \quad A < 9 \tag{22}$$

In (22), A is the atomic weight of the nucleus.

The equation for phase (13) was solved by the fifth-order Runge-Kutta implicit method with step control [15]. We used the Ila scheme to obtain the three-stage Rado IIA method.

From two possible variants of implementation of condition (21), we will fulfil it, like in [11], using equality

$$G(\rho_N) = 0. \tag{23}$$

Some reason for this is known smallness of function $G(\rho_N)$ in comparison with function $F(\rho)$ in nonrelativistic approximation of Dirac equation.

It follows from (23) that the condition for the phase equals

$$\Phi(\varepsilon, \kappa, Z) = k\frac{\pi}{2}, \quad k = \pm 1, \pm 3, \pm 5 \ldots \tag{24}$$

Tables 1–3 contain the values of energy levels for the hydrogen atom $Z = 1$, $A = 1$ obtained by numerical calculations of Eq. (14) with the boundary conditions (16), (24) for $\kappa = \pm 1, \pm 2, \pm 3$ and $n = 1 \div 11$.

The tables also present corresponding energy values obtained from (1) and relative deviations of calculated values from analytical ones in percent.

Table 1. Energy levels of the hydrogen atom for the $S_{\frac{1}{2}}$, $P_{\frac{1}{2}}$ -states $(\kappa = \pm 1)$.

n	$1-\varepsilon_{an}$	$1-\varepsilon_{num}$	$\delta(\%)$	Comment
1	2.6640E–05	2.6641E–05	–0.004	No solution available for $\kappa = +1$
2	6.6600E–06	6.6602E–06	–0.003	
3	2.9600E–06	2.9601E–06	–0.003	
4	1.6650E–06	1.6651E–06	–0.006	
5	1.0656E–06	1.0656E–06	0.000	
6	7.4000E–07	7.3999E–07	0.001	
7	5.4367E–07	5.4367E–07	0.000	
8	4.1625E–07	4.1624E–07	0.002	
9	3.2889E–07	3.2888E–07	0.002	
10	2.6640E–07	2.6639E–07	0.003	
11	2.2016E–07	2.2015E–07	0.006	

Table 2. Energy levels of the hydrogen atom for the $P_{\frac{3}{2}}$, $D_{\frac{3}{2}}$ - states $(\kappa = \pm 2)$.

n	$1-\varepsilon_{an}$	$1-\varepsilon_{num}$	$\delta(\%)$	Comment
2	6.6599E–06	6.6585E–06	0.022	No solution available for $\kappa = +2$
3	2.9600E–06	2.9603E–06	–0.009	
4	1.6650E–06	1.6653E–06	–0.016	
5	1.0656E–06	1.0656E–06	0.004	
6	7.3999E–07	7.3997E–07	0.004	
7	5.4367E–07	5.4367E–07	0.001	
8	4.1625E–07	4.1622E–07	0.007	
9	3.2889E–07	3.2887E–07	0.006	
10	2.6640E–07	2.6637E–07	0.012	
11	2.2016E–07	2.2017E–07	–0.001	

We can see that the calculated and analytical values of energy values are in close agreement to within hundredths of percent $\left(\delta = \dfrac{\varepsilon_{num.} - \varepsilon_{an.}}{\varepsilon_{an.}} \lesssim 10^{-4} \right)$.

Within the above accuracy, the calculations reproduce degeneration of the energy levels with the same total momentum j (the same value of $|\kappa|$) typical for the fine-structure formula (1).

Table 3. Energy levels of the hydrogen atom for the $D_{5/2}$, $F_{5/2}$ - states $(\kappa = \pm 3)$.

n	$1 - \varepsilon_{an}$	$1 - \varepsilon_{num}$	$\delta(\%)$	Comment
3	2.9600E–06	2.9597E–06	0.011	No solution available for $\kappa = +3$
4	1.6650E–06	1.6652E–06	–0.010	
5	1.0656E–06	1.0657E–06	–0.006	
6	7.3999E–07	7.3997E–07	0.004	
7	5.4367E–07	5.4367E–07	0.000	
8	4.1625E–07	4.1622E–07	0.007	
9	3.2889E–07	3.2887E–07	0.006	
10	2.6640E–07	2.6637E–07	0.012	
11	2.2016E–07	2.2017E–07	–0.001	

Next, energy levels of the one-electron atoms were calculated for the following nuclei: B $(Z = 5, A = 10)$, Ne $(Z = 10, A = 21)$, $Mn(Z = 25, A = 5)$, Sn $(Z = 50$, $A = 119)$, $U (Z = 92$, $A = 238)$, $(Z = 104$, $A = 261)$. For hypothesized nuclei, $Z > 104$, the ratio A/Z was chosen equal to 2.9.

The results of the calculations for three lower levels and for the values of $\kappa = \pm 1, \pm 2, \pm 3$ are shown in Figs. 1–6. For comparison, the same figures present some numerical results [8] and analytical values from the fine-structure formula (1). In the calculations [8] the nucleus radiuses were determined from the relationship $r_N = 1.2 \cdot 10^{-13} \cdot A^{1/3}$ cm.

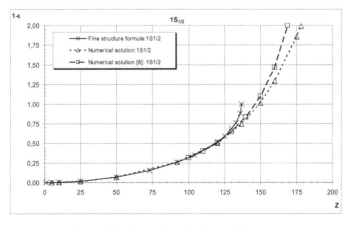

Fig. 1. The plots of $E(Z)$ for the $1S_{1/2}$-state.

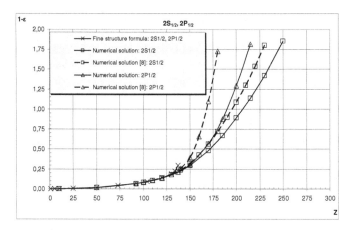

Fig. 2. The plots of $E(Z)$ for the $2S_{1/2}$, $2P_{1/2}$-states.

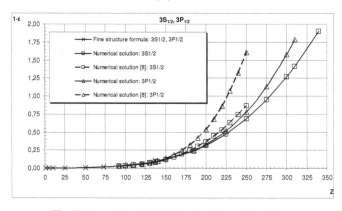

Fig. 3. The plots of $E(Z)$ for the $3S_{1/2}$, $3P_{1/2}$-states.

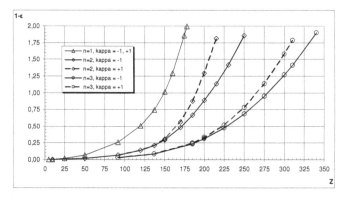

Fig. 4. Calculated plots of $E(Z)$ for the states with $n = 1, 2, 3$ and $\kappa = \pm 1$.

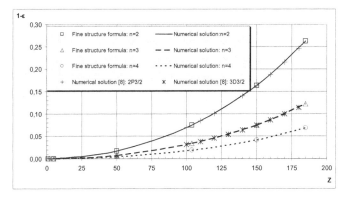

Fig. 5. The plots of $E(Z)$ for $P_{3/2}$, $D_{3/2}$-states and $n = 2, 3, 4$.

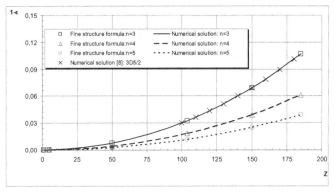

Fig. 6. The plots of $E(Z)$ for the $D_{5/2}$, $F_{5/2}$-states and $n = 3, 4, 5$.

These results indicate that the formula (1) is in a good agreement with the calculated values of energy levels for all the known elements of the periodic table.

For $\kappa = -1$ $\left(1S_{1/2}\right)$, any noticeable discrepancy for the lower level $(>1\%)$ occurs at $Z > 105$ (Fig. 1).

The calculated plots of $E(Z)$ are smooth and monotone.

The lower level $1S_{1/2}$ reaches the value of $\varepsilon = -1$ (the electron "drop" on a nuclei) at $Z_c \simeq 178$.

If the level $1S_{1/2}$ reaches the lower continuum $\varepsilon = -1$ at $Z > 178$, one must move from single-body quantum mechanics to many-body quantum field theory [7].

In this paper, the plots of $E(Z)$ for $Z > 178$ are shown in Figs. 2–4 for methodological reasons. These plots have no singularities and are qualitatively similar to the plots of $E(Z)$ for the lower energy level $1S_{1/2}$.

In accordance with the results obtained in [7], [8] in Figs. 2–4 one can see that energy levels with the same j are no more degenerate for $Z > 137$.

As the values of n and κ grow, the values of Z, at which energy levels with the same j begin to differ, get higher. It follows from Figs. 5, 6 that the levels $P_{3/2}, D_{3/2}$ and $D_{5/2}, F_{5/2}$ coincide up to $Z_c = 178$. For these levels, one can also see good agreement with the fine-structure formula.

As a result of effective accounting of the finite size of nuclei using the boundary condition for the Dirac wave functions (21), (24), energy levels for $Z \leq 105$ practically coincide with the fine-structure formula (1) and with the results in [7], [8] using effective nucleus potentials (4), (6).

It means an absence of appreciable effect of a values of electron location probability in a nucleus on the energy spectrum (maximum probability — in the calculations with the use of the singular Coulomb potential; smaller probability — in the calculations with the use of the finite electrostatic potentials of nuclei; zero probability — in the calculation of this paper with the use of the boundary condition (21)).

For $Z > 105$, the plots of $E(Z)$ based on the results of this work are less steep (see Figs. 1–6). This leads to a somewhat higher value of $Z_c = 178$ compared to the values of $Z_c \approx 170$ in [7], [8]. The difference between the plots of $E(Z)$ decreases as the quantum numbers n and κ grow.

A single-body quantum-mechanical consideration becomes more approximate when Z increases. It is necessary to take into account the effects of quantum electrodynamics and using of a many-body relativistic quantum theory of heavy and superheavy nuclei. Considering this fact, a value $Z_c = 178$ derived in this paper with the boundary condition (21), which provides the zero probability of electron location within a nucleus, one should consider as the upper limit of a true value Z_c. In [10] there is a formulation of conditions which must be fulfilled for determination Z_c in future experiments.

5. Conclusions

The calculations to determine energy levels of hydrogen-like atoms with effective accounting (21) of the finite size of nuclei allow us to draw the following conclusions:

1. Calculations with $Z = 1, A = 1$ reproduce the fine-structure formula (1) for the hydrogen atom to within $\sim 10^{-4}$.

2. The calculations are in a good agreement with the fine-structure formula for all the known nuclei of the periodic table. For the lower level, any noticeable discrepancy occurs at $Z > 105$.

3. The calculated plots of $E(Z)$ are smooth and monotone.

4. The lower level $1S_{1/2}$ reaches the value of $\varepsilon = -1$ ($E = -mc^2$ is the electron "drop" on a nuclei) at $Z_c = 178$.

5. To account of the finite size of nuclei, the boundary condition (21), which shows well for the one-electron case, can be easily applied to calculations of many-electron atoms using solutions of the Dirac equation for radial wave functions.

Acknowledgments

We thank our colleagues professors P.P. Fiziev, M.A. Vronsky and A.A. Sadovy for fruitful discussions, and A.L. Novoselova for her significant technical help in the preparation of the manuscript.

References

1. A.Sommerfeld, Ann. D. Phys 51, 1 (1916).
2. P.A.M.Dirac, Proc. Roy. Soc. A 117, 610; 118, 341 (1928).
3. C.G.Darwin, Proc. Roy. Soc. Lond. Ser. A 118, 654 (1928).
4. W.Z.Gordon, Z. Phys. 48, 11 (1928).
5. I.Ya.Pomeranchuk, Ya.A.Smorodinsky, Jour. Phys. USSR 9, 97 (1945).
6. Ya.B.Zeldovich, Solid-State Physics, v. 1, no.11, pp. 1637–1641 (1959) (in Russian).
7. Ya.B.Zeldovich, V.S. Popov, UFN, v.105, no. 3, 1971 (in Russian).
8. W.Pieper, W.Greiner, Zs. Phys. 218, 327 (1969).
9. D.Andrae, Physics Reports 336, 413–525 (2000).

10. R.Ruffini, G.Vereshchagin, She-Sheng Xue, Physics Reports 487, 1–140 (2010).
11. M.V. Gorbatenko, N.S. Kolesnikov, V.P. Neznamov, E.V. Popov, I.I. Safronov, M.A. Vronsky, arxiv: 1301.7595 (gr-qc).
12. D.R. Brill, J.A. Wheeler, Rev. of Modern Physics 29, 465–479 (1957).
13. S.R. Dolan, Trinity Hall and Astrophysics Group, Cavendish Laboratory. Dissertation, 2006.
14. C.Colijn and E.R.Vrscay, Foundations of Physics Letters, Vol.16, No. 4 (2003).
15. E.Hairer, G.Wanner, Solving Ordinary Differential Equations II. Stiff and Differential-Algebraic Problems Second Revised Edition, Springer-Verlag 1991, 1996.
16. W.R.Johnson, G.Soff, At. Data Nucl. Nables 33, 405–446 (1985).

TWO-DIMENSIONAL SIGMA MODELS RELATED TO NON-ABELIAN STRINGS IN SUPER-YANG-MILLS

M. SHIFMAN

William I. Fine Theoretical Physics Institute, University of Minnesota, Minneapolis, MN 55455, USA

A. YUNG

William I. Fine Theoretical Physics Institute, University of Minnesota, Minneapolis, MN 55455, USA
Petersburg Nuclear Physics Institute, Gatchina, St. Petersburg 188300, Russia

We review diverse two-dimensional models emerging on the world sheet of non-Abelian strings in the low-energy limit. Non-Abelian strings are supported in a class of four-dimensional bulk theories with or without supersymmetry. In supersymmetric bulk theories we are mostly interested in BPS-saturated strings. Some of these two-dimensional models, in particular, heterotic models, were scarcely studied in the past, if at all.

1. Introduction

Isaak Yakovlevich Pomeranchuk, the founder of the ITEP Theory Department, died in December of 1966, only six years before the advent of revolutionary changes in high-energy physics. His work with Landau[1] (see also[2]) on the so-called Moscow zero charge (currently known as infrared freedom in Abelian gauge theories), shaped the subsequent research on gauge theories which culminated in 1973, with the discovery of asymptotic freedom in non-Abelian gauge theories.[3] Non-Abelian gauge theories proved to be the basis of the modern theory.[*]

Asymptotic freedom is just one aspect of these theories. Another aspect is a unique behavior in the infrared domain, at strong coupling, known as *confinement*, or, sometimes, color confinement. Despite four decades of vigorous efforts analytic understanding of the phenomenon of color confine-

---

[*]Pomeranchuk witnessed the discovery of a two-dimensional asymptotically free field theory as early as in 1958,[4] but at that time due attention was not paid to this work.

ment in quantum chromodynamics is still incomplete. At the same time significant advances occurred in 1994 when Seiberg and Witten solved $\mathcal{N} = 2$ super-Yang-Mills theory.[5]

In the mid-1970s Nambu, 't Hooft, and Mandelstam (independently) put forward an idea[6] of a "dual Meissner effect" as the underlying mechanism for color confinement. Within their conjecture, in appropriate Yang-Mills theories chromomagnetic "monopoles" condense leading to formation of "chromoelectric flux tubes" between the probe quarks. At that time the Nambu-'t Hooft-Mandelstam paradigm was not even a physical scenario, rather a dream, since people had no clue as to the main building blocks such as non-Abelian flux tubes.

The Seiberg-Witten solution[5] triumphantly demonstrated the emergence of the confining strings as a result of a small $\mathcal{N} = 1$ deformation $\mathcal{N} = 2$ super-Yang-Mills theory.

However, although these strings appear in the non-Abelian theory they turned out to be Abelian in their structure,[7] in essence identical to the Abrikosov-Nielsen-Olesen (ANO) strings.[8]

Just like the fundamental string in string theory, the ANO string (at low excitation energies) is fully characterized by the position of its center in the perpendicular plane Đ the so-called translational moduli. The orientation of the magnetic flux in the string core is rigidly fixed in the SW solution. Say, for the SU(2) gauge group it can be aligned along the third axis in the color space. Shortly after the SW discovery it was realized that for QCD-like theories, in which there are no preferred directions in the color space, it would be more appropriate to have the flux in the string core fluctuating freely "inside" the non-Abelian group. In other words for QCD strings it is desirable to have additional orientational moduli on the string world sheet. Such strings became known as non-Abelian.

The search for genuinely non-Abelian strings started in the end of 1990s and culminated in their discovery[9] in 2003. Dynamics of the extra – orientational – moduli on the string world sheet was demonstrated to be described by CP$(N - 1)$ model, where N is the number of colors in the bulk theory. Since then a large variety of non-Abelian strings became known; some of them support two-dimensional theories that had been known from long ago, others exhibit nontrivial and largely unexplored sigma models on the string world sheet. This review is devoted to two-dimensional sigma models which came into the limelight in connection with the non-Abelian strings. The review is by necessity brief and represents, in a sense, a travel guide in this subject.

Historically nontrivial sigma models on the string world sheet first emerged in the context of supersymmetric bulk theories. Now it is clear that supersymmetry is not necessary, nonsupersymmetric bulk theories can support them too.[10,11] Due to the fact that we will mainly focus on least explored world-sheet theories – heterotic two-dimensional sigma models – our discussion will be tied up with supersymmetry. A significant part of this review is devoted to results which we obtained after 2008. For a review before 2009 see.[12]

2. How world-sheet models appear: the simplest example

The simplest and historically the first model supporting non-Abelian strings is[9] $\mathcal{N} = 2$ super-Yang–Mills theory with the number of colors equal to the number of flavors (i.e. if the gauge group is SU(2), to which we will limit ourselves in this section, we introduce two (s)quark flavors). Moreover, we add a U(1) factor to the gauge group, so that, in fact, the gauge group is U(2). We endow this U(1) factor with the Fayet–Iliopoulos term ξ.[13] The latter is needed to make non-Abelian strings BPS-saturated. BPS saturation is *not* a necessary condition. However, it simplifies calculations.

The bosonic part of the basic U(2) theory with two flavors has the form[9] (in the Euclidean space)

$$\mathcal{L} = \frac{1}{4g_2^2} \left(F_{\mu\nu}^a \right)^2 + \frac{1}{4g_1^2} \left(F_{\mu\nu} \right)^2 + \frac{1}{g_2^2} \left| D_\mu a^a \right|^2 + \frac{1}{g_1^2} \left| \partial_\mu a \right|^2$$

$$+ \left| \nabla_\mu q^A \right|^2 + \left| \nabla_\mu \bar{\tilde{q}}^A \right|^2 + V(q^A, \tilde{q}_A, a^a, a) . \tag{1}$$

Here D_μ is the covariant derivative in the adjoint representation of SU(2), and

$$\nabla_\mu = \partial_\mu - \frac{i}{2} A_\mu - i A_\mu^a T^a , \qquad T^a = \frac{1}{2} \tau^a , \tag{2}$$

where τ^a are the Pauli matrices acting in the color SU(2) group. The coupling constants g_1 and g_2 correspond to the U(1) and SU(2) sectors, respectively. With our conventions, the U(1) charges of the fundamental matter fields are $\pm 1/2$. Two squark fields are denoted by q^A and $\tilde{q}_A$, respectively (the flavor index $A = 1, 2$). The doubling of the (s)quark fields is required by $\mathcal{N} = 2$ supersymmetry. In addition to the flavor index A the the (s)quark fields carry SU(2) doublet index too; therefore, they can be viewed as a 2×2 matrix. Moreover, a^c ($c = 1, 2, 3$) is the complex scalar field in the adjoint representation of SU(2), the superpartner of the SU(2) gauge bosons, while

a without the superscript is the superpartner of the U(1) gauge boson. For brevity we will refer to these fields as to "adjoints."

The potential $V(q^A, \tilde{q}_A, a^a, a)$ in the Lagrangian (1) is a sum of D and F terms,

$$V(q^A, \tilde{q}_A, a^a, a) = \frac{g_2^2}{2} \left(\frac{i}{g_2^2} \varepsilon^{abc} \bar{a}^b a^c + \bar{q}_A T^a q^A - \tilde{q}_A T^a \bar{\tilde{q}}^A \right)^2$$

$$+ \frac{g_1^2}{8} \left(\bar{q}_A q^A - \tilde{q}_A \bar{\tilde{q}}^A \right)^2 + 2g_2^2 \left| \tilde{q}_A T^a q^A \right|^2 + \frac{g_1^2}{2} \left| \tilde{q}_A q^A - \xi \right|^2$$

$$+ \frac{1}{2} \sum_{A=1}^{N} \left\{ \left| (a + \sqrt{2} m_A + 2T^a a^a) q^A \right|^2 + \left| (a + \sqrt{2} m_A + 2T^a a^a) \bar{\tilde{q}}^A \right|^2 \right\}. \quad (3)$$

Here m_A are the (s)quark mass terms, and the sum over the repeated flavor indices $A = 1, 2$ is implied.

Let us discuss the vacuum structure of this model. Nonvanishing of the Fayet-Iliopoulos term $\xi \neq 0$ implies an isolated vacuum with the maximal possible value of condensed (s)quarks – two. The vacua of the theory (1) are determined by the zeros of the potential (3). The adjoint fields develop the following vacuum expectation values (VEVs):

$$\langle \Phi \rangle = -\frac{1}{\sqrt{2}} \begin{pmatrix} m_1 & 0 \\ 0 & m_2 \end{pmatrix}, \quad (4)$$

where we defined the scalar adjoint matrix as

$$\Phi \equiv \frac{1}{2} a + T^a a^a. \quad (5)$$

If $m_1 = m_2$ and $\xi = 0$ the SU(2)×U(1) gauge group remains classically unbroken, since in this case m can be absorbed in a. Alternatively, we can set $m = 0$ from the beginning. However, if $m_1 \neq m_2$ SU(2) is broken down to U(1). Furthermore, if $\xi \neq 0$ we must take into account the squark VEVs which results in Higgsing of all gauge bosons.

In the vacuum the squark VEVs have a peculiar color-flavor locked form

$$\langle q^{kA} \rangle = \langle \bar{\tilde{q}}^{kA} \rangle = \sqrt{\frac{\xi}{2}} \begin{pmatrix} 1 & 0 \\ 0 & 1 \end{pmatrix}, \quad k = 1, 2, \quad A = 1, 2. \quad (6)$$

The potential (3) vanishes if Φ and q are chosen according to (4) and (6), respectively. (ξ is assumed to be large, $\xi \gg \Lambda^2$, to warrant quasi classical treatment.)

The vacuum field (6) results in the spontaneous breaking of both gauge and flavor SU(2)'s. A diagonal global SU(2) survives, however,

$$U(2)_{\text{gauge}} \times SU(2)_{\text{flavor}} \to SU(2)_{C+F} \,. \tag{7}$$

Thus, a color-flavor locking takes place in the vacuum.

Why does the model described above support a novel type of strings, non-Abelian?

The conventional ANO string corresponds to a U(1) winding of the phase of all squark fields in the plane, perpendicular to the string axis,

$$q^{kA} \longrightarrow \sqrt{\frac{\xi}{2}} \, e^{i\alpha} \begin{pmatrix} 1 & 0 \\ 0 & 1 \end{pmatrix}, \tag{8}$$

where α is the polar angle in the perpendicular plane (see Fig. 1). Its topological stability is due to $\pi_1(U(1)) = Z$. Now we have more options, however, due to the fact that $SU(2)_{C+F}$ has center. Usually people say that $\pi_1(SU(2))$ is trivial and, therefore there are no other topologically stable strings.

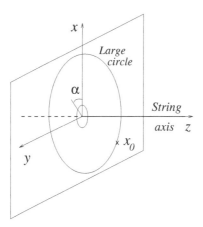

Fig. 1. Geometry of the string.

This is not quite the case in the model at hand. Observe that the center of the SU(2) group, Z_2, belongs to the U(1) factor too. This means that we can split the 2π windings in two halves: the first (from 1 to -1) is carried out in U(1), while the second, from -1 to 1 in SU(2) (e.g. around the third axis). This is clearly, a topologically stable configuration, albeit the stability is of the Z_2 type. Correspondingly, the winding ansatz takes the form

$$q^{kA} \longrightarrow \sqrt{\frac{\xi}{2}} \begin{pmatrix} e^{i\alpha} & 0 \\ 0 & 1 \end{pmatrix} \quad \text{or} \quad q^{kA} \longrightarrow \sqrt{\frac{\xi}{2}} \begin{pmatrix} 1 & 0 \\ 0 & e^{i\alpha} \end{pmatrix}, \tag{9}$$

depending on whether we use the combination of generators $T_{U(1)} + T^3_{SU(2)}$ or $T_{U(1)} - T^3_{SU(2)}$.

It is clear, that the ansatz (9) breaks the color-flavor locked SU(2) down to U(1). The particular way of embedding is unimportant. Instead of T^3 we could have chosen any other generator of SU(2). In other words, the existence of the string (9) implies the existence of the whole family of strings parametrized by SU(2)/U(1) moduli. The theory of the moduli fields on the string world sheet is the sigma model on the SU(2)/U(1) coset space. This is the celebrated CP(1) model. It is asymptotically free in the UV and strongly coupled in the IR. Since the bulk theory has eight supercharges and the string is 1/2 BPS saturated, the world-sheet model has four supercharges. In other words, its supersymmetry is $\mathcal{N} = (2,2)$. The existence of the orientational moduli means that the flux through the string does not have a preferred orientation inside SU(2). This is a genuinely non-Abelian string. Non-Abelian strings are formed if all non-Abelian bulk degrees of freedom participate are equally operative at the scale of string formation.

Since the CP(1) model is equivalent to O(3) (see e.g.[14]), the orientational moduli can be represented as a unit vector (in the "isospace") attached to every point of the string and allowed to fluctuate freely see Fig. 2.

If the tension of the ANO string is $4\pi\xi$, the tension of the non-Abelian string is $2\pi\xi$, in the U(2) bulk theory. Thus, the ANO string is, in a sense, composite.

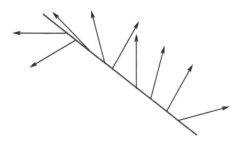

Fig. 2.   O(3) sigma model on the string world sheet.

For bulk theories with the U(N) gauge group the world sheet theory on the non-Abelian string is given by CP($N - 1$) models.[9]

3. Supersymmetry in the bulk and BPS strings

The degree of supersymmetry of BPS-saturated[15] non-Abelian strings is determined by supersymmetry in the bulk. Thus, $\mathcal{N} = 2$ bulk theories support $\mathcal{N} = (2, 2)$ models on their world sheet,[9] while $\mathcal{N} = 0$ bulk theories give rise to non-supersymmetric non-Abelian strings (e.g.[10]). The most interesting type of strings – heterotic – appear in the $\mathcal{N} = 1$ bulk theories.[16–18] In this case the world-sheet Lagrangian possesses $\mathcal{N} = (0, 2)$ supersymmetry (it can be minimal or nonminimal), which is usually further spontaneously broken due to an appropriate Goldstino field on the world sheet.

Nonsupersymmetric and (2,2) supersymmetric two-dimensional sigma models are thoroughly studied, see e.g. the review.[19] As far as heterotic (0,2) models are concerned, till recently only some general aspects have been discussed.[20–25] However, emergence of these theories on the string world sheet gave a strong impetus for further studies, see e.g.[26–31]

4. Basic models

In the vast majority of examples studied so far, the world-sheet theories on non-Abelian strings are various versions of $CP(N - 1)$ models: with or without twisted masses, with or without extra fields, with or without supersymmetry, and their extensions such as the so-called zn and weighted $CP(N, M)$ models. All these models have two (sometimes even three) distinct representations. In this section we will briefly discuss these representation using the simplest example: non-supersymmetric and (2,2) supersymmetric $CP(N - 1)$ without twisted mass.

4.1. *Geometric formulation*

A generic Lagrangian of any sigma-model with the Kählerian target space is

$$\mathcal{L}_{CP(N-1)} = G_{i\bar{j}}\, \partial^{\mu}\bar{\phi}^{\bar{j}}\, \partial_{\mu}\phi^{i}\,, \tag{10}$$

where $G_{i\bar{j}}$ is the Kähler metric,

$$G_{i\bar{j}} = \frac{\partial^2 K(\phi, \bar{\phi})}{\partial\phi^i \partial\bar{\phi}^{\bar{j}}}$$

and $K(\phi, \bar{\phi})$ is the Kähler potential. For the $CP(N–1)$ model one can choose the following Kähler potential:

$$K = \frac{2}{g_0^2} \log \left(1 + \sum_{i,\bar{j}=1}^{N-1} \bar{\phi}^{\bar{j}} \delta_{\bar{j}i} \phi^{i} \right), \tag{11}$$

corresponding to the so-called round Fubini-Study metric. The bare coupling constant is denoted by g_0^2.

It is not difficult to supersymmetrize the model (10) and (11). Its $\mathcal{N} = (2, 2)$ generalization can be written as[32]

$$\mathcal{L}_{\mathcal{N}=(2,2)} = \int d^4\theta K(\Phi, \bar{\Phi}) = G_{i\bar{j}} \left[\partial^\mu \bar{\phi}^{\bar{j}} \, \partial_\mu \phi^i + i\bar{\psi}^{\bar{j}} \gamma^\mu \mathcal{D}_\mu \psi^i \right]$$

$$- \frac{1}{2} R_{i\bar{j}k\bar{l}} (\bar{\psi}^{\bar{j}} \psi^i)(\bar{\psi}^{\bar{l}} \psi^k), \tag{12}$$

where Φ^i and $\bar{\Phi}$ are the chiral and antichiral superfields

$$\Phi^i(x^\mu + i\bar{\theta}\gamma^\mu\theta), \qquad \bar{\Phi}^{\bar{j}}(x^\mu - i\bar{\theta}\gamma^\mu\theta) \tag{13}$$

of which the lowest components are ϕ^i and $\bar{\phi}$ (see e.g.[14]), $R_{i\bar{j}k\bar{l}}$ is the Riemann tensor,

$$\mathcal{D}_\mu \psi^i = \partial_\mu \psi^i + \Gamma^i_{kl} \partial_\mu \phi^k \psi^l \tag{14}$$

is the covariant derivative, Γ^i_{kl} are the Christoffel symbols, and we use the notation $\bar{\theta} = \theta^\dagger\gamma^0$, $\bar{\psi} = \psi^\dagger\gamma^0$ for the fermion objects. The γ matrices are chosen as

$$\gamma^0 = \gamma^t = \sigma_2, \qquad \gamma^1 = \gamma^z = i\sigma_1, \qquad \gamma_5 \equiv \gamma^0\gamma^1 = \sigma_3. \tag{15}$$

For $CP(N-1)$ target space, as for any symmetric manifold, the Ricci-tensor $R_{i\bar{j}}$ is proportional to the metric, see Eq. (18) below. Both versions of this model – supersymmetric and non-supersymmetric are asymptotically free.[33] In the former case the β function is one-loop exact,

$$\beta_{\mathcal{N}=(2,2)} \equiv \frac{\partial g_0^2}{\partial \log M_{\mathrm{uv}}} = -\frac{g^4 N}{4\pi}. \tag{16}$$

Only bosons contribute at first loop. In non-supersymmetric $CP(N - 1)$ model[14,34]

$$\beta_{CP(N-1)} = -\frac{g^4 N}{4\pi} \left(1 + \frac{g^2}{2\pi} + ... \right), \tag{17}$$

where ellipses stand for the third and higher loops.

For completeness, concluding this section let us add a few extra useful expressions,

$$G_{i\bar{j}} = \frac{2}{g^2}\left(\frac{\delta_{i\bar{j}}}{\chi} - \frac{\bar{\phi}^i\phi^{\bar{j}}}{\chi^2}\right), \qquad G^{i\bar{j}} = \frac{g^2}{2}\chi\left(\delta^{i\bar{j}} + \phi^i\bar{\phi}^{\bar{j}}\right),$$

$$\Gamma^i_{kl} = -\frac{\delta^i_k\bar{\phi}^{\bar{l}} + \delta^i_l\bar{\phi}^{\bar{k}}}{\chi}, \qquad \Gamma^{\bar{i}}_{\bar{k}\bar{l}} = -\frac{\delta^{\bar{i}}_{\bar{k}}\phi^l + \delta^{\bar{i}}_{\bar{l}}\phi^k}{\chi},$$

$$R_{i\bar{j}k\bar{l}} = -\frac{g^2}{2}\left(G_{i\bar{j}}G_{k\bar{l}} + G_{k\bar{j}}G_{i\bar{l}}\right), \qquad R_{i\bar{j}} = -G^{k\bar{j}}R_{i\bar{j}k\bar{l}} = \frac{g^2 N}{2}G_{i\bar{j}},$$

$$\chi \equiv 1 + \sum_{m}^{N-1}\bar{\phi}^{\bar{m}}\phi^m. \tag{18}$$

4.2. Gauged formulation

An alternative formulation – the so-called gauged formulation – was suggested by Witten.[35,36] Being completely equivalent to the geometric formulation it is more convenient for the large-N solution of the model.

The CP$(N-1)$ target space is the coset SU(N)/(SU$(N-1) \times$ U(1)). In the gauged formulation we build the Lagrangian $\mathcal{L}_{\mathrm{CP}(N-1)}$ starting from an N-plet of complex bosonic fields n^i where $i = 1, 2, ..., N$. The fields n^i are scalar (i.e. spin-0), and are subject to the constraint

$$\bar{n}_i\, n^i = 1, \tag{19}$$

The Lagrangian takes the form

$$\mathcal{L}_{\mathrm{CP}(N-1)} = \frac{2}{g^2}\left|\mathcal{D}_\mu n^i\right|^2 - D\left(n^\dagger_i\, n^i - 1\right), \tag{20}$$

where the covariant derivative $\mathcal{D}_\mu$ is defined as

$$\mathcal{D}_\mu n^i \equiv (\partial_\mu - iA_\mu)\, n^i. \tag{21}$$

Here A_μ is an auxiliary vector field implementing U(1) gauge invariance, while D is an auxiliary real scalar field implementing the constraint (19). Neither A_μ nor D have kinetic terms in the Lagrangian (20).

Sometimes it is convenient to rescale the n and D fields as follows:

$$\mathcal{L}_{\mathrm{CP}(N-1)} = \left|\mathcal{D}_\mu n^i\right|^2 - D\left(\bar{n}_i\, n^i - 2\beta\right),$$

$$\beta \equiv \frac{1}{g^2}, \tag{22}$$

making the kinetic term canonic. The vacuum expectation value of D will then play the role of the n-field mass squared.

From Sec. 4.1 we see that the $CP(N-1)$ target space is parametrized by $2N - 2$ real degrees of freedom. There are $2N$ real degrees of freedom in the n^i fields. The constraint (19) reduces this number to $2N - 1$, while the U(1) gauge invariance further reduces it to $2N - 2$.

The fields ϕ^i of the geometric formulation can be related to n^i (on a particular patch) by singling out one of the components of n^i, say, n^N, and defining

$$\phi^i = \frac{n^i}{n^N}, \quad i = 1, 2, ..., N - 1. \tag{23}$$

The easiest way to extend the above formalism to $\mathcal{N} = (2,2)$ supersymmetry is to start from $\mathcal{N} = 1$ SQED in four dimensions with N flavors of chiral matter superfields (with one and the same charge), plus the Fayet-Iliopoulos term $\tilde{\xi}$,

$$\mathcal{L} = \left\{ \frac{1}{4e^2} \int d^2\theta\, W^2 + \text{H.c.} \right\} + \int d^4\theta \sum_{i=1}^{N} \left(\bar{Q}_i e^V Q^i \right) - \tilde{\xi} \int d^4\theta\, V, \tag{24}$$

where

$$Q^i = n^i + \sqrt{2}\,\theta\xi^i + \theta^2 F^i.$$

This theory does not exist in four dimensions due to the chiral anomaly. However, we will use it only as a starting point, with the intention of reducing it to two dimensions. In two dimensions it becomes well-defined. The $\mathcal{N} = (2,2)$ $CP(N-1)$ model is obtained in the limit $e^2 \to \infty$. In this limit both the photon and photino kinetic terms can be dropped, and we obtain (in components)

$$\mathcal{L}_{\mathcal{N}=(2,2)} = \left| \mathcal{D}_\mu n^i \right|^2 - 2|\sigma|^2 \left| n^i \right|^2 - D \left(|n^i|^2 - 2\beta \right)$$

$$+ \bar{\xi}_{jR}\, i\mathcal{D}_L \xi_R^j + \bar{\xi}_{jL}\, i\mathcal{D}_R \xi_L^j$$

$$+ \left[\sqrt{2}\sigma \bar{\xi}_{jR} \xi_L^j + \sqrt{2}\bar{n}_j \left(\lambda_R \xi_L^j + \lambda_L \xi_R^j \right) + \text{H.c.} \right]. \tag{25}$$

Here $\sigma = (A_x + iA_y)/\sqrt{2}$ is a part of the superfield V (in the Wess-Zumino gauge), along with $A_{t,z}$, D, and $\lambda_{R,L}$ (for geometrical conventions see Fig. 1). All these fields enter in the Lagrangian (25) without kinetic terms. The latter will be generated, however, at one-loop level, dynamically. The spinor fields $\lambda_{R,L}$ implement the constraint $\bar{n}_j \xi^j = 0$. The constraint $|n^i|^2 = 2\beta$ is implemented by the auxiliary D field. The covariant derivative is defined in (21). Finally, the Fayet-Iliopoulos term in (24) is related to 2β, namely, $\tilde{\xi} \to 2\beta$ (note that in two dimensions the Fayet-Iliopoulos term is dimensionless).

4.3. CP(1): a special case

The case $N = 2$, when we deal with the CP(1) model, is special. Indeed, the CP(1) target space is isomorphic to O(3), implying that the CP(1) model can be formulated in terms of a triplet of real fields S^a. The O(3) model, in turn, opens the series of the O(N) models. For $N > 3$ the O(N) target space is not Kählerian. Thus, supersymmetrization of the O(N) models with $N > 3$ results in $\mathcal{N} = (1,1)$ supersymmetry.

To explicitly pass from CP(1) to O(3) one needs expressions relating the $\vec{S}$ fields to the n^i fields. Given the fact that in this case n^i's are spinors of SU(2) while $\vec{S}$ is the O(3) vector one can write

$$S^a = \bar{n}\,\tau^a\,n\,, \quad a = 1, 2, 3, \tag{26}$$

where τ^a are the Pauli matrices which satisfy the Fierz transformation formula

$$\vec{\tau}_{\alpha\beta}\,\vec{\tau}_{\delta\gamma} = 2\delta_{\alpha\gamma}\,\delta_{\delta\beta} - \delta_{\alpha\beta}\,\delta_{\delta\gamma}\,. \tag{27}$$

Making use of (27) one concludes that

$$\vec{S}^2 = (\bar{n}\,n)^2 = 1\,. \tag{28}$$

Thus,

$$\mathcal{L}_{O(3)} = \frac{1}{2g^2}\,\partial_\mu S^a \partial^\mu S^a\,, \quad S^a S^a = 1\,, \quad a = 1, 2, 3. \tag{29}$$

Supergeneralization of (29) is straightforward.[37] One introduces a triplet of real superfields $\sigma^a(x, \theta)$,

$$N^a(x, \theta) = S^a + \bar{\theta}\chi^a + \frac{1}{2}\,\bar{\theta}\theta\,F^a\,, \tag{30}$$

where $\vec{S}$ and $\vec{F}$ are bosonic fields while $\vec{\chi}$ denotes two-component Majorana fields (the θ coordinate is also Majorana).

Then the supersymmetric Lagrangian takes the form

$$\mathcal{L}_{O(3)} = \frac{1}{g^2}\int d^2\theta \left(\frac{1}{2}\bar{D}_\alpha N^a D_\alpha N^a\right)$$

$$= \frac{1}{2g^2}\left\{\partial^\mu S^a \partial_\mu S^a + \frac{i}{2}\,\bar{\chi}^a \gamma^\mu \overset{\leftrightarrow}{\partial}_\mu \chi^a + \vec{F}^{\,2}\right\}\,. \tag{31}$$

with the constraint

$$N^a(x,\,\theta)\,N^a(x,\,\theta) = 1\,, \tag{32}$$

which replaces the nonsupersymmetric version of this constraint $\vec{S}^2 = 1$. Decomposing (32) in components we get

$$\vec{S}^2 = 1\,, \qquad \vec{S}\vec{\chi} = 0\,, \qquad \vec{F}\vec{S} = \frac{1}{2}\,(\bar{\chi}^a\chi^a)\,. \tag{33}$$

As usual, the F term enters with no derivatives. Eliminating F by virtue of the equations of motion one obtains[37,38]

$$\mathcal{L} = \frac{1}{2g^2}\left\{\partial^\mu S^a\,\partial_\mu S^a + \frac{i}{2}\,\bar{\chi}^a\gamma^\mu \overset{\leftrightarrow}{\partial}_\mu \chi^a + \frac{1}{4}\,(\bar{\chi}^a\chi^a)^2\right\}\,, \tag{34}$$

plus the first two constraints in Eq. (33).

The global O(3) symmetry is explicit in this Lagrangian. Moreover, $(1,1)$ supersymmetry is built in. In fact, supersymmetry of this model is $\mathcal{N} = (2,2)$ due to the Kählerian nature of the target space 2-sphere.[39]

A special nature of the CP(1) target space manifests itself in the fact that for the CP($N-1$) with $N > 2$ minimal heterotic $(0,2)$ models do not exist,[40] while it does exists for CP(1), see Sec. 9.2.1.

5. Witten's large-N solution

In this section we will briefly discuss large-N solutions of the simplest two-dimensional models emerging on the world sheet of non-Abelian strings. Massless non-supersymmetric and $\mathcal{N} = (2,2)$ supersymmetric CP($N-1$) models were solved at large N by Witten.[35]

5.1. Non-supersymmetric CP($N-1$)

Let us turn to (22) rescaling the coupling constant à la 't Hooft to make explicit the N dependence,

$$\mathcal{L}_{\mathrm{CP}(N-1)} = (\partial_\alpha + iA_\alpha)\,\bar{n}_i\,(\partial^\alpha - iA^\alpha)\,n^i - D\left(\bar{n}_in^i - \frac{N}{\lambda}\right)^2\,, \tag{35}$$

where

$$\lambda \equiv \frac{g^2N}{2}\,. \tag{36}$$

First, we study the vacuum structure of this model. Note that the Lagrangian (35) is quadratic in the n^i fields; therefore these fields can be integrated out,

$$Z = \int \mathcal{D}A_\alpha \, \mathcal{D}D \exp\left\{ -N\mathrm{Tr}\ln\left[-(\partial_\alpha - iA_\alpha)^2 - D \right] + i\frac{N}{\lambda} \int d^2x\, D \right\}.$$
(37)

The Lorentz invariance of the theory tells us that if the saddle point exists it must be achieved at an x independent value of D. Hence we can treat D as a constant, vary with respect to D, and require the result to vanish. The same Lorentz invariance tells us that at the saddle point $A_\alpha = 0$. In this way we arrive at the following equation:

$$\frac{i}{\lambda} + \int \frac{d^2k}{4\pi^2} \frac{1}{k^2 - D} = 0, \qquad \frac{1}{\lambda} - \frac{1}{4\pi} \log \frac{M_{\mathrm{uv}}^2}{D}.$$
(38)

The integral in (38) is logarithmic and diverges in the ultraviolet, therefore we cut it off at M_{uv}^2. In this way, starting from (38), we arrive at the equation

$$D_{\mathrm{vac}} \equiv m^2 = M_{\mathrm{uv}}^2\, e^{-4\pi/\lambda} = M_{\mathrm{uv}}^2\, e^{-8\pi/Ng^2} \equiv \Lambda^2.$$
(39)

The assumption of existence of the saddle point is confirmed *a posteriori*. The n-quanta mass m is a physical parameter. Therefore, the right-hand side of (39) is renormalization-group invariant, Λ^2, the dynamical scale parameter of the $\mathrm{CP}(N-1)$ model. This is in full agreement with the first coefficient of the β function in (17). The second coefficient is invisible to the leading order in $1/N$.

Integrating the second equation in (38) over D one readily reconstructs the effective potential as a function of D,

$$V_{\mathrm{eff}} = \frac{N}{4\pi} D \log \frac{D}{e\, m^2}, \quad e = 2.718....$$
(40)

From the large-N solution one can see that the constraint $\bar{n}_i n^i = 1$ is lifted and the massive n particles form a full N-plet. The n-mass is given in (39). Moreover, it is not difficult to see that the field A_μ acquires kinetic terms and become dynamical. Expanding the effective action (37) around the saddle point, one can easily check that cubic and higher orders in D and A are suppressed by powers of $1/\sqrt{N}$. The linear term of expansion vanish. This is the essence of Eq. (38). We will focus on the quadratic terms of expansion.

It is not difficult to check (see e.g.[14]) that the cross term of the DA type also vanish (see Fig. 3). Therefore, we need only consider the terms quadratic in A, see Fig. 4. A straightforward computation yields for the A^2

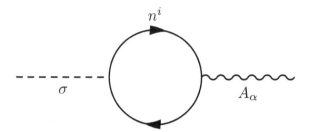

Fig. 3.   The vanishing of the $D\,A_\alpha$ mixing term in the effective Lagrangian.

term [†]

$$\frac{N}{12\pi\,m^2}\left(-g_{\mu\nu}k^2 + k_\mu k_\nu\right)\left(1 + O(k^2/m_n^2)\right).\qquad(41)$$

This expression is automatically transversal, as expected given the U(1) gauge invariance of (35). The $O(k^2)$ term in (41) represents the standard kinetic term $F_{\mu\nu}^2$ of the photon field, more exactly,

$$-\frac{N}{48\pi\,m^2}F_{\mu\nu}F^{\mu\nu}.\qquad(42)$$

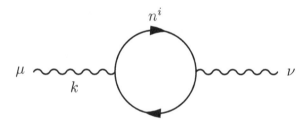

Fig. 4. $O(A^2)$ terms in the effective Lagrangian.

It is convenient to rescale the A field to make its kinetic term (42) canonically normalized. Upon this rescaling the effective Lagrangian takes the form

$$\mathcal{L}_{\mathrm{eff}} = -\frac{1}{4}F_{\mu\nu}^2 + \left(\partial_\alpha + ie_n A_\alpha\right)\bar{n}_i\left(\partial^\alpha - ie_n A^\alpha\right)n^i - m^2\,\bar{n}_i n^i,\qquad(43)$$

[†]The $O(k^4)$, $O(k^4)$, and so terms can be ignored since they have no impact on the position of the pole at $k^2 = 0$ of the photon Green's function.

where the electric charge of the n quanta e_n is

$$e_n \equiv m \sqrt{\frac{12\pi}{N}} \, . \tag{44}$$

It has dimension of mass, which is the correct dimension of the electric charge in two-dimensional theories. Moreover, one should stress that at large N the electric charge becomes small, $e_n/m \ll 1$, which implies, in turn, weak coupling.

Emergence of the massless gauge U(1) field ensures the presence of the Coulomb potential between charges states. In two dimensions static Coulomb potential is a linear rising potential. It leads to the confinement of kinks which carry electric charge.[35] Therefore this phase of the theory is called Coulomb/confining phase.

5.2. Supersymmetric CP(N − 1)

It is easy to generalize the large-N solution of Sec. 5.1 to $\mathcal{N} = (2,2)$ model.[35,36] The Lagrangian (25) is quadratic in both, the n fields and their fermion superpartners ξ. Therefore, they can be integrated out exactly. Note that the auxiliary fields A, σ and $\lambda_{L,R}$ form a supermultiplet. Hence, it is sufficient to find the kinetic term and mass for one of them in order to determine them all, provided that supersymmetry is unbroken. As we will see momentarily, it is indeed unbroken.

As in (38) we set $A_\mu = 0$, and then integrate out n^i and ξ^i. This yields

$$\frac{\mathrm{Det}\left(-\partial_\alpha^2 - 2|\sigma|^2\right)^N}{\mathrm{Det}\left(-\partial_\alpha^2 - D - 2|\sigma|^2\right)^N} \, . \tag{45}$$

The denominator comes from the boson loop while the numerator from the fermion loop. It is obvious that supersymmetric vacuum (with $E_{\mathrm{vac}} = 0$) is attained at $D = 0$, when the ratio of the determinants in (45) reduces to unity.

The above conclusion is confirmed by an explicit calculation of the effective potential, the analog of (64),

$$V_{\mathrm{eff}} = \frac{N}{4\pi}\left[\left(D + 2|\sigma|^2\right)\log\frac{D + 2|\sigma|^2}{m^2} - D - 2|\sigma|^2\log\frac{2|\sigma|^2}{m^2}\right], \tag{46}$$

where we carried out renormalization using the analog of (39),

$$2|\sigma_{\mathrm{vac}}|^2 \equiv m^2 = M_{\mathrm{uv}}^2 \, e^{-8\pi/Ng^2} \, . \tag{47}$$

The vacuum values of D and $|\sigma|$ are obtained through minimization, i.e. by differentiating V_{eff} in (46) over D and $2|\sigma|^2$,

$$\log \frac{D + 2|\sigma|^2}{m^2} = 0\,,$$

$$\log \frac{D + 2|\sigma|^2}{m^2} - \log \frac{2|\sigma|^2}{m^2} = 0\,. \tag{48}$$

As was mentioned, the mass of the ξ field is the same as as that of n, due to supersymmetry.

The kinetic term of the A_μ field and its superpartners is dynamically generated in much the same way as in Sec. 5.1. A crucial difference is that now the photon field A_μ acquires a nonvanishing (albeit small) mass through the Schwinger mechanism: the massless fermion loop shifts the pole in the photon propagator away from zero. This was noted already in 1979.[35] Needless to say, all superpartners of the photon field receive the same mass.

Consequences of massless vs. massive photon in two dimensions are radically different. Massless photons in two-dimensions (non-supersymmetric $\text{CP}(N-1)$) lead to confinement of charged particles, while massive photons (supersymmetric $\text{CP}(N-1)$) do not confine. In one-to-one correspondence with this is the existence of N degenerate vacua in the non-confining case. In the confining case (i.e. massless photon) one of these vacua remains genuine while the remaining $N-1$ are uplifted and become quasistable states.

6. Twisted masses

The so-called twisted masses is the only mass deformation of the $\mathcal{N} = (2,2)$ model which preserves supersymmetry. The essence of this deformation is as follows.[41] One starts from four-dimensional $\text{CP}(N-1)$ model and couples $N-1$ conserved U(1) currents of this model to background gauge four-potential A_μ. Then one reduces the model to two dimensions t and z simultaneously declaring the background fields A_x and A_y (Fig. 1) to be nonvanishing constants. The twisted masses μ and $\bar\mu$ are proportional to $A_x \pm iA_y$.

In the geometric formulation of Sec. 4.1 the formal procedure can be described as follows. The theory (10) can be interpreted as an $\mathcal{N} = 1$ theory of $N-1$ chiral superfields in four dimensions. The theory possesses $N-1$ U(1) isometries parametrized by t^a, $a = 1,\ldots,N-1$. The Killing vectors of the isometries can be expressed via derivatives of the Killing potentials

$$D^a(\phi, \phi^\dagger),$$

$$\frac{d\phi^i}{dt_a} = -iG^{i\bar{j}}\frac{\partial D^a}{\partial \bar{\phi}^{\bar{j}}}, \qquad \frac{d\bar{\phi}^{\bar{j}}}{dt_a} = iG^{i\bar{j}}\frac{\partial D^a}{\partial \phi^i}. \qquad (49)$$

This defines the U(1) Killing potentials up to additive constants.

The isometries are evident from the expression (11) for the Kähler potential,

$$\delta\phi^i = -i\delta t_a (T^a)^i_k(\phi)^k, \qquad \delta\bar{\phi}^{\bar{j}} = i\delta t_a (T^a)^{\bar{j}}_{\bar{l}}\bar{\phi}^{\bar{l}}, \qquad a = 1,\ldots,N-1, \qquad (50)$$

(together with the similar variation of fermionic fields), where the generators T^a have a simple diagonal form,

$$(T^a)^i_k = \delta^i_a \delta^a_k, \qquad a = 1,\ldots,N-1. \qquad (51)$$

The explicit form of the Killing potentials D^a in CP(N–1) with the Fubini–Study metric is

$$D^a = \frac{2}{g_0^2}\frac{\bar{\phi}T^a\phi}{1+\bar{\phi}\phi}, \qquad a = 1,\ldots,N-1. \qquad (52)$$

Here we use the matrix notation implying that ϕ is a column ϕ^i and $\bar{\phi}$ is a row $\bar{\phi}^{\bar{j}}$.

The isometries allow us to introduce an interaction with $N-1$ *distinct* background U(1) gauge superfields V_a by modifying the Kähler potential (11) in a gauge invariant way,

$$K(\Phi,\bar{\Phi}) \to \tilde{K}(\Phi,\bar{\Phi},V) = \frac{2}{g_0^2}\log\left(1 + \bar{\Phi}\,e^{V_a T^a}\,\Phi\right). \qquad (53)$$

where

$$V_a = -\mu_a\bar{\theta}(1+\gamma_5)\theta - \bar{\mu}_a\bar{\theta}(1-\gamma_5)\theta. \qquad (54)$$

Thus, in our notation the complex masses m_a are linear combinations of the constant U(1) gauge potentials,

$$m_a = A^a_y + iA^a_x, \qquad \bar{m}_a = m^*_a = A^a_y - iA^a_x. \qquad (55)$$

Passing to two dimensions we assume, of course, that there is no dependence on x and y in the chiral fields. It gives us the Lagrangian with the twisted masses included,[41]

$$\mathcal{L}_m = \int d^4\theta\, K_m(\Phi,\Phi^\dagger,V) = G_{i\bar{j}}\, g_{MN}\left[\mathcal{D}^M\phi^{\dagger\bar{j}}\,\mathcal{D}^N\phi^i + i\bar{\psi}^{\bar{j}}\gamma^M\mathcal{D}^N\psi^i\right]$$

$$- \frac{1}{2}R_{i\bar{j}k\bar{l}}\,(\bar{\psi}^{\bar{j}}\psi^i)(\bar{\psi}^{\bar{l}}\psi^k), \qquad (56)$$

where summation over M includes, besides $M = \alpha = 0, 1$, also $M = +, -$. The metric g_{MN} and extra γ matrices are

$$g_{MN} = \begin{pmatrix} 1 & 0 & 0 & 0 \\ 0 & -1 & 0 & 0 \\ 0 & 0 & 0 & -\frac{1}{2} \\ 0 & 0 & -\frac{1}{2} & 0 \end{pmatrix}, \qquad \gamma^+ = -i(1 + \gamma_5), \quad \gamma^- = i(1 - \gamma_5). \quad (57)$$

The gamma-matrices satisfy the following algebra:

$$\bar{\Gamma}^M \Gamma^N + \bar{\Gamma}^N \Gamma^M = 2g^{MN}, \quad (58)$$

where the set $\bar{\Gamma}^M$ differs from Γ^M by interchanging of the $+, -$ components, $\bar{\Gamma}^\pm = \Gamma^\mp$. The gauge covariant derivatives $\mathcal{D}^M$ are defined as

$$\mathcal{D}^\alpha \phi = \partial^\alpha \phi, \qquad \mathcal{D}^\alpha \bar{\phi} = \partial^\alpha \bar{\phi},$$

$$\mathcal{D}^+ \phi = -\bar{\mu}_a T^a \phi, \qquad \mathcal{D}^- \phi = \mu_a T^a \phi,$$

$$\mathcal{D}^+ \bar{\phi} = \bar{\phi} T^a \bar{\mu}_a, \qquad \mathcal{D}^- \bar{\phi} = -\bar{\phi} T^a \mu_a, \quad (59)$$

and similarly for $\mathcal{D}^M \psi$, while the general covariant derivatives $D^M \psi$'s are

$$D^M \psi^i = \mathcal{D}^M \psi^i + \Gamma_{kl}^i \, \mathcal{D}^M \phi^k \, \psi^l. \quad (60)$$

In the geometrical formulation we have $N - 1$ complex twisted mass parameters. Introduction of the twisted masses in the gauged formulation will be discussed in Part II, Secs. 7.1 and 7.2. In the gauged formulation there are N complex twisted mass parameters m_i related to μ^a,

$$\mu^i = m_i - m_N, \qquad i = 1, 2, ..., N - 1. \quad (61)$$

(see (23)) and subject to the constraint

$$\sum_{i=1}^{N} m_i = 0. \quad (62)$$

One of our tasks in what follows is the study of the phase diagram of the two-dimensional model on the string world sheet. To this end it is convenient to have a discrete symmetry. A Z_N symmetry is guaranteed if the mass parameters are adjusted as

$$m_j = m_0 \exp\left(\frac{2\pi i j}{N}\right), \qquad j = 1, 2, ..., N. \quad (63)$$

Such a choice is referred to as Z_N symmetric. It is always assumed in what follows if not stated to the contrary.

Note that m_0 can be chosen to be real and positive. Then m_N is real and positive too. Alternatively, if $m_0 = |m_0| \exp(-2\pi i/N)$, then m_1 is real and positive.

7. Large-N solutions with twisted masses

In this section we will briefly review those two-dimensional sigma models that are in the limelight ever since the discovery of the non-Abelian strings.

7.1. $CP(N-1)$ with Z_N symmetric masses

As a world-sheet model in the non-supersymmetric context, $CP(N-1)$ with twisted masses was discussed in,[10] and its large-N solution in the Z_N symmetric case was found in.[42]

In the gauged formulation the Lagrangian has the form

$$\mathcal{L} = |\mathcal{D}_\alpha n^i|^2 + D\left(|n^i|^2 - 2\beta\right) + \sum_{i=1}^{N} \left|(\sigma - m_i)n^i\right|^2 , \qquad (64)$$

were $\mathcal{D}_\alpha = \partial_\alpha - iA_\alpha$, the mass parameters m_i are defined in Eq. (63), and 2β is the bare coupling constant, see Eq. (22).

Intuitively it is clear that the structure of the solution depends on the ratio of m and the dynamical scale Λ generated in this theory. As we will see below, there are two distinct cases – the Higgs and the Coulomb/confining phases – in this theory at large and small $|m_0/\Lambda|$, respectively.

In the Higgs phase the field n^{i_0} develops a VEV. One can always choose $i_0 = 1$ and denote $n^{i_0} = n^1 \equiv n$. There are N equivalent choices, N vacua. This corresponds to the spontaneous breaking of Z_N. Setting the background A_α field to zero, as in Sec. 5.1, and integrating out all n_i except $n^{i_0} = n$ we arrive at

$$\mathcal{L}_{\text{eff}} = |\partial_\alpha n|^2 + \left(D + |\sigma - m_1|^2\right)|n|^2$$

$$+ \frac{1}{4\pi} \sum_{i=2}^{N} \left(D + |\sigma - m_i|^2\right) \left[1 - \log \frac{D + |\sigma - m_i|^2}{\Lambda^2}\right]$$

$$+ \frac{1}{4\pi} c \sum_{i=2}^{N} |\sigma - m_i|^2 \qquad (65)$$

where

$$c = \frac{1}{N} \sum_{i=2}^{N} \left(1 - \frac{m_i}{m_1}\right) \log \frac{|m_i - m_1|^2}{\Lambda^2} , \qquad (66)$$

and we used the renormalization condition

$$\frac{2}{g_0^2} = \frac{N}{4\pi} \ln \frac{M_{uv}^2}{\Lambda^2} \,. \tag{67}$$

This condition introduces the dynamical scale Λ through dimensional transmutation, just like in Sec. 5.1.

Minimizing this effective potential with respect to D, n and σ we determine the vacuum values of these parameters. It is not surprising that n_{vac} turns out to be exactly as it follows from the renormalized constraint $|n^i|^2 = 2\beta$ in the Higgs phase,

$$n_{\text{vac}} = \left(\frac{N}{2\pi} \ln \left| \frac{m_0}{\Lambda} \right| \right)^{1/2} \tag{68}$$

while

$$D_{\text{vac}} = 0 \,, \qquad \sigma_{\text{vac}} = m_1 \,. \tag{69}$$

It is also obvious that there are N vacua corresponding to cyclic permutation. In each of them the Z_N symmetry is spontaneously broken.

Substituting (68) and (69) in (65) we obtain the vacuum energy density,

$$E_{\text{Higgs vac}} = \frac{N}{2\pi} m_0^2 \,, \tag{70}$$

where the parameter m_0 is assumed to be real and positive (see the bold line in Fig. 5). This formula is valid at

$$m_0 \geq \Lambda \,. \tag{71}$$

The Higgs phase has a clear-cut meaning at large m_0. The above result is compatible with intuition. We will see momentarily that the lower bound of the allowed domain, $m_0 = \Lambda$, is the phase transition point (presumably, the phase transition is of the second order).

Now let us discuss the Coulomb/confining phase. At small $|m_0|$

$$\sigma_{\text{vac}} = 0 \,, \qquad (n^i)_{\text{vac}} = 0 \text{ for all } i, \ i = 1, 2, ..., N \,, \tag{72}$$

and

$$D_{\text{vac}} = \Lambda^2 - m_0^2 \,. \tag{73}$$

The Z_N symmetry remains unbroken. Hence, we deal with a unique vacuum. Inspecting Eq. (64) we conclude that in the saddle point the mass of all n^i quanta is Λ, independent of the value of the mass deformation parameter m_0.

The vacuum energy in the Coulomb phase is obtained by substituting the vacuum values (72) and (73) in (65) and using expression (66) for the value of the constant c. In this way one arrives at

$$E_{\text{Coulomb vac}} = \frac{N}{4\pi} \left\{ \Lambda^2 + m_0^2 + m_0^2 \log \frac{m_0^2}{\Lambda^2} \right\} , \tag{74}$$

(see the solid line in Fig. 5).

At the point of the phase transition at $m_0 = \Lambda$ the energy densities in the both phases coincide. Moreover, their first derivatives with respect to m_0^2 at this point coincide too. The dashed line corresponds to a formal extrapolation of the Higgs and Coulomb/confinement vacuum energies to "forbidden" values of m_0^2 below and above the phase transition point.

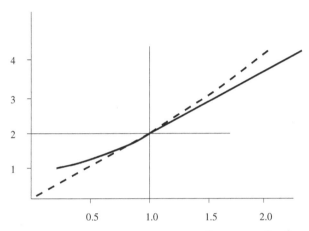

Fig. 5. Normalized vacuum energies $(4\pi E_{\text{vac}}/N \Lambda^2)$ versus m_0^2/Λ^2. The solid line shows the actual vacuum energy, while dashed lines correspond to a formal extrapolation of the Higgs and Coulomb/confinement vacuum energies to unphysical values of m below and above the phase transition point, respectively.

To reiterate, at $m_0 \geq \Lambda$, at weak coupling, we have N strictly degenerate vacua; the Z_N symmetry is broken. At $m_0 \leq \Lambda$ the Z_N symmetry is unbroken, and the vacuum is unique. The order parameter which marks these vacua is the VEV of n^i.

Introduction of an additional axion field in this model is discussed in.[43]

7.2. Supersymmetric $CP(N-1)$ with $\mathcal{N} = (2, 2)$

Two-dimensional $CP(N-1)$ models with twisted masses and with $\mathcal{N} = (2, 2)$ supersymmetry[44] emerge as effective low-energy theories on the world sheet of non-Abelian strings in a class of four-dimensional $\mathcal{N} = 2$ gauge theories with unequal (s)quark masses,[9] for a complete derivation see.[18] In the gauged formulation the $\text{CP}(N-1)$ Lagrangian with the twisted masses (replacing the zero mass limit (25)) is (see e.g.[18,27])

$$
\mathcal{L} = \left[|\mathcal{D}n|^2 + 2 \left| \sigma - \frac{m^l}{\sqrt{2}} \right|^2 |n^l|^2 + iD \left(|n^l|^2 - 2\beta \right) \right.
$$

$$
+ \bar{\xi}_R \, i\mathcal{D}_L \xi_R + \bar{\xi}_L \, i\mathcal{D}_R \xi_L + i\sqrt{2} \left(\sigma - \frac{m^l}{\sqrt{2}} \right) \bar{\xi}_{Rl} \xi_L^l + i\sqrt{2} \left(\bar{\sigma} - \frac{\bar{m}^l}{\sqrt{2}} \right) \bar{\xi}_{Ll} \xi_R^l
$$

$$
+ \left. \left(i\sqrt{2}\, \bar{\xi}_R \lambda_L \; n - i\sqrt{2}\, \bar{n} \, \lambda_R \, \xi_L + \text{H.c.} \right) \right]. \tag{75}
$$

To solve the model in the large-N limit one can basically repeat the derivation of Sec. 5.2 since the fields n and ξ enter in the Lagrangian bilinearly. Integrating them out yields

$$
\frac{\prod_{i=2}^{N} \det \left(-\partial_k^2 + \left| \sqrt{2}\sigma - m_i \right|^2 \right)}{\prod_{i=2}^{N} \det \left(-\partial_k^2 + iD + \left| \sqrt{2}\sigma - m_i \right|^2 \right)}, \tag{76}
$$

and we obtain[27] an analog of the nonsupersymmetric formula (65) for the vacuum structure. Above we integrated over $N-1$ fields n^i and ξ^i with $i > 1$. The resulting effective action is to be considered as a functional of $n^1 \equiv n$, D and σ. We will again assume that the twisted masses are Z_N symmetric, see (63).

The ensuing effective Lagrangian is

$$
\mathcal{L} = \sum_{i=2}^{N} \frac{1}{4\pi} \left\{ \left(iD + \left| \sqrt{2}\sigma - m_i \right|^2 \right) \left(\log \frac{M_{\text{uv}}^2}{iD + \left| \sqrt{2}\sigma - m_i \right|^2} + 1 \right) \right.
$$

$$
\left. - \left| \sqrt{2}\sigma - m_i \right|^2 \left(\log \frac{M_{\text{uv}}^2}{\left| \sqrt{2}\sigma - m_i \right|^2} + 1 \right) \right\}, \tag{77}
$$

Using (67) for the bare coupling constant we can eliminate M_{uv} in a usual way. Then the effective potential as a function of n, D and σ fields takes the form

$$V_{\text{eff}} = \left(iD + \left| \sqrt{2}\sigma - m_1 \right|^2 \right) |n|^2$$

$$- \frac{1}{4\pi} \sum_{i=2}^{N} \left(iD + \left| \sqrt{2}\sigma - m_i \right|^2 \right) \log \frac{iD + \left| \sqrt{2}\sigma - m_i \right|^2}{\Lambda^2}$$

$$+ \frac{1}{4\pi} \sum_{i=2}^{N} \left| \sqrt{2}\sigma - m_i \right|^2 \log \frac{\left| \sqrt{2}\sigma - m_i \right|^2}{\Lambda^2} + \frac{1}{4\pi} iD (N-1) .$$

$$(78)$$

Minimization of (78) gives two solutions: either

$$iD + \left| \sqrt{2}\sigma - m_1 \right|^2 = 0 \tag{79}$$

or

$$n = 0 . \tag{80}$$

These two distinct solutions correspond to the weak and strong-coupling regimes of the theory, respectively. They are analogous to two phases we observed in Sec. 7.1. In the case at hand supersymmetry is unbroken but in both weak and strong-coupling regimes the Z_N symmetry is spontaneously broken, and there are N distinct vacua. At strong coupling, in the regime $n_{\text{vac}} = 0$, the order parameter is $\bar{\xi}_R \xi_L$ and its Hermitian conjugate.

As usual, supersymmetry suppresses phase transitions. The passage from weak (large $|m_0|$) and strong (small $|m_0|$) regimes presents a crossover rather than a phase transition. Supersymmetry is preserved in both regimes.

At large $|m_0|$

$$D = 0, \qquad \sqrt{2}\sigma_{\text{vac}} = m_1, \qquad |n_{\text{vac}}|^2 = \frac{N}{2\pi} \log \frac{m}{\Lambda} . \tag{81}$$

This is similar to the Higgs phase in Sec. 7.1.

For small $|m_0|$ we have

$$D = 0, \qquad n_{\text{vac}} = 0, \tag{82}$$

while the vacuum equation on σ can be written as

$$\prod_{i=1}^{N} \left| \sqrt{2}\sigma - m_i \right| = \Lambda^N . \tag{83}$$

For the Z_N-symmetric masses Eq. (83) can be solved. Say, for even N one can rewrite this equation in the form

$$\left| \left(\sqrt{2}\sigma \right)^N - m^N \right| = \Lambda^N , \qquad (84)$$

due to the fact that with the masses given in (63)

$$\sum m_i = 0 ,$$

$$\sum_{i,j;\, i \neq j} m_i m_j = 0 ,$$

$$\cdots$$

$$\sum_{i_1, i_2, \ldots, i_{N-1}} m_{i_1} m_{i_2} \ldots m_{i_{N-1}} = 0 , \qquad (i_1 \neq i_2 \neq \ldots \neq i_{N-1}) . \qquad (85)$$

Equation (84) has N solutions (i.e. N distinct vacua),

$$\sqrt{2}\sigma = \left(\Lambda^N + m_0^N \right)^{1/N} \exp\left(\frac{2\pi i k}{N} \right) , \quad k = 1, \ldots, N, \qquad (86)$$

The crossover occurs at $m_0 = \Lambda$. The width of the crossover domain is not seen in the leading order in $1/N$. In fact, it is exponentially small in N. The transition from weak to strong coupling is depicted in Fig. 6.

Summarizing, in both regimes – weak and strong coupling – supersymmetry is unbroken and there is no confinement of charged particles due to the fact that the photon (which becomes dynamical) acquires a mass. The spontaneous breaking of Z_N implies N degenerate vacua.

7.3. Curves of marginal stability in (2.2) CP(N − 1) with Z_N twisted masses

An exact twisted superpotential of the Veneziano-Yankielowicz type[45] is known to describe the $\mathcal{N} = (2,2)$ supersymmetric CP(N − 1) model.[36,44,46–48] Integrating out the fields n^P and ρ^K we obtain the following exact twisted superpotential:

$$\mathcal{W}_{\text{CP}(N-1)}(\sigma) = \frac{1}{4\pi} \left\{ \sum_{l=1}^{N} \left(\sqrt{2}\sigma - m_l \right) \ln \frac{\sqrt{2}\sigma - m_l}{\Lambda} - N \sqrt{2}\sigma \right\} , \qquad (87)$$

where we use one and the same notation σ for the twisted superfield[36] and its lowest scalar component. Minimizing this superpotential with respect to

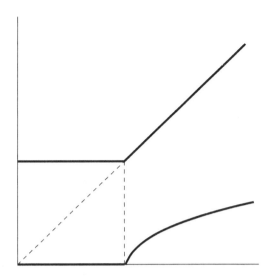

Fig. 6. Plots of n and σ VEVs (thick lines) vs. m_0 in the $\mathcal{N} = (2,2)$ CP($N-1$) model with the Z_N-symmetric twisted masses.

σ we get the equation for the σ-field VEVs,

$$\prod_{l=1}^{N}(\sqrt{2}\sigma - m_l) = \Lambda^N . \tag{88}$$

This equation has N roots σ_p ($p = 1, ..., N$) associated with N vacua of the CP($N-1$) model. Note that this *exact* equation is a holomorphic version of Eq. (83) which appears in the large-N solution. It takes into account that chiral U(1)$_R$ symmetry is broken by chiral anomaly down to discrete Z_{2N} symmetry. This is the reason for the presence of N distinct vacua.

The masses of the BPS kinks interpolating between the vacua σ_p and $\sigma_{p'}$ are given by the appropriate differences of the superpotential (87) calculated at distinct roots,[44,48,49]

$$M_{pp'}^{\text{BPS}} = 2 \left| \mathcal{W}_{\text{CP}(N-1)}(\sigma_{p'}) - \mathcal{W}_{\text{CP}(N-1)}(\sigma_p) \right| , \qquad p, p' = 1, ..., N . \tag{89}$$

In addition to kinks, the BPS spectrum of the model contains elementary excitations with masses given by $|m_l - m_p|$ ($l = 1, ..., N$ and $p = 1, ..., N$).

Due to the presence of branches in the logarithmic functions in (87) each kink comes together with a tower of dyonic kinks carrying global U(1) charges. The dyonic kinks are reflected in (89) through terms

$$\text{integer} \times i\, m_l \tag{90}$$

with different l which appear from the logarithm branches. We stress that all these kinks with the imaginary part (90) in the mass formula (89) interpolate between the same pair of vacua: p and p'.

Generically there are way too many choices in (89). Not all of them are realized. Moreover, the kinks present in the quasiclassical domain (i.e. at large $|m_0|$) decay on the curves of marginal stability (CMS) or form new bound states. Therefore, the quasiclassical spectrum outside CMS and the quantum spectrum inside CMS (i.e. at small $|m_0|$) are different. This phenomenon is referred to as "wall crossing." There exists a general procedure[50] which allows one to determine the full BPS spectrum starting from the strong coupling spectrum inside CMS. However, this procedure is rather cumbersome. In certain cases one can use a simpler approach based on analysis of various limits. Below we will briefly review the BPS spectra and CMS in CP(1) and CP(2) with the Z_N twisted masses.[48,51,52]

These CMS were obtained by matching the weak coupling BPS spectrum found using semiclassical considerations with the strong coupling spectrum found from mirror representation of the CP($N - 1$) model.[53] The later spectrum includes only N kinks which become massless at strong coupling.

The strong coupling spectrum of the CP(1) model includes two BPS states with the following masses:

$$M_0 = |m_D^{\text{CP}(1)} + i m_1|, \qquad M_1 = |m_D^{\text{CP}(1)} + i m_2|, \tag{91}$$

where

$$m_D^{\text{CP}(1)} = \frac{1}{\pi}\left[2\sqrt{m_0^2 + \Lambda^2} - m_0 \log \frac{\sqrt{m_0^2 + \Lambda^2} + m_0}{\sqrt{m_0^2 + \Lambda^2} - m_0} \right], \tag{92}$$

while m_1, m_2 are given in (63) with $N = 2$, namely, $m_1 = -m_0$, $m_2 = m_0$.

The weak coupling spectrum of the CP(1) model includes the tower of dyonic kinks

$$M_n = \left| m_D^{\text{CP}(1)} + i m_1 + i n(m_2 - m_1) \right|, \tag{93}$$

where n is an integer. The two states of the strong coupling spectrum (91) belong to this tower with $n = 0, 1$. Other states from this tower as well as elementary (i.e. non-kink) states decay on the closed single CMS around the origin in the m_0^2 complex plane.[51]

Now let us briefly discuss a more contrived situation, the CP(2) model with the Z_3 twisted masses. For simplicity we will restrict ourselves to kinks interpolating between the third and first vacua ($\sqrt{2}\sigma_3 \approx m_3$ and $\sqrt{2}\sigma_1 \approx m_1$ in the large mass limit). The strong coupling spectrum of the CP(2) model consists of three states

$$M_k^{13} = \left| m_D^{CP(2)} + im_k \right|, \qquad k = 1, 2, 3, \tag{94}$$

where

$$m_D^{CP(2)} = -\frac{1}{2\pi} \left(e^{2\pi i/3} - 1 \right) \left\{ 3 \sqrt[3]{m_0^3 + \Lambda^3} + \sum_j m_j \log \frac{\sqrt[3]{m_0^3 + \Lambda^3} - m_j}{\Lambda} \right\} \tag{95}$$

and the mass terms m_k are given in (63) with $N = 3$.

The weak coupling spectrum of the CP(2) model includes two towers of dyonic kinks

$$M_{n_1}^{13} = \left| m_D^{CP(2)} + im_3 + in_1(m_1 - m_3) \right| \text{ and}$$

$$M_{n_2}^{13} = \left| m_D^{CP(2)} + im_3 + in_2(m_1 - m_3) + i(m_3 - m_2) \right|, \tag{96}$$

plus elementary states. Here n_1 and n_2 are integers. The two states of the strong coupling spectrum (94) with $k = 3, 1$ both belong to the first tower. The CP(2) model with the Z_3 masses has several CMS where all other states except these two decay,[52] see Fig 7. The third kink of the strong coupling spectrum (with $k = 2$) does not make it to the weak coupling domain. It decays on the most inner curve in Fig. 7.

8. Weighted CP(N, M) models and zn model

Considering $\mathcal{N} = 2$ bulk theories of the type discussed in Sec. 2 with $N_f > N$ (here N_f is the number of flavors) we arrive at the so-called semilocal non-Abelian strings.[54–57] Instead of N_f we can introduce a positive number M,

$$N_f = N + M. \tag{97}$$

The semilocal string solutions on the Higgs branches (typical for multiflavor theories) usually are not fixed-radius strings, but, rather, possess radial moduli ρ^k, also known as the size moduli (see[58] for a review of the Abelian semilocal strings).

As previously, the orientational moduli of the semilocal non-Abelian string can be described by a complex vector n^P (here $P = 1, ..., N$), while

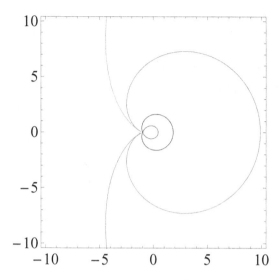

Fig. 7.   The decay curves of CP^2 in m_0^3 plane.

its size moduli are parametrized by a complex vector ρ^K (here $K = N + 1, ..., N + M$).

Originally it was conjectured[54] (on the basis of string theory arguments) that the effective two-dimensional sigma model describing low-energy dynamics on the semilocal string is the so-called weighted $CP(N, M)$ model. This turned out to be not quite correct. The world-sheet theory of the moduli fields was derived in[57,59] and is known as the zn model. Its Lagrangian is

$$\mathcal{L}_{zn} = \left|\partial_\alpha (n^P \rho^K)\right|^2 + \left|\mathcal{D}_\alpha n^P\right|^2 + |m_K - m_P|^2 \left|n^P\right|^2 \left|\rho^K\right|^2$$

$$+ \left|\sqrt{2}\sigma - m_P\right|^2 \left|n^P\right|^2 + iD\left(|n^P|^2 - 2\beta\right)^2,$$

$$P = 1, ..., N, \qquad K = N + 1, ..., N + M, \tag{98}$$

The zn model so far remains largely unexplored. We refer the reader to the original papers for a brief discussion.

The zn model is similar but not identical to the weighted $CP(N, M)$ model. However, it was demonstrated[57,59] that its vacuum structure and BPS spectrum coincide with those of the $\mathcal{N} = 2$ weighted $CP(N, M)$ model. Moreover, at $N \to \infty$ the zn model and the weighted $CP(N, M)$ model coincide. Thus, the difference between them lies in the non-BPS sector at finite N.

Technically, it seems more convenient to work with the weighted $CP(N, M)$ model. The bosonic part of its Lagrangian is

$$\mathcal{L}_{\text{WCP}} = \left|\mathcal{D}_\alpha n^P\right|^2 + \left|\tilde{\mathcal{D}}_\alpha \rho^K\right|^2 + \left|\sqrt{2}\sigma - m_P\right|^2 \left|n^P\right|^2$$

$$+ \left|\sqrt{2}\sigma - m_K\right|^2 \left|\rho^K\right|^2 + iD\left(|n^P|^2 - |\rho^K|^2 - 2\beta\right)^2,$$

$$P = 1, ..., N, \qquad K = N+1, ..., N+M, \tag{99}$$

where

$$\mathcal{D}_\alpha = \partial_\alpha - iA_\alpha, \qquad \tilde{\mathcal{D}}_\alpha = \partial_\alpha = iA_\alpha. \tag{100}$$

The mass terms M_K and M_P in (99) are viewed as generic in this section.

The fields n^P and ρ^K have the opposite charges, $+1$ and -1, with respect to the auxiliary U(1) gauge field. This seemingly insignificant detail is crucial. Strictly speaking, the name 'weighted CP' model is misleading since the geometry of the target-space following from (99) has nothing to do with the $CP(N-1)$ geometry in which all target space covariant quantities reduce to the metric, see (18). The weighted CP models are not even renormalizable in the usual sense of this word. Nevertheless, the large-N solution exists and is unique.[60] We will discuss it in more detail in Sec. 9.6.

9. Heterotic models

Heterotic two-dimensional models we will discuss below have two chiral supercharges, say, Q_L and $\bar{Q}_L$ with the defining anticommutator

$$\{Q_L, \bar{Q}_L\} = 2(H - P). \tag{101}$$

They are known as $\mathcal{N} = (0, 2)$ supersymmetric sigma models.[‡] Previously they were studied mainly from the mathematical perspective.[20,21,23,24] They can be divided into two classes: the so-called minimal and nonminimal models. This classification is in a sense similar to pure Yang-Mills theories and Yang-Mills theories with matter. Later we will explain the difference between these two classes in more detail. In particular, the minimal CP(1) model was considered in.[22] This minimal model cannot be extended to

---

[‡]In Sec. 9.6 we will briefly comment on a $\mathcal{N} = (0, 1)$ model.

$CP(N-1)$ with $N > 2$. The general hypercurrent structure in $\mathcal{N} = (0,2)$ was analyzed in.[25] In what follows we will focus on those heterotic two-dimensional models that are obtained on the world sheet of non-Abelian strings.

9.1. *How heterotic models appear*

If the bulk four-dimensional theory has $\mathcal{N} = 2$ and supports 1/2-BPS strings, then the low-energy theory on its world sheet has four supercharges and, thus, possesses $\mathcal{N} = (2,2)$ supersymmetry. Now, if we slightly deform the bulk theory breaking $\mathcal{N} = 2$ down to $\mathcal{N} = 1$ we will have four supercharges in the bulk. For small deformations BPS saturation remains intact and so does the the target space of the two-dimensional sigma model. Now, the world-sheet model must have two, not four supercharges. However, Zumino's theorem tells us that given a Kähler target space any supersymmetric nonchiral model is automatically uplifted to $\mathcal{N} = (2,2)$, i.e. four supercharges.

A way out was suggested by Edalati and Tong[16] who conjectured a nonminimal $\mathcal{N} = (0,2)$ model on the string world sheet in the case of nonvanishing $\mathcal{N} = 2$ breaking deformation in the bulk (see also[61]). This nonminimal theory was derived by Shifman and Yung[17] from the analysis of the string solution. The nonminimal theory, as it emerged on the string world sheet, has no twisted masses. In fact, even today we do not know which bulk theory might result in the nonminimal heterotic model with twisted masses. However, the inclusion of the twisted masses is straightforward in the two-dimensional model *per se*, without any reference to the bulk theory. This is the model to be discussed below too.

Large-N solutions of the heterotic models generically exhibit spontaneous breaking of supersymmetry. For nonvanishing twisted masses this breaking occurs at the tree level.

9.2. *Minimal vs. nonminimal $CP(N-1)$ models with $\mathcal{N} = (0,2)$ supersymmetry*

9.2.1. *Geometric formulation*

The minimal model can be obtained from (12) by keeping only left-handed fermions and discarding all right-handed fermions,

$$\mathcal{L}_{\mathcal{N}=(0,2)} = G_{i\bar{j}} \left[\partial^\mu \phi^{\dagger \bar{j}} \partial_\mu \phi^i + i\bar{\psi}_L^{\bar{j}} \mathcal{D}_R \psi_L^i \right], \qquad (102)$$

where

$$\mathcal{D}_R \psi_L^i = \partial_R \psi_L^i + \Gamma_{kl}^i (\partial_R \phi^k) \psi_L^l, \tag{103}$$

and

$$\partial_R \equiv \partial_t - \partial_z. \tag{104}$$

The fields ϕ and ψ_L form an $(0,2)$ supermultiplet. In terms of $\mathcal{N} = (0,2)$ superfields[20] one can act as follows. Introduce a superfield

$$A = \phi(x_R + 2i\theta^\dagger\theta, x_L) + \sqrt{2}\,\theta\,\psi_L(x_R + 2i\theta^\dagger\theta, x_L), \tag{105}$$

where θ is a single (right-handed) complex Grassmann variable on the $(0,2)$ superspace, and

$$x_L = t - z \equiv x^0 - x^1, \qquad x_R = t + z \equiv x^0 + x^1. \tag{106}$$

Then

$$\mathcal{L}_{\min} = \frac{1}{2} \int d\theta d\bar{\theta} \left[K_i(A, A^\dagger) i\partial_R A^i + \text{H.c.} \right]$$

$$= -\frac{1}{4} \int d\theta\, G_{i\bar{j}}(A, A^\dagger)(\bar{D}A^{\dagger\bar{j}})\,i\partial_R A^i + \text{H.c.} \tag{107}$$

Warning: Due to an anomaly pointed out in[40] the heterotic minimal model is self-consistent only for CP(1) (see also[31]). Minimal CP(N − 1) models with N > 2 do not exist. However, minimal heterotic O(N) models exist for any N. For N > 3 they have $(0,1)$ supersymmetry. For N = 3 we have O(3) = CP(1). Nonminimal models presented in (109) exist for CP(N − 1) at any N.

Alternatively, one can start from Eq. (25) and discard all terms containing ξ_R.

One last remark is in order here concerning the minimal CP(1) model presented in (107). This is a strongly coupled theory. Since large-N expansion is unavailable, we cannot solve it by virtue of the large-N expansion (we will apply it, however, to nonminimal heterotic CP(N − 1)). Nevertheless, one feature of this model is known. As was shown in,[25] current algebra in this model allows for a nonperturbative Schwinger term (see Eq. (5.7) in[25]), namely, $C \sim \Lambda^2 \sim M_{\text{uv}}^2 \exp\left(-\frac{4\pi}{g_0^2}\right)$. This Schwinger term is saturated by a single instanton due to the fact that in the model at hand it has just two fermion zero modes. The occurrence of this Swinger term implies spontaneous supersymmetry breaking. The interpolating field for Goldstino is

$$g \sim R_{i\bar{j}} \left(\partial_R \phi^i \right) \bar{\psi}_L^{\bar{j}} \,. \tag{108}$$

Spontaneous breaking of supersymmetry will be explicit in the large-N solution of the nonminimal heterotic CP($N-1$).

The bulk theories supporting non-Abelian strings are usually obtained by deforming $\mathcal{N}=2$ theories by a mass term of the adjoint superfield which breaks bulk supersymmetry down to $\mathcal{N}=1$. In this case the moduli fields on the string include all those inherent to the $\mathcal{N}=(2,2)$ CP($N-1$) model plus an extra $\mathcal{N}=(0,2)$ supermultiplet with a peculiar interaction. The heterotic model obtained in this way is to be referred to as nonminimal. In the geometric formulation its Lagrangian is

$$\mathcal{L} = G_{i\bar{j}} \left[\partial_R \phi^{\dagger\bar{j}} \partial_L \phi^i + \psi_L^{\dagger\bar{j}} \, i\mathcal{D}_R \, \psi_L^i + \psi_R^{\dagger\bar{j}} \, i\mathcal{D}_L \psi_R^i \right] + R_{i\bar{j}k\bar{l}} \, \psi_L^{\dagger\bar{j}} \psi_L^i \, \psi_R^{\dagger\bar{l}} \psi_R^k$$

$$+\zeta_R^\dagger \, i\partial_L \, \zeta_R + \left[\kappa \, \zeta_R \, G_{i\bar{j}} \left(i \, \partial_L \phi^{\dagger\bar{j}} \right) \psi_R^i + \text{H.c.} \right] + |\kappa|^2 \zeta_R^\dagger \, \zeta_R \left(G_{i\bar{j}} \, \psi_L^{\dagger\bar{j}} \psi_L^i \right)$$

$$- |\kappa|^2 \left(G_{i\bar{j}} \psi_L^{\dagger\bar{j}} \psi_R^i \right) \left(G_{k\bar{l}} \psi_R^{\dagger\bar{l}} \psi_L^k \right) \,. \tag{109}$$

Here $\mathcal{D}_{L,R}$ are covariant derivatives,

$$\mathcal{D}_{L,R} \, \psi_{R,L}^i = \partial_{L,R} \, \psi_{R,L}^i + \Gamma_{kl}^i \, \partial_{L,R} \, \phi^k \, \psi_{R,L}^l \,. \tag{110}$$

The first line in (109) coincides with the (2,2) Lagrangian in Eq. (12). The second and third lines present a heterotic deformation. The right-handed fermion field ζ_R is absent in the (2,2) model.

In terms of superfields the nonminimal heterotic model can be written as follows:

$$\mathcal{L} = -\frac{1}{2} \int d\theta \left[\frac{1}{2} \, G_{i\bar{j}}(A, A^\dagger)(\bar{D}A^{\dagger\bar{j}}) \, i\partial_R A^i - \kappa \, \mathcal{B} \, G_{i\bar{j}}(A, A^\dagger)(\bar{D}A^{\dagger\bar{j}}) B^i + \text{H.c.} \right]$$

$$+\frac{1}{2} \int d^2\theta \left[G_{i\bar{j}}(A, A^\dagger) \, B^{\dagger\bar{j}} B^i + \mathcal{B}^\dagger \mathcal{B} \right] , \tag{111}$$

where κ is the deformation parameter, and the extra (compared to the minimal model) (0,2) superfields are

$$\mathcal{B} = \zeta_R(x_R + 2i\theta^\dagger\theta, \, x_L) + \sqrt{2} \, \theta \, F_\zeta(x_R + 2i\theta^\dagger\theta, \, x_L) \text{ and}$$

$$B = \psi_R(x_R + 2i\theta^\dagger\theta, \, x_L) + \sqrt{2} \, \theta F_\psi(x_R + 2i\theta^\dagger\theta, \, x_L) \,. \tag{112}$$

On mass shell both $\mathcal{B}$ and B contain one fermion degree of freedom, ζ_R and ψ_R, respectively.

9.2.2. Gauged formulation

The gauged formulation is most convenient for large-N solution. The Lagrangian is obtained by adding to (25) the following deformation:

$$\Delta\mathcal{L}_{(0,2)} = \bar{\zeta}_R \, i\partial_L \zeta_R + \left[\frac{\omega}{\sqrt{2}\,\beta} \, (i\partial_L \bar{n}_j) \, \xi_R^j \zeta_R + \text{H.c.} \right]$$
$$- 4|\omega|^2 \, |\sigma|^2 \,, \tag{113}$$

where the deformation constant ω is related to κ in (109) as follows:

$$\omega = \sqrt{2}\,\beta\kappa \,. \tag{114}$$

The deformation parameter ω is renormalization-group invariant, see Sec. 9.3. In the large-N solution we will see that physical effects are determined by an N-independent deformation parameter,

$$u = \frac{8\pi}{N}|\omega|^2 = \frac{16\pi}{Ng^2}\frac{\kappa^2}{g^2} \,. \tag{115}$$

Both constants, κ^2 and g^2 scale with N as $1/N$.

9.2.3. Twisted masses

Twisted masses were added in.[18,27] The corresponding expressions are quite bulky. The interested reader is referred to the original publications. A novel element worth noting is as follows. In the absence of the heterotic deformation the CP($N-1$) model has $N-1$ complex twisted mass parameters, see Sec. 6. With $\kappa \neq 0$ the number of independent complex mass parameters generally speaking increases. In the generic case the nonminimal (0,2) model will have N independent mass parameters.

9.3. Beta functions

All models under consideration in this review paper are asymptotically free. As was noted by Polyakov in 1975[33] at one loop only the bosonic fields contribute to the β functions. Fermion contribution shows up at the two-loop level.

The exact all-loop β function in the minimal CP(1) model was found in.[30] It has the form

$$\beta_{g\,(0,2)\,\mathrm{min}} = -\frac{g^4}{2\pi}\left(1 - \frac{g^2}{4\pi}\right)^{-1}, \tag{116}$$

where g^2 is the coupling constant in the Kähler potential and metric.

Its structure is perfectly analogous to that of the NSVZ β function in four-dimensional $\mathcal{N} = 1$ gluodynamics.[62,63] In fact, the above two-dimensional model is the closest analog of $\mathcal{N} = 1$ four-dimensional theories one can think of. One can show[30] that the analogy extends further than Eq. (116) and is maintained when one introduces "matter" fields. Then the β function (116) acquires a numerator typical of the NSVZ β function in the presence of matter.

In the nonminimal model one deals with two coupling constants, g^2 appearing in the metric, and the deformation parameter κ. At one loop the corresponding β functions were calculated in,[29] while the two-loop loop corrections and an exact relation were found in[31] (see also[29]),

$$\beta_{g\,\mathrm{nonmin}} = -\frac{g^2}{4\pi}\frac{Ng^2 + h^2\gamma}{1 + (h^2/4\pi)},$$

$$\beta_{h\,\mathrm{nonmin}} = -\frac{h^2}{1 + (h^2/4\pi)}\left[\frac{Ng^2}{2\pi} - \gamma\left(1 - \frac{h^2}{4\pi}\right)\right]. \tag{117}$$

Here

$$h^2 = \frac{|\kappa|^2}{Z\mathcal{Z}}, \tag{118}$$

where Z and $\mathcal{Z}$ are filed renormalization constants for ψ_R and ζ_R respectively, and γ is the corresponding anomalous dimension. At one loop[29]

$$\gamma = \frac{Nh^2}{2\pi}. \tag{119}$$

Both formulas in (117) are exact all-order results. Equation (119) implies that the two-loop β functions are

$$\beta_{g\,\mathrm{nonmin}} = -\frac{Ng^2}{4\pi}\frac{g^2 + (h^4/2\pi)}{1 + (h^2/4\pi)},$$

$$\beta_{h\,\mathrm{nonmin}} = -\frac{Nh^2}{1 + (h^2/4\pi)}\left[\frac{g^2}{2\pi} - \frac{h^2}{2\pi}\left(1 - \frac{h^2}{4\pi}\right)\right]. \tag{120}$$

There is another consequence from Eqs. (117) and (119). In the limit $N \to \infty$ the constant h^2 scales as $1/N$, implying that $\beta_{g\,\mathrm{nonmin}}$ reduces to one loop and becomes exactly the same as in the undeformed $\mathcal{N} = (2,2)$ $CP(N-1)$ model. This is in full agreement with the large-N solution of the nonminimal heterotic model to be presented below. The combination κ^2/g^4 is renormalization-group invariant,

$$\frac{\kappa^2}{g^4} = \text{RGI}, \tag{121}$$

cf. Eq. (114).

9.4. Large-N solution of nonminimal CP(N − 1)

This model was solved with the Z_N symmetric twisted masses[27] and arbitrary value of the mass parameter m_0. This solution includes of course the massless heterotic model[26] as a limiting case $m_0 = 0$. Therefore, we will pass directly to the nonminimal model with the Z_N symmetric twisted masses.

One brief remark is in order before this passage. At small values of u, vanishing mass parameter m_0, and *arbitrary* (i.e. not necessarily large) N it is easy to find both the Goldstino and the vacuum energy,

$$g \sim \omega \left\langle R_{i\bar{j}} \, \bar{\psi}_R^{\bar{j}} \psi_L^i \right\rangle_{\text{vac}} \zeta_R \,, \tag{122}$$

where the vacuum averaging is performed in the undeformed $(2,2)$ massless $CP(N-1)$ model, see Eqs. (6.26) and (6.27) in.[17] The extra right-handed field ζ_R plays the role of Goldstino.

Now, let us diñcuss the solution found in.[27] Conceptually, the strategy of solving this model at large N is similar to that described in Sec. 7. Since in the model at hand we have two parameters, u and m_0, we discover a rather rich and not quite trivial phase diagram, in which we observe phases with broken or unbroken Z_N symmetry. If $u \neq 0$ we have two phases with the broken Z_N symmetry, on the left and on the right in Fig. 8. The first Z_N phase is strongly coupled, the second (the Higgs phase) is weakly coupled. In the middle lies the phase of unbroken Z_N symmetry, in which the vacuum is unique, the photon does not acquire a mass, and the corresponding dynamical regime is that of charge confinement.

Analytical solution for the vacuum structure is easier to obtain at large deformations, $u \gg 1$.

9.4.1. *Strong coupling phase with broken Z_N*

This phase occurs at very small masses, namely,

$$m_0 \leq \Lambda \, e^{-u/2} \,, \qquad u \gg 1 \,. \tag{123}$$

u

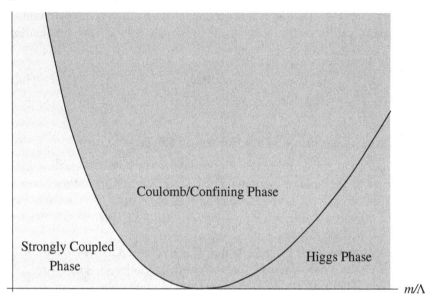

Fig. 8. The phase diagram of the twisted-mass deformed heterotic $CP(N-1)$ theory in the plane u and m_0 where m_0 is assumed to be real. The parameter u denotes the amount of deformation, $u = \frac{8\pi}{N}|\omega|^2$.

In this phase we have

$$|n| = 0, \qquad iD \approx \Lambda^2, \tag{124}$$

while the vacuum value of the σ field is

$$\sqrt{2}\,\langle\sigma\rangle_{\text{vac}} = e^{\frac{2\pi i}{N}k}\,\Lambda\,e^{-u/2}, \qquad k = 1, ..., (N). \tag{125}$$

The vacuum value of σ is exponentially small at large u. The bound $m_0 < |\sqrt{2}\sigma|$ translates into the condition (123) for m_0. For simplicity we will assume in this section m_0 to be real and positive.

We have N degenerate vacua in this phase. The chiral Z_{2N} symmetry is broken down to Z_2, the order parameter is $\langle\sigma\rangle$. Moreover, the absolute value of σ in these vacua does not depend on m. This solution essentially coincides with one obtained in[26] in the massless case. In this aspect the situation is quite similar to the strong coupling phase of the $\mathcal{N} = (2,2)$ model. The difference is that the absolute value of σ depends now on u and becomes exponentially small in the limit $u \gg 1$.

The vacuum energy is positive (see Fig. 9). Supersymmetry is spontaneously broken.

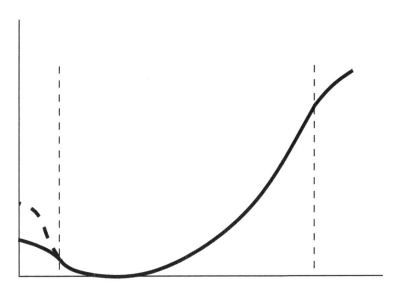

Fig. 9. Vacuum energy density vs. m_0. The dashed line shows the behavior of the energy density (128) extrapolated into the strong coupling region.

9.4.2. Coulomb/confining phase

Now we increase m_0 above the bound (123). The exponentially small σ_{vac} solution no longer exists. The only solution is

$$\langle \sigma \rangle_{\text{vac}} = 0 \,. \tag{126}$$

In addition, Eq. (124) implies

$$|n| = 0, \qquad iD = \Lambda^2 - m^2 \,. \tag{127}$$

This solution describes a single Z_N symmetric vacuum. All other vacua are lifted and become quasivacua (metastable at large N). This phase is quite similar to the Coulomb/confining phase of nonsupersymmetric CP($N - 1$) model without twisted masses.[35] The presence of small splittings between quasivacua produces a linear rising confining potential between kinks that interpolate between, say, the true vacuum and the lowest quasivacuum,[10] see also the review.[12] Alternatively, this is a Coulomb interaction between charged particles due to a massless photon that results in confinement.

There is a phase transition (most likely of the second order) that separates these phases. As a rule, one does not have phase transitions in supersymmetric theories. However, in the model at hand supersymmetry is broken (in fact, it is broken already at the classical level[18]); therefore, the emergence of a phase transition is not too surprising.

One can calculate the vacuum energy explicitly to see the degree of supersymmetry breaking. Substituting (126) and (127) in the effective potential one gets

$$E_{\text{vac}}^{\text{Coulomb}} = \frac{N}{4\pi}\left[\Lambda^2 - m_0^2 + m_0^2 \ln \frac{m_0^2}{\Lambda^2}\right].$$ (128)

see Fig. 9. At $m_0 = \Lambda$ the vacuum energy vanishes in the large-N limit implying a supersymmetry restoration. Most likely, this vanishing will be lifted by $1/N$ corrections, so that supersymmetry is always spontaneously broken.

9.4.3. Higgs phase

The Higgs (weakly coupled) phase takes place in the model under consideration at large m_0,

$$m_0 > \sqrt{u}\Lambda, \qquad \text{if } u \gg 1.$$ (129)

In this phase $|n|$ develops a VEV, which is a clear-cut signal of the Z_N symmetry breaking. Thus, we conclude that

$$|n|_{\text{vac}}^2 = \frac{N}{4\pi} \ln \frac{\sqrt{2}\sigma\, m_0}{\Lambda^2} \sim \frac{N}{4\pi} \ln \frac{m_0^2}{u\,\Lambda^2}$$ (130)

in each of the N vacua in the Higgs phase, where

$$\sqrt{2}|\sigma|_{\text{vac}} = \left(\frac{8\pi}{N}\right)\frac{m_0}{u},$$ (131)

We have N degenerate vacua again, as in the strongly coupled phase. In each of them $|\sigma|$ is small ($\sim m_0/u$) but nonvanishing. The Z_N chiral symmetry is broken. Clearly, the Higgs phase is separated form the Coulomb/confining phase (where Z_N is unbroken) by a phase transition.

9.4.4. Goldstino

In this section we limit ourselves to the large-N solution of the massless heterotic model (25), (113) obtained in.[26] Due to the spontaneous supersymmetry breaking we have a massless Goldstino fermion in the world-sheet theory. To check this explicitly one can analyze the one-loop effective Lagrangian calculated in.[26] The appropriate fermionic part of the effective

Lagrangian is

$$\mathcal{L}_{\text{eff}}^{\text{ferm}} = \frac{1}{e_\lambda^2} \bar{\lambda}_R \, i \, \partial_L \, \lambda_R + \frac{1}{e_\lambda^2} \bar{\lambda}_L \, i \, \partial_R \, \lambda_L + \frac{1}{2} \bar{\zeta}_R \, i \, \partial_L \, \zeta_R$$

$$+ \left[i\sqrt{2}\,\Gamma\,\bar{\sigma}\,\bar{\lambda}_L\lambda_R + \sqrt{2}\,i\,\omega\,\bar{\lambda}_L\,\zeta_R + \text{H.c.} \right], \tag{132}$$

where the one-loop couplings e_λ and Γ were calculated in.[26]

First, we diagonalize the mass matrix for the ζ_R, λ_R and λ_L fermions. Equating the determinant of this matrix to zero produces the following equation for the mass eigenvalues m:

$$m^3 - m \left(2|\sigma|^2\,\Gamma^2\,e_\lambda^4 + 4\,\omega^2\,e_\lambda^2 \right) = 0. \tag{133}$$

For any ω we have a vanishing eigenvalue corresponding to a massless Goldstino. Clearly, at small ω this fermion coincides with ζ_R (with an $O(\omega)$ admixture from the λ fermions).

At large u

$$e_\lambda \sim \Lambda \text{ and } \Gamma \sim u/\Lambda^2,$$

while σ is given by (125). Thus, the last term in the second line in (132) dominates, giving masses to ζ_R, and λ_L. The role of Goldstino is assumed by the λ_R fermion field.

9.5. Large N in nonminimal heterotic weighted $CP(N, M)$ model

The unperturbed $\mathcal{N} = (2,2)$ model was discussed in Sec. 8. It is obtained on the world sheet of semilocal strings supported in the bulk $\mathcal{N} = 2$ theories if $N_f > N$.[54-57] In this case there are two distinct types of the moduli fields, ρ and n, (scale and orientation moduli, respectively), and we arrive at the so-called zn model on the world sheet. Hanany and Tong suggested[54] the weighted $CP(N, M)$ for the same purpose. Later it was shown that these two models lead to identical predictions in the large-N limit.

If one introduces a $\mu \, \text{Tr} \mathcal{A}^2$ deformation in the bulk theory breaking $\mathcal{N} = 2$ down to $\mathcal{N} = 1$ one arrives at a heterotically deformed model on the world sheet. As far as we know, no explicit derivation of the deformation term in two dimensions starting from the deformed bulk theory has ever been carried out. A conjecture that this deformation term is identical to that emerging in the $N_f = N$ case was formulated in.[60] Then the two-dimensional model obtained in this way was further generalized to include twisted masses of two types, corresponding to two types of the moduli fields, namely, the scale and orientational moduli. To reduce the number of adjustable parameters, it was assumed that the first set of the twisted masses is Z_M symmetric, while the second is Z_N symmetric (cf. (63)). As a result, there are two mass parameters m_0 and μ_0 plus two dimensionless parameters

$$\alpha = \frac{M}{N} \text{ and } u. \tag{134}$$

The limit $N \to \infty$ was assumed. The large-N analysis of the vacuum structure and the spectrum of the model is very similar to that discussed in Sec. 9.4. Under these conditions the model was solved[60] and a rich structure discovered on the phase diagrams, including two distinct Higgs phases and two distinct Coulomb phases and various patterns of the $Z_{N,M}$ breaking. An interesting phenomenon was observed on a two-dimensional subspace of mass parameter space on which a discrete Z_{N-M} symmetry is preserved. As was expected, supersymmetry is spontaneously broken for generic values of adjustable parameters. However, on a special curve in the parameter space we have the same phenomenon as at $m_0 = \Lambda$ in Fig. 9. Supersymmetry seems to be restored at $N \to \infty$. A new branch opens up for special values of m_0 and μ_0. In much the same way as in Sec. 9.4.2 one can expect that the vacuum energy on this curve will be lifted in a subleading order in $1/N$.

9.6. Large N in heterotic $O(N)$ model

To begin with, a few words about the minimal $(0, 1)$ $O(N)$ model will be in order. Assuming $N \geq 4$ it is easy to obtain this model by truncating the standard $(1, 1)$ model,[37] for a review see.[38] To this end we introduce the $(0, 1)$ superfield

$$N^a = S^a(x) + \theta_R \psi_L^a(x), \qquad a = 1, 2, ..., N, \tag{135}$$

with the following Lagrangian[20] (plus the standard constraint)

$$\mathcal{L}_{(0,1) \text{ min}} = \frac{1}{2g^2} \int d\theta_R (D_L N^a)(i \partial_R N^a), \qquad N^a N^a - 1 = 0, \tag{136}$$

where ψ_L^a is a Weyl-Majorana field, $\partial_R = \partial_t - \partial_z$ as usual, and

$$D_L = \frac{\partial}{\partial \theta_R} - i\theta_R \partial_L.$$

The constraint in (136) can be implemented by adding an appropriate Lagrange multiplier term

$$\Delta \mathcal{L}_{(0,1) \text{ min}} = \int d\theta_R X (N^a N^a - 1), \tag{137}$$

where

$$X = \frac{1}{2g^2} \left(-\lambda_R + \theta_R D \right) . \tag{138}$$

Note that, in contradistinction with the $\text{CP}(N-1)$ case, the minimal $O(N)$ model exists at all N. The large-N solution of the model (136) is constructed in much the same way as that for nonsupersymmetric $O(N)$ model.[38] Supersymmetry is spontaneously broken, the constraint $S^a S^a = 1$ is lifted, all S^a fields acquire a mass while the ψ_L fields remain massless. The field χ_R acquires a kinetic term.

To construct a nonminimal heterotic model we will follow the same line of reasoning as in Sec. 9.2.1. In fact, in the geometric formulation one can use the Lagrangian (111) with the replacement of the Kähler metric of $\text{CP}(N-1)$ by a real metric of the N-dimensional sphere, and assuming that the parameter θ in the definition of the superfields (105) and (112) is real.

. A slightly different formulation is more convenient for the large-N analysis, however. In addition to the $(0,1)$ superfield (135) let us introduce two right-handed "matter" superfields (both with one physical degree of freedom),

$$\mathcal{B} = \zeta_R(x) + \theta_R \, F_\zeta(x) , \qquad B^a = \psi_R^a(x) + \theta_R F_\psi^a(x) . \tag{139}$$

The Lagrangian of the model can be written as

$$\mathcal{L}_{(0,1)} = \int d\theta_R \left\{ \frac{1}{2g^2} \left[(D_L N^a) \, (i \, \partial_R N^a) + (D_L \, B^a) \, B^a \right] + \frac{1}{2} \, (D_L \, \mathcal{B}) \, \mathcal{B} \right. $$

$$\left. - \frac{\kappa}{g^2} \, (D_L N^a) \, B^a \mathcal{B} - X \, (N^a N^a - 1) - \tilde{X} \, (N^a B^a) \right\} , \tag{140}$$

where the last two terms implement the constraints $S^a S^a = 1$ and $S^a \psi_{L,R}^a = 0$ (plus the standard relation for F_ψ, see[38]) and $\tilde{X}$ is an auxiliary field analogous to (138), namely,

$$\tilde{X} = \frac{1}{g^2} \, (\sigma + \theta_R \lambda_L) . \tag{141}$$

In components (after eliminating the auxiliary fields $F_{\zeta, \psi}$ and a rescaling needed to make kinetic terms canonical) the Lagrangian takes the form[64]

$$\mathcal{L}_{(0,1)} = \frac{1}{2}\partial_L S^a \partial_R S^a + \frac{i}{2}\psi_L^a \partial_R \psi_L^a + \frac{i}{2}\psi_R^a \partial_L \psi_R^a + \frac{i}{2}\zeta_R \partial_L \zeta_R$$

$$+ \beta_L \psi_R^a S^a + \chi_R \psi_L^a S^a - \frac{1}{2}\left(D + \sigma^2\right) S^a S^a + \frac{1}{2}\frac{D}{g^2}$$

$$+ \sigma \psi_L^a \psi_R^a + \kappa\left(i\partial_L S^a\right)\psi_R^a \zeta_R + \frac{1}{2}\kappa^2 \sigma^2 . \tag{142}$$

It is not difficult to calculate the effective potential as a function of D and σ,[64]

$$V_{\text{eff}} = \frac{N}{8\pi}\left[D \log \frac{\Lambda^2}{D + \sigma^2} + \sigma^2 \log \frac{\sigma^2}{\sigma^2 + D} + D + u\sigma^2\right], \tag{143}$$

where

$$u = \frac{4\pi\kappa^2}{g^4 N} . \tag{144}$$

Minimizing the potential with respect to D and σ one finds two distinct vacua of the theory

$$\sigma_0 = \pm\Lambda e^{-\frac{u}{2}}, \qquad D = \Lambda^2 - \sigma^2, \tag{145}$$

which present continuations of two distinct vacua inherent to the super-symmetric $(1,1)$ limit of the model. The ensuing vacuum energy is

$$E_{\text{vac}} = \frac{N}{8\pi}\Lambda^2\left(1 - e^{-u}\right) . \tag{146}$$

Any nonvanishing value of u results in the spontaneous breaking of super-symmetry. The spectrum of the model and, in particular, the Goldstino composition can be readily found too.

10. In the uncharted waters

In Sec. 2 we outlined the simplest prototype bulk theory supporting non-Abelian strings which, in turn, give rise to the observed wealth of two-dimensional sigma models in Secs. 4 – 9. Extending the bulk theory one can expect do derive novel sigma models on the string world sheet. In this section we will briefly discuss an extended construction resulting in the $(0,2)$ two-dimensional model which has never been discussed previously. Moreover, its geometric formulation is not yet known.

For brevity of presentation we will stick to $N = 2$ and $N_f \geq 2$, referring the reader to the original papers[65,66] for the case of generic N. Unlike Sec.

2 we will switch off the Fayet-Iliopoulos D term (i.e. $\xi = 0$ in the last term in Eq. (3)) but, instead, switch on the mass term for the adjoint fields $\mathcal{A}$,

$$\mathcal{W}_{\text{def}} = \mu \operatorname{Tr} \Phi^2, \qquad \Phi \equiv \frac{1}{2} \mathcal{A} + T^a \, \mathcal{A}^a \tag{147}$$

in addition to non-vanishing mass terms for the bulk (s)quark fields.[65,66] The deformation (147) breaks bulk supersymmetry down to $\mathcal{N} = 1$, generally speaking.

This leads to the following modification of the bulk potential (3). The last two F terms in the second line in (3) responsible for the squark condensation are replaced by

$$2g_2^2 \left| \tilde{q}_A T^a q^A + \frac{\mu}{\sqrt{2}} a^a \right|^2 + \frac{g_1^2}{2} \left| \tilde{q}_A q^A + \frac{N}{\sqrt{2}} \mu a \right|^2 . \tag{148}$$

Since VEVs of the adjoint fields a and a^a are determined by squark masses (cf. (4)) this leads to the breaking of the color-flavor symmetry. The quark VEVs are no longer degenerate. Instead of (6) the quark VEVs take the form

$$\langle q^{kA} \rangle = \langle \bar{\tilde{q}}^{kA} \rangle = \frac{1}{\sqrt{2}} \begin{pmatrix} \sqrt{\xi_1} & 0 & 0 \dots 0 \\ 0 & \sqrt{\xi_2} & 0 \dots 0 \end{pmatrix},$$

$$k = 1, 2, \qquad A = 1, ..., N_f . \tag{149}$$

The parameters $\xi_{1,2}$ in (149) in the quasiclassical approximation are

$$\xi_{1,2} \approx 2 \, \mu \, m_{1,2} . \tag{150}$$

These parameters can be made large in the large-m limit even if μ is small, to ensure that the bulk theory is at weak coupling.

The squark condensation leads to the string formation. If $m_1 \neq m_2$ these strings have nondegenerate tensions. The U(2) gauge group is broken down to U(1)×U(1) by the quark mass difference. To the leading order in μ each U(1) gauge factor supports it own BPS string. The string tensions of two strings under consideration are[65]

$$T_{1,2} = 2\pi |\xi_{1,2}| . \tag{151}$$

If $|m_1 - m_2| \ll |m_{1,2}|$ these two strings can still be promoted to non-Abelian strings with a shallow potential in the world-sheet theory. As was mentioned, now $\mathcal{N} = (2,2)$ supersymmetry is broken down to $\mathcal{N} = (0,2)$ even to the leading order in μ. For the single-trace deformation (147) the bosonic part of the world-sheet theory becomes[65,66]

$$\mathcal{L} = \mathcal{L}_{(2,2)} + V_{\text{def}}(\sigma), \tag{152}$$

where $\mathcal{L}_{(2,2)}$ is the Lagrangian of the $\mathcal{N} = (2,2)$ supersymmetric model (99) while the deformation potential is

$$V_{\text{def}}(\sigma) = 4\sqrt{2}\pi \, |\mu\sigma| \, . \tag{153}$$

The deformation (153) respects only $(0,2)$ superalgebra.

This potential is radically different from the $|\sigma|^2$ potential in the heterotic deformation (113). The latter potential arises on the non-Abelian string in the massless bulk theory with the Fayet-Iliopoulos D-term deformed by the superpotential (147).

The total scalar potential is given by the sum of the twisted mass potential in (99) and deformation (153). Its minima correspond to tensions of two elementary non-Abelian strings,

$$V(\sigma_{1,2})_{\text{def}} = T_{1,2} \, . \tag{154}$$

To see that this is the case we note that at small μ the vacuum values of σ are still determined by the squark masses $\sqrt{2}\sigma_{1,2} \approx m_{1,2}$ in the quasiclassical approximation. Then (154) follows [§] from (150).

If $m_1 \neq m_2$ the minima are nondegenerate. Only the lowest-lying vacuum is stable. The stability of the lowest vacuum in two dimensions means that only the lightest non-Abelian string is stable, the other one is metastable. Moreover, since generically the string tensions do not vanish, $\mathcal{N} = (0,2)$ supersymmetry is broken spontaneously already at the classical level.[65]

To conclude this section let us mention that at the generic quark masses the deformation (147) leads to the emergence of a whole set of isolated vacua in the bulk theory, the so-called r vacua, $r \leq N$. In each r vacuum r quarks and $(N - r - 1)$ monopoles condense. The vacuum in (149) correspond to the $r = N$ vacuum, with the maximal number of condensed quarks. The simplest example of $r < N$ vacuum, namely, an $r = N - 1$ vacuum (with $r = N - 1$ condensed squarks and no monopoles) was considered in.[67] This vacuum also supports non-Abelian strings. However, in contradistinction with the $r = N$ vacuum, the two-dimensional theory on the string world-sheet receives in this case nonperturbative corrections from the bulk, through the bulk gaugino condensate. Nonperturbative bulk effects deforming the theory on the string world sheet were found in[67] by virtue of the method of resolvents suggested by Gaiotto, Gukov and Seiberg for surface defects.[68]

[§]This statement is valid beyond the quasiclassical approximation (to all orders in $\Lambda/m_{1,2}$). In this case the σ VEVs are determined[65] by the roots of the equation (88).

11. Conclusions

Forty years ago A. Polyakov emphasized that asymptotically free two-dimensional sigma models could present the best laboratory for the four-dimensional Yang-Mills theories. This prophecy came true in various aspects – even more than it was anticipated. First and foremost, a remarkable 2D-4D correspondence was detected in supersymmetric theories (see[12,67] and references therein): the BPS spectrum of the sigma models on the string world sheet proves to be in one-to-one correspondence with that in the bulk four-dimensional theory. Moreover, diverse two-dimensional sigma models *per se* exhibit nontrivial dynamical features which, quite unexpectedly, proved to be in close parallel with some features of four-dimensional Yang-Mills. Novel models continue to appear in the limelight. Today the task of their exploration is highly challenging. This path is fruitful.

Acknowledgments

This work is supported in part by DOE grant DE-FG02-94ER40823. The work of A.Y. was supported by FTPI, University of Minnesota, by RFBR Grant No. 13-02-00042a and by Russian State Grant for Scientific Schools RSGSS-657512010.2.

References

1. L. D. Landau and I. Ya. Pomeranchuk, On point-like interaction in Quantum Electrodynamics, Dokl. Akad. Nauk SSSR **102**, 489 (1955). [Reprinted in *L.D. Landau's Collected Papers*, (Nauka, Moscow, 1969), Vol. 2, p. 247].
2. I. Ya. Pomeranchuk, Doklady Akad. Nauk USSR **103**, 1005 (1955); I. Ya. Pomeranchuk, V. V. Sudakov and K. A. Ter-Martirosyan, ÊÊPhys. Rev. **103**, 784 (1956); ÊÊI. Ya. Pomeranchuk, Doklady Akad. Nauk USSR **104**, 51 (1955); Doklady Akad. Nauk USSR **105**, 461 (1955); Nuovo Cim. **3**, 1186 1956.
3. D. J. Gross and F. Wilczek, Phys. Rev. Lett. **30**, 1343 (1973); H. D. Politzer, Phys. Rev. Lett. **30**, 1346 (1973).
4. A. A. Anselm, Sov. Phys. JETP **9 (36)**, 608 (1959) [reprinted in M. Shifman (Ed.) *Under the Spell of Landau*, (World Scientific, Singapore, 2013), p. 526].
5. N. Seiberg and E. Witten, Nucl. Phys. B **426**, 19 (1994), (E) B **430**, 485 (1994) [hep-th/9407087]; Nucl. Phys. B **431**, 484 (1994) [hep-th/9408099].
6. Y. Nambu, Phys. Rev. D **10**, 4262 (1974);
G. 't Hooft, Gauge theories with unified weak, electromagnetic and strong interactions, in Proc. of the E.P.S. Int. Conf. on High Energy Physics, Palermo, 23-28 June, 1975, ed. A. Zichichi (Editrice Compositori, Bologna, 1976); S. Mandelstam, Phys. Rept. **23**, 245 (1976).

7. M. R. Douglas and S. H. Shenker, Nucl. Phys. B **447**, 271 (1995) [hep-th/9503163]; A. Hanany, M. J. Strassler and A. Zaffaroni, Nucl. Phys. B **513**, 87 (1998) [hep-th/9707244].

8. A. Abrikosov, Sov. Phys. JETP **32** 1442 (1957) [Reprinted in *Solitons and Particles*, Eds. C. Rebbi and G. Soliani (World Scientific, Singapore, 1984), p. 356]; H. Nielsen and P. Olesen, Nucl. Phys. B **61** 45 (1973) [Reprinted in *Solitons and Particles*, Eds. C. Rebbi and G. Soliani (World Scientific, Singapore, 1984), p. 365].

9. A. Hanany and D. Tong, JHEP **0307**, 037 (2003) [hep-th/0306150]; R. Auzzi, S. Bolognesi, J. Evslin, K. Konishi and A. Yung, Nucl. Phys. B **673**, 187 (2003) [hep-th/0307287]; M. Shifman, A. Yung, Phys. Rev. D **70**, 045004 (2004). [hep-th/0403149]; A. Hanany and D. Tong, JHEP **0404**, 066 (2004) [hep-th/0403158].

10. A. Gorsky, M. Shifman and A. Yung, Phys. Rev. D **71**, 045010 (2005) [hep-th/0412082].

11. M. Shifman and A. Yung, Phys. Rev. Lett. **110**, 201602 (2013) [arXiv:1303.7010 [hep-th]]; S. Monin, M. Shifman and A. Yung, Phys. Rev. D **88**, 025011 (2013) [arXiv:1305.7292 [hep-th]].

12. M. Shifman and A. Yung, *Supersymmetric Solitons* (Cambridge University Press, 2009).

13. P. Fayet and J. Iliopoulos, Phys. Lett. B **51**, 461 (1974).

14. M. Shifman, *Advanced Topics in Quantum Field Theory*, (Cambridge University Press, 2012).

15. E. B. Bogomol'nyi, *Sov. J. Nucl. Phys.* **24**, 449 (1976), reprinted in *Solitons and Particles*, eds. C. Rebbi and G. Soliani (World Scientific, Singapore, 1984) p. 389. M. K. Prasad and C. M. Sommerfield, *Phys. Rev. Lett.* **35**, 760 (1975), reprinted in *Solitons and Particles*, eds. C. Rebbi and G. Soliani (World Scientific, Singapore, 1984) p. 530.

16. M. Edalati and D. Tong, JHEP **0705**, 005 (2007) [hep-th/0703045].

17. M. Shifman and A. Yung, Phys. Rev. D **77**, 125016 (2008) [Erratum-ibid. D **79**, 049901 (2009)] [arXiv:0803.0158 [hep-th]].

18. P. A. Bolokhov, M. Shifman and A. Yung, Phys. Rev. D **79**, 085015 (2009) [arXiv:0901.4603 [hep-th]]; Phys. Rev. D **81**, 065025 (2010) [arXiv:0907.2715 [hep-th]].

19. A. M. Perelomov, Phys. Rept. **146**, 135 (1987); Phys. Rept. **174**, 229 (1989).

20. P. West, *Introduction to Supersymmetry and Supergravity*, Second Edition, (World Scientific, Singapore, 1990), Chapter 23.

21. E. Witten, Adv. Theor. Math. Phys. **11** (2007) [hep-th/0504078].

22. M. -C. Tan and J. Yagi, Lett. Math. Phys. **84**, 257 (2008) [arXiv:0801.4782 [hep-th], arXiv:0805.1410 [hep-th]].

23. J. Yagi, Adv. Theor. Math. Phys. **16**, 1 (2012) [arXiv:1001.0118 [hep-th]].

24. B. Jia, E. Sharpe and R. Wu, Notes on non-Abelian (0,2) theories and dualities, arXiv:1401.1511 [hep-th].

25. T. T. Dumitrescu and N. Seiberg, JHEP **1107**, 095 (2011) [arXiv:1106.0031 [hep-th]].

26. M. Shifman and A. Yung, Phys. Rev. D **77**, 125017 (2008) [Erratum-ibid. D **81**, 089906 (2010)] [arXiv:0803.0698 [hep-th]].
27. P. A. Bolokhov, M. Shifman and A. Yung, Phys. Rev. D **82**, 025011 (2010) [arXiv:1001.1757 [hep-th]].
28. X. Cui and M. Shifman, Phys. Rev. D **82**, 105022 (2010) [arXiv:1009.4421 [hep-th]].
29. X. Cui and M. Shifman, Phys. Rev. D **84**, 105016 (2011) [arXiv:1105.5107 [hep-th]].
30. X. Cui and M. Shifman, Phys. Rev. D **85**, 045004 (2012) [arXiv:1111.6350 [hep-th]].
31. J. Chen, X. Cui, M. Shifman, and A. Vainshtein, work in progress.
32. J. Wess and J. Bagger, *Supersymmetry and Supergravity*, Second Edition, Princeton University Press, 1992.
33. A. M. Polyakov, Phys. Lett. B **59**, 79 (1975).
34. D. Friedan, Phys. Rev. Lett. **45**, 1057 (1980); L. Alvarez-Gaumé, D. Z. Freedman and S. Mukhi, Annals Phys. **134**, 85 (1981).
35. E. Witten, Nucl. Phys. B **149**, 285 (1979).
36. E. Witten, Nucl. Phys. B **403**, 159 (1993) [hep-th/9301042].
37. E. Witten, Phys. Rev. D **16**, 2991 (1977); P. Di Vecchia and S. Ferrara, Nucl. Phys. B **130**, 93 (1977).
38. V. A. Novikov, M. A. Shifman, A. I. Vainshtein and V. I. Zakharov, Phys. Rept. **116**, 103 (1984).
39. B. Zumino, Phys. Lett. B **87**, 203 (1979).
40. G. W. Moore and P. C. Nelson, Commun. Math. Phys. **100**, 83 (1985); A. Manohar, G. W. Moore and P. C. Nelson, Phys. Lett. B **152**, 68 (1985).
41. L. Alvarez-Gaumé and D. Z. Freedman, Commun. Math. Phys. **91**, 87 (1983); S. J. Gates, Nucl. Phys. B **238**, 349 (1984); S. J. Gates, C. M. Hull and M. Roček, Nucl. Phys. B **248**, 157 (1984).
42. A. Gorsky, M. Shifman and A. Yung, Phys. Rev. D **73**, 065011 (2006) [hep-th/0512153].
43. A. Gorsky, M. Shifman and A. Yung, Phys. Rev. D **73**, 125011 (2006) [hep-th/0601131].
44. A. Hanany and K. Hori, Nucl. Phys. B **513**, 119 (1998) [arXiv:hep-th/9707192].
45. G. Veneziano and S. Yankielowicz, Phys. Lett. B **113**, 231 (1982).
46. A. D'Adda, A. C. Davis, P. Di Vecchia and P. Salomonson, Nucl. Phys. B **222**, 45 (1983).
47. S. Cecotti and C. Vafa, Commun. Math. Phys. **158**, 569 (1993) [hep-th/9211097].
48. N. Dorey, JHEP **9811**, 005 (1998) [hep-th/9806056].
49. N. Dorey, T. J. Hollowood and D. Tong, JHEP **9905**, 006 (1999) [arXiv:hep-th/9902134].
50. M. Kontsevich and Y. Soibelman, Stability structures, motivic Donaldson-Thomas invariants and cluster transformations, arXiv:0811.2435 [math.AG]. Wall-crossing structures in Donaldson-Thomas invariants, integrable systems and Mirror Symmetry, arXiv:1303.3253 [math.AG]; D. Gaiotto, G. W. Moore

and A. Neitzke, Commun. Math. Phys. **299**, 163 (2010) [arXiv:0807.4723 [hep-th]], D. Gaiotto, G. W. Moore and A. Neitzke, Framed BPS States, [arXiv:1006.0146 [hep-th]].

51. M. Shifman, A. Vainshtein and R. Zwicky, J. Phys. A **39**, 13005 (2006) [hep-th/0602004].

52. P. A. Bolokhov, M. Shifman and A. Yung, Phys. Rev. D **84**, 085004 (2011) [arXiv:1104.5241 [hep-th]], Phys. Rev. D **85**, 085028 (2012) [arXiv:1202.5612 [hep-th]]; N. Dorey and K. Petunin, JHEP **1205**, 085 (2012) [arXiv:1202.5595 [hep-th]].

53. K. Hori and C. Vafa, Mirror symmetry, arXiv:hep-th/0002222.

54. A. Hanany and D. Tong, JHEP **0307**, 037 (2003) [hep-th/0306150]. JHEP **0404**, 066 (2004) [hep-th/0403158].

55. M. Shifman and A. Yung, Phys. Rev. D **73**, 125012 (2006) [arXiv:hep-th/0603134].

56. M. Eto, J. Evslin, K. Konishi, G. Marmorini, M. Nitta, K. Ohashi, W. Vinci, N. Yokoi, Phys. Rev. D **76**, 105002 (2007) [arXiv:0704.2218 [hep-th]].

57. M. Shifman, W. Vinci and A. Yung, Phys. Rev. D **83**, 125017 (2011) [arXiv:1104.2077 [hep-th]].

58. A. Achucarro and T. Vachaspati, Phys. Rept. **327**, 347 (2000) [hep-ph/9904229].

59. P. Koroteev, M. Shifman, W. Vinci and A. Yung, Phys. Rev. D **84**, 065018 (2011) [arXiv:1107.3779 [hep-th]].

60. P. Koroteev, A. Monin and W. Vinci, Phys. Rev. D **82**, 125023 (2010) [arXiv:1009.6207 [hep-th]].

61. D. Tong, JHEP **0709**, 022 (2007) [hep-th/0703235 [hep-th]].

62. V. A. Novikov, M. A. Shifman, A. I. Vainshtein and V. I. Zakharov, Nucl. Phys. B **229**, 381 (1983); Phys. Lett. B **166**, 329 (1986)

63. M. A. Shifman and A. I. Vainshtein, Nucl. Phys. B **277**, 456 (1986).

64. P. Koroteev and A. Monin, Phys. Rev. D **81**, 105001 (2010) [arXiv:1003.2645 [hep-th]].

65. M. Shifman and A. Yung, Phys. Rev. D **82**, 066006 (2010) [arXiv:1005.5264 [hep-th]].

66. P. A. Bolokhov, M. Shifman and A. Yung, Phys. Rev. D **88**, 085016 (2013) [arXiv:1308.4494 [hep-th]].

67. M. Shifman and A. Yung, Quantum Deformation of the Effective Theory on Non-Abelian string and 2D-4D correspondence, arXiv:1401.1455 [hep-th].

68. D. Gaiotto, S. Gukov and N. Seiberg, JHEP **1309**, 070 (2013) [arXiv:1307.2578].

VACUUM STRUCTURE
IN 3d SUPERSYMMETRIC GAUGE THEORIES

A. V. SMILGA*

SUBATECH, Université de Nantes,
4 rue Alfred Kastler, BP 20722, Nantes 44307, France[†]
* E-mail: smilga@subatech.in2p3.fr

This minireview (written on the basis of the talk that the author delivered at the Pomeranchuk memorial conference in ITEP in June 2013 and of the original papers[1-3]) is devoted to the problem of vacuum dynamics in 3d supersymmetric Yang-Mills-Chern-Simons theories with and without extra matter multiplets. By analyzing the effective Born-Oppenheimer Hamiltonian in a small spatial box, we calculate the number of vacuum states (the Witten index) for these theories and analyze their structure. The results coincide with those derived by other methods.

Keywords: Supersymmetric Gauge Theories, Chern-Simons Theories, Witten index.

1. Introduction

Probably, the most known scientific achievement of Isaak Yakovlich Pomeranchuk was the concept of the vacuum Regge pole that is called nowadays *pomeron*. I did not have a chance to meet Pomeranchuk personally — I came to ITEP when he was already gone. But I heard many times from his colleagues and collaborators that Isaak Yakovlich attributed a great significance to studying vacuum properties and even used to joke about an urgent need for the ITEP theory group to buy a powerful pump for that purpose.

Pomeranchuk did not know that, with the advent of supersymmetry, the problems of vacuum structure and vacuum counting would acquire a special interest. The existence of supersymmetric vacua (ground states of the Hamiltonian annihilated by the action of supercharges and having zero energy) is a signature that supersymmetry stays intact, while the absence of such states signalizes spontaneous breaking of supersymmetry. The crucial

[†]On leave of absence from ITEP, Moscow, Russia

quantity to be studied in this respect is *Witten index*, the difference between the numbers of bosonic and fermionic vacuum states, which also can be represented as

$$I = \text{Tr}\{(-1)^F e^{-\beta H}\} \tag{1.1}$$

where H is the Hamiltonian and F is the fermion charge operator. Due to supersymmetry, non-vacuum contributions in the trace cancel out. It is important that the quantity (1.1) represents an *index*, a close relative of the Atiyah-Singer index and other topological invariants, which is invariant under smooth Hamiltonian deformations. The latter circumstance allows one to evaluate the Witten index for rather complicated theories — it is sufficient to find out a proper simplifying deformation.

My talk will be devoted exactly to that. I will study the vacuum dynamics in a particular class of theories — supersymmetric 3-dimensional gauge theories involving the Chern-Simons term. Such theories attracted recently a considerable attention in view of newly discovered dualities between certain $\mathcal{N} = 8$ and $\mathcal{N} = 6$ versions of these theories and string theories on $AdS_4 \times S^7$ or $AdS_4 \times \mathbb{CP}^3$ backgrounds, respectively.[4,5] [a] Note, however, that the field theories dual to string theories are conformal and do not involve a mass gap. In such theories, the conventional Witten (alias, toroidal) index we are interested in here is not well defined and the proper tool to study them is the so called superconformal (alias, spherical) index.[9]

I calculate the index by deforming the theory, with putting it in a small spatial box and studying the dynamics of the Hamiltonian thus obtained in the framework of the Born–Oppenheimer (BO) approximation. The results coincide with those obtained by other methods.

Let us discuss first the simplest such theory — the $\mathcal{N} = 1$ supersymmetric Yang-Mills-Chern-Simons theory with the Lagrangian

$$\mathcal{L} = \frac{1}{g^2} \left\langle -\frac{1}{2} F_{\mu\nu}^2 + i\bar{\lambda}\slashed{D}\lambda \right\rangle + \kappa \left\langle \epsilon^{\mu\nu\rho} \left(A_\mu \partial_\nu A_\rho - \frac{2i}{3} A_\mu A_\nu A_\rho \right) - \bar{\lambda}\lambda \right\rangle . \tag{1.2}$$

The conventions are: $\epsilon^{012} = 1$, $D_\mu \mathcal{O} = \partial_\mu \mathcal{O} - i[A_\mu, \mathcal{O}]$ (such that A_μ is Hermitian), λ_α is a 2-component Majorana $3d$ spinor belonging to the adjoint representation of the gauge group, and $\langle \ldots \rangle$ stands for the color trace. We choose

$$\gamma^0 = \sigma^2, \quad \gamma^1 = i\sigma^1, \quad \gamma^2 = i\sigma^3 . \tag{1.3}$$

---

[a]More known is the Maldacena duality between the 4d $\mathcal{N} = 4$ SYM and string theory on $AdS_5 \times S^5$.[6,7] A nice review of the latter was recently published in Uspekhi.[8]

This is a $3d$ theory and the gauge coupling constant g^2 carries the dimension of mass. The physical boson and fermion degrees of freedom in this theory are massive,

$$m = \kappa g^2 . \tag{1.4}$$

In three dimensions, a nonzero mass brings about parity breaking. The requirement for e^{iS} to be invariant under certain large (non-contractible) gauge transformations (see e.g. Ref.[11] for a nice review) leads to the quantization condition

$$\kappa = \frac{k}{4\pi} . \tag{1.5}$$

with integer or sometimes (see below) half-integer $level$ k.

The index (1.1) was evaluated in[12] with the result

$$I(k, N) = [\text{sgn}(k)]^{N-1} \binom{|k| + N/2 - 1}{N - 1} . \tag{1.6}$$

for $SU(N)$ gauge group. This is valid for $|k| \geq N/2$. For $|k| < N/2$, the index vanishes and supersymmetry is broken. In the simplest $SU(2)$ case, the index is just

$$I(k, 2) = k . \tag{1.7}$$

For $SU(3)$, it is

$$I(k, 3) = \frac{k^2 - 1/4}{2} . \tag{1.8}$$

We can notice now that, for the index to be integer, the level k should be half-integer rather than integer for $SU(3)$ and for all unitary groups with odd N. The explanation is that, in these cases, a large gauge transformation mentioned above not only shifts the classical action, but also brings about the extra factor (-1) due to the modification of the fermion determinant.[13,14]

The result (1.6) was derived in[12] by the following reasoning. Consider the theory in a $large$ spatial volume, $g^2 L \gg 1$. Consider then the functional integral for the index (1.1) and mentally perform a Gaussian integration over fermionic variables. This gives an effective bosonic action that involves the CS term, the Yang-Mills term and other higher-derivative gauge-invariant terms. After that, the coefficient of the CS term is renormalized, [b]

$$k \rightarrow k - \frac{N}{2}.$$ (1.9)

At large β, the sum (1.1) is saturated by the vacuum states of the theory depending thus on the low-energy dynamics of the corresponding effective Hamiltonian. The latter is determined by the term with the lowest number of derivatives, i.e. the Chern-Simons term, the effects due to the YM term and still higher derivative terms being suppressed at small energies and large spatial volume. Basically, the spectrum of vacuum states coincides with the full spectrum in the topological pure CS theory. The latter was determined some time ago

- by establishing a relationship between the pure $3d$ CS theories and $2d$ WZNW theories.[15]
- by canonical quantization of the CS theory and direct determination of the wave functions annihilated by the Gauss law constraints.[16,17]

Then the index (1.6) is determined as the number of states in pure CS theory with the shift (1.9). For example, in the $SU(2)$ case, the number of CS states is $k + 1$, which gives (1.7) after the shift.

In Sec. 2, 3, we will rederive the result (1.6) using another method. We choose the spatial box to be *small* rather than large, $g^2 L \ll 1$, and study the dynamics of the corresponding BO Hamiltonian.

This method was developed in[18] and applied there to $4d$ SYM theories. Let us explain how it works. Take the simplest $SU(2)$ theory. If imposing *periodic* boundary conditions for all fields,[c] the slow variables in the effective BO Hamiltonian are just the zero Fourier modes of the spatial components of the Abelian vector potential and its superpartners,

$$C_j = A_j^{(0)3}, \qquad \lambda_\alpha = \lambda_\alpha^{(0)3}.$$ (1.10)

(In the $4d$ case, the spatial index j takes three values, $j = 1, 2, 3$. λ_α is the Weyl 2-component spinor describing the gluino field.). The motion in the field space $\{C_j\}$ is actually finite because a shift

$$C_j \rightarrow C_j + 4\pi n_j / L$$ (1.11)

---

[b]This is for $k > 0$. In the following, k will be assumed to be positive by default though the results for negative k will also be mentioned.

[c]We stick to this choice here though, in a theory involving only adjoint fields, one could also impose the so called *twisted* boundary conditions. In $4d$ theories, this gives the same value for the index,[18] but, in $3d$ theories, the result turns out to be different.[19]

with an integer n_j amounts to a *contractible* (this is a non-Abelian specifics) gauge transformation with respect to which the wave functions are invariant. To the leading BO order, the effective Hamiltonian is nothing but the Laplacian

$$H^{\text{eff}} = \frac{g^2}{2L^2} P_j^2 , \qquad (1.12)$$

where P_j is the momentum conjugate to C_j. The vacuum wave function is thus just a constant which can be multiplied by a function of holomorphic fermionic variables λ_α. We seem to obtain four vacuum wave functions of fermion charges 0, 1, and 2:

$$\Psi^{F=0} = 1 , \qquad \Psi_\alpha^{F=1} = \lambda_\alpha , \qquad \Psi^{F=2} = \epsilon^{\alpha\beta} \lambda_\alpha \lambda_\beta . \qquad (1.13)$$

However, the fermion wave functions are not allowed in this case. The matter is that the wave functions in the original theory should be invariant with respect to gauge transformations. For the effective wave functions, this translates into invariance with respect to *Weyl reflections*. In the $SU(2)$ case, the latter amount just to the sign flip,

$$C_j \to -C_j, \qquad \lambda_\alpha \to -\lambda_\alpha . \qquad (1.14)$$

The functions $\Psi^{F=1}$ in (1.13) are not invariant under (1.14) and are thus not allowed. We are left with 2 bosonic vacuum functions giving the value $I = 2$ for the index. A somewhat more complicated analysis (it is especially nontrivial for orthogonal and exceptional groups[20–22]) allows one to evaluate the index for other groups. It coincides with the adjoint Casimir eigenvalue c_V (another name for it is the dual Coxeter number $h^\vee$). For $SU(N)$, $I = N$.

The analysis of the $3d$ SYMCS theories along the same lines turns out to be more complicated:

- The tree level effective Hamiltonian is not just a free Laplacian, but involves an extra homogeneous magnetic field.
- The effective wave functions are not just invariant with respect to the shifts (1.11), but are multiplied by certain phase factors.[23]
- It is *not* enough to analyze the effective Hamiltonian to the leading BO order, but one-loop corrections should also be taken into account.

In Sec. 2, we will perform an accurate BO analysis at the tree level. In Sec. 3, we discuss the loop corrections. Sec. 4 is devoted to the SYMCS theories with matter. We discuss both $\mathcal{N} = 1$ theories and $\mathcal{N} = 2$ theories. For the latter, we reproduce the results of,[24] but derive them in a more transparent and simple way.

2. Pure $\mathcal{N} = 1$ SYMCS theory: the leading BO analysis

2.1. $SU(2)$

Consider $SU(2)$ theory first. As was explained above, we impose the periodic boundary conditions on all fields. In the $3d$ case, we are left with two bosonic slow variables $C_{j=1,2} = A_j^{(0)3}$ and one holomorphic fermion slow variable $\lambda = \lambda_1^{(0)3} - i\lambda_2^{(0)3}$. The tree-level effective BO supercharges and Hamiltonian describe the motion in a homogeneous magnetic field proportional to the Chern-Simons coupling and take the form

$$Q^{\text{eff}} = \frac{g}{L}\lambda(P_- + \mathcal{A}_-)$$
$$\bar{Q}^{\text{eff}} = \frac{g}{L}\bar{\lambda}(P_+ + \mathcal{A}_+),\tag{2.1}$$

$$H^{\text{eff}} = \frac{g^2}{2L^2}\left[(P_j + \mathcal{A}_j)^2 + \mathcal{B}(\lambda\bar{\lambda} - \bar{\lambda}\lambda)\right],\tag{2.2}$$

where

$$\mathcal{A}_j = -\frac{\kappa L^2}{2}\epsilon_{jk}C_k,\tag{2.3}$$

$P_\pm = P_1 \pm iP_2$, $\mathcal{A}_\pm = \mathcal{A}_1 \pm i\mathcal{A}_2$, and $\mathcal{B} = \frac{\partial \mathcal{A}_2}{\partial C_1} - \frac{\partial \mathcal{A}_1}{\partial C_2}$.

The effective vector potential (2.3) depends on the field variables $\{C_1, C_2\}$ and has nothing to do, of course, with $A_j^a(\vec{x})$. It is defined up to a gauge transformation

$$\mathcal{A}_j \rightarrow \mathcal{A}_j + \partial_j f(\vec{C}).\tag{2.4}$$

Indeed, the particular form (2.3) follows from the CS terms $\sim \epsilon_{jk}A_j\dot{A}_k$ in the Lagrangian (1.2), but one can always add to the latter a total time derivative, which adds a gradient to the canonical momentum P_j and to the effective vector potential.

Similarly to what we had in the $4d$ case, the motion in the space $\{C_1, C_2\}$ is finite. However, as was mentioned, the wave functions are not just invariant under the shifts along the cycles of the dual torus, but acquire extra phase factors,

$$\Psi(X + 1, Y) = e^{-2\pi ikY}\Psi(X, Y),$$
$$\Psi(X, Y + 1) = e^{2\pi ikX}\Psi(X, Y),\tag{2.5}$$

where $X = C_1 L/(4\pi), Y = C_2 L/(4\pi)$.

Let us explain where these factors come from. As was mentioned, the shifts $X \rightarrow X + 1$ and $Y \rightarrow Y + 1$ represent contractible gauge transformations. In the $4d$ theories, wave functions are invariant under such transformations. But the YMCS theory is special in this respect. Indeed, the Gauss law constraint in the YMCS theory (and in SYMCS theories) is not just $D_j \Pi_j^a$, but has the form

$$G^a = \frac{\delta \mathcal{L}}{\delta A_0^a} = D_j \Pi_j^a + \frac{\kappa}{2} \epsilon_{jk} \partial_j A_k^a \,,$$

where $\Pi_j^a = F_{0j}^a / g^2 + (\kappa/2)\epsilon_{jk} A_k^a$ are the canonical momenta. The second term gives rise to the phase factor associated with an infinitesimal gauge transformation $\delta A_j^a(\vec{x}) = D_j \alpha^a(\vec{x})$ (do not confuse the spatial coordinates $\vec{x}$ with the rescaled vector potentials X, Y),

$$\Psi[A_j^a + D_j \alpha^a] = \exp\left\{ -\frac{i\kappa}{2} \int d\vec{x}\, \epsilon_{kl}\, \partial_k \alpha^a A_l^a \right\} \Psi[A_j^a] \,. \tag{2.6}$$

This property holds also for the finite contractible gauge transformations $\alpha^a = (4\pi x/L)\delta^{a3}$ or $\alpha^a = (4\pi y/L)\delta^{a3}$ implementing the shifts $C_{1,2} \to C_{1,2} + 4\pi/L$. The phase factors $\mathcal{E}_{1,2}(X, Y)$ thus obtained coincide with those quoted in Eq. (2.5); they are nothing but the holonomies $\mathcal{E}_1 = \exp\left\{ i \int_{\gamma_1} \mathcal{A}_1 dC_1 \right\}$ and $\mathcal{E}_2 = \exp\left\{ i \int_{\gamma_2} \mathcal{A}_2 dC_2 \right\}$, with $\gamma_{1,2}$ being two cycles of the torus attached to the point (X, Y). The factors $\mathcal{E}_{1,2}$ satisfy the property

$$\mathcal{E}_1(X, Y)\mathcal{E}_2(X + 1, Y)\mathcal{E}_1^{-1}(X + 1, Y + 1)\mathcal{E}_2^{-1}(X, Y + 1) = e^{4\pi i k} = 1 \,. \tag{2.7}$$

The phase $4\pi k$ that one acquires going around the sequence of two direct and two inverse cycles is nothing that $2\pi\Phi$, with Φ being the magnetic flux. For the wave functions to be uniquely defined, the latter must be quantized.

Note that, if choosing another gauge for $\mathcal{A}_j$, the holonomies $\mathcal{E}_{1,2}$ would be different, but the property (2.7) would of course be still there.

The eigenfunctions of the Hamiltonian (2.2) satisfying the boundary conditions (2.5) represent elliptic functions — a variety of theta functions. There are $2k$ ground state wave functions. For $k > 0$, their explicit form is

$$\Psi_{\text{tree}}^{\text{eff}}(X, Y) \propto e^{-\pi k \bar{z} z} e^{\pi k \bar{z}^2} Q_m^{2k}(\bar{z}), \tag{2.8}$$

where $z = X + iY$, $m = 0, \ldots, 2k - 1$, and the functions Q_m^q are defined in the Appendix. For negative k, the functions have the same form, but with the interchanged z and $\bar{z}$ and with the extra fermionic factor λ.

The index $I = 2k$ of the effective Hamiltonian (2.2) coincides with the flux of the effective magnetic field on the dual torus divided by 2π.[25,26]

Note now that not all $2|k|$ states are admissible. We have to impose the additional Weyl invariance condition (following from the gauge invariance of the original theory). For $SU(2)$, this amounts to [d] $\Psi^{\text{eff}}(-C_j) = \Psi^{\text{eff}}(C_j)$, which singles out $|k| + 1$ vacuum states, bosonic for $k > 0$ and fermionic for $k < 0$.

When $k = 0$, the effective Hamiltonian (2.2) describes free motion on the dual torus. There are two zero energy ground states, $\Psi^{\text{eff}} = \text{const}$ and $\Psi^{\text{eff}} = \text{const} \cdot \lambda$ (we need not to bother about Weyl oddness of the factor λ by the same reason as above). The index is zero. We thus derive

$$I_{SU(2)}^{\text{tree}} = (|k| + 1)\text{sgn}(k) \,. \tag{2.9}$$

2.2. Higher unitary groups

The effective Hamiltonian for the group $SU(N)$ involves $2r = 2(N-1)$ slow bosonic and $r = N - 1$ slow fermionic variables $\{C_j^a, \lambda^a\}$ belonging to the Cartan subalgebra of $SU(N)$ (r is the rank of the group). It has the form

$$H = \frac{g^2}{2L^2}\left[(P_j^a + \mathcal{A}_j^a)^2 + \mathcal{B}^{ab}(\lambda^a\bar{\lambda}^b - \bar{\lambda}^b\lambda^a)\right] \,, \tag{2.10}$$

where

$$\mathcal{A}_j^a = -\frac{\kappa L^2}{2}\epsilon_{jk}C_k^a \,,$$
$$\mathcal{B}^{ab} = \kappa L^2\delta^{ab} \,, \tag{2.11}$$

$a = 1, \ldots, r$. By the same token as in the $SU(2)$ case, the motion is finite and extends over the manifold $T \times T$, with T being the maximal torus of the group. For $SU(3)$, the latter is depicted in Fig. 1. Each point in Fig. 1 is a *coweight* $\{w^3, w^8\}$ such that the group element brought on the maximal torus is $g^{\text{torus}} = \exp\{4\pi i(w^3 t^3 + w^8 t^8)\}$. The meaning of the dashed lines and of special points marked by the box and triangle will be shortly explained.

The index of the effective Hamiltonian can be evaluated semiclassically by reducing the functional integral for (1.1) to an ordinary one.[27] The latter represents a *generalized* magnetic flux (this is nothing but that the

---

[d]Note that, in contrast to what should be done in 4 dimensions, we did not include here the Weyl reflection of the fermion factor λ entering the effective wave function for negative k. The reason is that the conveniently defined *fast* wave function (to which the effective wave function depending only on C_j and λ should be multiplied) involves, for negative k, a Weyl-odd factor $C_1 + iC_2$. This oddness compensates the oddness of the factor λ in the effective wave function.[1]

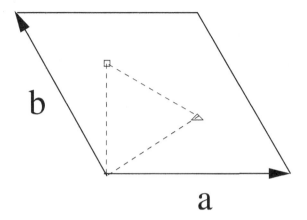

a

Fig. 1. Maximal torus and Weyl alcove for $SU(3)$. **a** and **b** are the simple coroots. The points $\square$ and $\triangle$ are fundamental coweights.

r-th Chern class of the $U(1)$ bundle over $T \times T$ with the connection $\mathcal{A}_j^a$),

$$I = \frac{1}{(2\pi)^r} \int_{T \times T} \prod_{ja} dC_j^a \, \det\|\mathcal{B}^{ab}\| \, . \tag{2.12}$$

In the case of $SU(N)$,

$$I^{SU(N)} = Nk^{N-1} \, . \tag{2.13}$$

Let us find the explicit expressions for the $3k^2$ ground state wave functions in the case of $SU(3)$. They represent generalized theta functions defined on the coroot lattice of $SU(3)$. They satisfy the boundary conditions

$$\Psi(\mathbf{X} + \mathbf{a}, \mathbf{Y}) = e^{-2\pi i k \mathbf{a} \mathbf{Y}} \Psi(\mathbf{X}, \mathbf{Y}) \, ,$$
$$\Psi(\mathbf{X} + \mathbf{b}, \mathbf{Y}) = e^{-2\pi i k \mathbf{b} \mathbf{Y}} \Psi(\mathbf{X}, \mathbf{Y}) \, ,$$
$$\Psi(\mathbf{X}, \mathbf{Y} + \mathbf{a}) = e^{2\pi i k \mathbf{a} \mathbf{X}} \Psi(\mathbf{X}, \mathbf{Y}) \, ,$$
$$\Psi(\mathbf{X}, \mathbf{Y} + \mathbf{b}) = e^{2\pi i k \mathbf{b} \mathbf{X}} \Psi(\mathbf{X}, \mathbf{Y}) \, , \tag{2.14}$$

where $\mathbf{X} = 4\pi \mathbf{C}_1/L$, $\mathbf{Y} = 4\pi \mathbf{C}_2/L$ and $\mathbf{a} = (1,0)$, $\mathbf{b} = (-1/2, \sqrt{3}/2)$ are the simple coroots. When $k = 1$, there are 3 such states:

$$\Psi_0 = \sum_{\mathbf{n}} \exp\left\{ -2\pi(\mathbf{n} + \mathbf{Y})^2 - 2\pi i \mathbf{X} \mathbf{Y} - 4\pi i \mathbf{X} \mathbf{n} \right\} \, ,$$

$$\Psi_\triangle = \sum_{\mathbf{n}} \exp\left\{ -2\pi(\mathbf{n} + \mathbf{Y} + \triangle)^2 - 2\pi i \mathbf{X} \mathbf{Y} - 4\pi i \mathbf{X}(\mathbf{n} + \triangle) \right\} \, ,$$

$$\Psi_\square = \sum_{\mathbf{n}} \exp\left\{ -2\pi(\mathbf{n} + \mathbf{Y} + \square)^2 - 2\pi i \mathbf{X} \mathbf{Y} - 4\pi i \mathbf{X}(\mathbf{n} + \square) \right\} \, , \tag{2.15}$$

where the sums run over the coroot lattice, $\mathbf{n} = m_a \mathbf{a} + m_b \mathbf{b}$ with integer $m_{a,b}$. Now, $\triangle$, $\square$ are certain special points on the maximal torus (fundamental coweights) satisfying

$$\triangle \mathbf{a} = \square \mathbf{b} = 1/2, \quad \square \mathbf{a} = \triangle \mathbf{b} = 0 .$$

The group elements that correspond to the points $0, \triangle$, and $\square$ belong to the center of the group,

$$
\begin{aligned}
U_0 &= \mathrm{diag}(1,1,1) , \\
U_\square &= \mathrm{diag}(e^{2i\pi/3}, e^{2i\pi/3}, e^{2i\pi/3}) , \\
U_\triangle &= \mathrm{diag}(e^{4i\pi/3}, e^{4i\pi/3}, e^{4i\pi/3}) .
\end{aligned}
\tag{2.16}
$$

They are obviously invariant with respect to Weyl symmetry, which permutes the eigenvalues.[e] Thus, all three states (2.15) at the level $k = 1$ are Weyl invariant. But for $k > 1$, the number of invariant states is less than $3k^2$. For an arbitrary k, the wave functions of all $3k^2$ eigenstates can be written in the same way as in (2.15),

$$\Psi_n = \sum_{\mathbf{n}} \exp\left\{ -2\pi(\mathbf{n} + \mathbf{Y} + \mathbf{w}_n)^2 - 2\pi i \mathbf{X}\mathbf{Y} - 4\pi i \mathbf{X}(\mathbf{n} + \mathbf{w}_n) \right\} , \tag{2.17}$$

where $\mathbf{w}_n$ are coweights whose projections on the simple coroots $\mathbf{a}$, $\mathbf{b}$ represent integer multiples of $1/(2k)$. Only the functions (2.17) with $\mathbf{w}_n$ lying in the vertices of the Weyl alcove are Weyl invariant. For all other $\mathbf{w}_n$, one should construct Weyl invariant combinations

$$\Psi = \sum_{\hat{x} \in W} \hat{x} \Psi_{\mathbf{w}_n} . \tag{2.18}$$

As a result, the number of Weyl invariant states is equal to the number of the coweights $\mathbf{w}_n$ lying within the Weyl alcove (including the borders). For example, in the case $k = 4$, there are 15 such coweights shown in Fig. 2 and, correspondingly, 15 vacuum states.

For a generic k, the number of the states is

$$I^{\mathrm{tree}}_{SU(3)}(k > 0) = \sum_{m=1}^{k+1} m = \frac{(k+1)(k+2)}{2} . \tag{2.19}$$

The analysis for $SU(4)$ is similar. The Weyl alcove is the tetrahedron with the vertices corresponding to the center elements of $SU(4)$. A pure geometric counting gives

---

[e]For a generic coweight, the Weyl group elements permuting the eigenvalues $1 \leftrightarrow 2$, $1 \leftrightarrow 3$ and $2 \leftrightarrow 3$ act as the reflections with respect to the dashed lines bounding the Weyl alcove (the quotient T/W in Fig. 1).

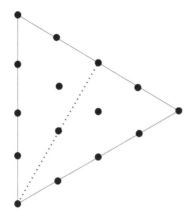

Fig. 2. $SU(3)$: 15 vacuum states for $k = 4$. The dotted line marks the boundary of the Weyl alcove for G_2.

$$I^{\text{tree}}_{SU(4)}(k > 0) = \sum_{m=1}^{k+1} \sum_{p=1}^{m} p = \frac{(k+1)(k+2)(k+3)}{6} . \qquad (2.20)$$

The generalization for an arbitrary N is obvious. It gives the result

$$I^{\text{tree}}(k, N) = \binom{k + N - 1}{N - 1} \qquad (2.21)$$

We also performed a similar analysis for the symplectic groups and for G_2. Let us dwell on G_2. The simple coroots for G_2 are $\mathbf{a} = (1, 0)$ and $\mathbf{b} = (-3/2, \sqrt{3}/2)$. The lattice of coroots and the maximal torus look exactly in the same way as for $SU(3)$ (see Fig. 3). Hence, before Weyl-invariance requirement is imposed, the index is equal to $3k^2$, as for $SU(3)$. The difference is that the Weyl group involves now 12 rather than 6 elements, and the Weyl alcove is two times smaller than for $SU(3)$. As a result, for $k = 4$, we have only 9 (rather than 15) Weyl-invariant states (see Fig. 2). The general formula is

$$I^{\text{tree}}_{G_2}(k) = \left\{ \begin{array}{ll} \frac{(|k|+2)^2}{4} & \text{for even } k \\ \frac{(|k|+1)(|k|+3)}{4} & \text{for odd } k \end{array} \right\} . \qquad (2.22)$$

3. Loop corrections

We will mostly discuss in this section the $SU(2)$ theory. An accurate generalization of all arguments to higher unitary groups should not be difficult, but it is yet to be done.

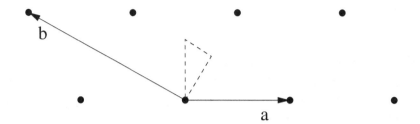

Fig. 3. Coroot lattice and Weyl alcove for G_2.

Fig. 4. Renormalization of the structure $\propto \epsilon_{\mu\nu\rho}\mathrm{Tr}\{A_\mu\partial_\nu A_\rho\}$ by a fermion loop.

3.1. Infinite volume

It is known since[28] that the CS coupling κ in the pure YMCS theory is renormalized at the 1-loop level. For $\mathcal{N} = 1, 2, 3$ SYMCS theories, the corresponding calculations have been performed in.[29] The effect can be best understood by considering the fermion loop contribution to the renormalization of the structure $\propto A\partial A$ in the Chern-Simons term (see Fig. 4). We obtain (remind that κ and k are assumed to be positive by default).

$$\Delta\kappa = -mc_V \int \frac{d^3 p_E}{(2\pi)^3} \frac{1}{(p_E^2 + m^2)^2} = -\frac{c_V}{8\pi}. \tag{3.1}$$

There is also a contribution coming from the gluon loop.[f] It is convenient[2] to choose the Hamilton gauge $A_0 = 0$, in which case the gluon propagator

$$D_{jk}^{ab}(\omega, \vec{p}) = \frac{ig^2 \delta^{ab}}{\omega^2 - \vec{p}^2 - m^2} \left[\delta_{jk} - \frac{p_j p_k}{\omega^2} - \frac{im}{\omega} \epsilon_{jk} \right] \quad (3.2)$$

involves only transverse degrees of freedom and there are no ghosts.

An accurate calculation gives

$$k_{\text{ren}} = k + c_V - \frac{c_V}{2}, \quad (3.3)$$

where the first term comes from the gluon loop and the second term from the fermion loop.

A legitimate question is whether also the second and higher loops bring about a renormalization of the level k. The answer is, however, negative. The proof is simple. Consider the case $k \gg c_V$. This is the perturbative regime where the loop corrections are ordered such that $\Delta k^{(1 \text{ loop})} \sim O(1)$, $\Delta k^{(2 \text{ loops})} \sim O(1/k)$, etc. But the corrections of order $\sim 1/k$ to k are not allowed. To provide for gauge invariance, k_{ren} should be integer. Thereby, all higher loop contributions in k_{ren} must vanish, and they do.

Note finally that the renormalization (3.3) refers to supersymmetric Yang-Mills-Chern-Simons theory — dynamical theory with nontrivial interactions. There is no such renormalization in the topological pure supersymmetric Chern-Simons theory where the fermions decouple. The number of states in this theory is the same as in the pure CS theory.

3.2. Finite volume

As was mentioned above, the coefficient κ (with the factor L^2) has the meaning of magnetic field on the dual torus for the effective finite volume BO Hamiltonian. Renormalization of κ translates into renormalization of this magnetic field. At the tree level, the magnetic field was constant. The renormalized field is not constant, however, but depends on the slow variables C_j. To find this dependence, we have to evaluate the effective Lagrangian in the slow Abelian background $C_j(t)$. The effective vector potential is extracted from the term $\sim \mathcal{A}_j(\vec{C}) \dot{C}_j$ in this Lagrangian. The latter can be evaluated in the background field approach. Up to certain technical fine points[1,2] that we will not discuss here, the result can be obtained by taking the same Feynman graphs that determined renormalization of κ in the

---

[f]On the other hand, spinless scalars (present in SYMCS theories with extended supersymmetries) do not contribute to the renormalization of κ.

infinite volume and replacing

$$\vec{p} \to \frac{2\pi\vec{n}}{L} - C_j \,, \qquad \int \frac{d^2p}{(2\pi)^2} \to \frac{1}{L^2} \sum_{n_j} \tag{3.4}$$

in the spatial integrals there. The shift $-\vec{C}$ in the momentum is due to replacing the usual derivative by the covariant one.

For the effective vector-potentials induced by the fermion and the gluon loop, we derive[g]

$$\mathcal{A}_j^F = \frac{\epsilon_{jk}}{2} \sum_{n_j} \frac{\left(C - \frac{2\pi n}{L}\right)_k}{\left(\vec{C} - \frac{2\pi\vec{n}}{L}\right)^2} \left[1 - \frac{m}{\sqrt{\left(\vec{C} - \frac{2\pi\vec{n}}{L}\right)^2 + m^2}} \right] \overset{m\to 0}{\longrightarrow}$$

$$\frac{\epsilon_{jk}}{2} \sum_{n_j} \frac{\left(C - \frac{2\pi n}{L}\right)_k}{\left(\vec{C} - \frac{2\pi\vec{n}}{L}\right)^2} \tag{3.5}$$

and

$$\mathcal{A}_j^B = -\frac{\epsilon_{jk}}{2} \sum_{n_j} \frac{\left(C - \frac{2\pi n}{L}\right)_k}{\left(\vec{C} - \frac{2\pi\vec{n}}{L}\right)^2} \left[2 - \frac{3m}{\sqrt{\left(\vec{C} - \frac{2\pi\vec{n}}{L}\right)^2 + m^2}} \right.$$

$$\left. + \frac{m^3}{\left[\left(\vec{C} - \frac{2\pi\vec{n}}{L}\right)^2 + m^2\right]^{3/2}} \right] \overset{m\to 0}{\longrightarrow} -\epsilon_{jk} \sum_{n_j} \frac{\left(C - \frac{2\pi n}{L}\right)_k}{\left(\vec{C} - \frac{2\pi\vec{n}}{L}\right)^2} \tag{3.6}$$

The corresponding induced magnetic fields are

$$\Delta\mathcal{B}^F(\vec{C}) = -\frac{m}{2} \sum_{n_j} \frac{1}{\left[\left(\vec{C} - \frac{2\pi\vec{n}}{L}\right)^2 + m^2\right]^{3/2}} \,, \tag{3.7}$$

and

$$\Delta\mathcal{B}^B(\vec{C}) = \frac{3m}{2} \sum_{n_j} \frac{\left(\vec{C} - \frac{2\pi\vec{n}}{L}\right)^2}{\left[\left(\vec{C} - \frac{2\pi\vec{n}}{L}\right)^2 + m^2\right]^{5/2}} \,. \tag{3.8}$$

---

[g]The sums (3.5) and (3.6) diverge at large $\|\vec{n}\|$. Their exact meaning will be clarified soon.

For most values of C_j, the corrections (3.7), (3.8) are of order $\sim mL^3 = \kappa g^2 L^3$, which is small compared to $\mathcal{B}^{\text{tree}} \sim \kappa L^2$ if $g^2 L \ll 1$, which we assume. There are, however, four special points (the "corners" of the torus)

$$C_j = 0, \ C_j = (2\pi/L, 0), \ C_j = (0, 2\pi/L), \ C_j = (2\pi/L, 2\pi/L), \quad (3.9)$$

at the vicinity of which the loop-induced magnetic field is *much larger* than the tree-level magnetic field. This actually means that the "Abelian" BO approximation, with an assumption that the energy scale associated with the slow variables $\{C_j, \lambda\}$ is small compared to the energy scale of the non-Abelian components and higher Fourier modes, breaks down in this region.

Disregarding this for a while, one can observe that the loop corrections bring about effective *flux lines* similar to Abrikosov vortices located at the corners. The width of these vortices is of order m. Gluon corrections generate the lines of unite flux $\Phi^B = \frac{1}{2\pi} \int \Delta \mathcal{B}^B(\vec{C}) \, d^2 C$, while the fermion loops generate the lines of flux $\Phi^F = -1/2$. Altogether, we have a line of flux $\Phi_{\text{line}} = 1/2$ in each corner.

Adding the induced fluxes to the tree-level flux, one obtains the total flux

$$\Phi^{\text{tot}} = 2k + 4 \cdot 1 - 4 \cdot \frac{1}{2} = 2k + 2 \quad (3.10)$$

suggesting the presence of $2k + 2$ vacuum states in the effective BO Hamiltonian (that is before the Weyl invariance requirement being imposed).

Not all these states are admissible, however. The wave functions of *four* such states turn out to be singular at the corners, and they should be dismissed.

Indeed, let us find the effective wave functions of all $2k + 2$ states in the Abelian valley far enough from the corners (3.9). The effective vector potential corresponding to one of the loop-induced flux lines can be chosen in the form

$$\mathcal{A}_j = -\frac{\epsilon_{jk} C_k}{2\vec{C}^2} F(m^2, \vec{C}^2), \quad (3.11)$$

where the core profile function deduced from (3.5) and (3.6),

$$F(m^2, \vec{C}^2) = 1 - \frac{2m}{\sqrt{\vec{C}^2 + m^2}} + \frac{m^3}{\left(\vec{C}^2 + m^2\right)^{3/2}} \quad (3.12)$$

vanishes at $C_j = 0$ and tends to 1 for large C_j.

Consider the effective supercharge Q^{eff} given by (2.1) at the vicinity of the origin, but outside the core of the vortex (3.11),

$$m \ll C_j \ll 4\pi/L \,. \tag{3.13}$$

We can then set $F(m^2, \vec{C}^2) = 1$, neglect the contribution of other flux lines as well as the homogeneous field contribution (2.3). The equation $Q^{\text{eff}}\chi^{\text{eff}} = 0$ for a vacuum effective wave function acquires the form

$$\left(\frac{\partial}{\partial z} + \frac{1}{4z} \right) \chi^{\text{eff}} = 0 \tag{3.14}$$

(we remind that $z = \frac{C_+ L}{4\pi}$). Its solution is

$$\chi^{\text{eff}}(z, \bar{z}) \sim \frac{F(\bar{z})}{z^{1/4}} \,. \tag{3.15}$$

The effective wave function on the whole torus can be restored from two conditions:

- It should behave as in (3.15) at the vicinity of each corner.
- It should satisfy the boundary conditions with the twist (2.7) corresponding to the total flux (3.10).

This gives the structure

$$\chi^{\text{eff}}_m(z, \bar{z}) \propto \frac{Q^{2k+2}_m(\bar{z})}{[\Pi(z)\Pi(\bar{z})]^{1/4}} \,, \tag{3.16}$$

where

$$\Pi(\bar{z}) = Q^4_3(\bar{z}) - Q^4_1(\bar{z}) \,. \tag{3.17}$$

is a θ function of level 4 having zeros at the "corners" (3.9).[h]

We can go back now to the sums (3.5), (3.6). The divergences can be regularized by subtracting from $\mathcal{A}_j$ a certain infinite pure gauge part $\sim \partial_j f(\vec{C})$ [as a side remark, this regularization breaks the apparent periodicity of (3.5), (3.6)]. After that, the massless limits of $\mathcal{A}^{F,B}_+$ and $\mathcal{A}^{F,B}_-$ represent meromorphic toric functions $P(\bar{z}), P(z)$ having simple poles at the corners (3.9). They are obviously expressed via $\Pi^{-1}(\bar{z})$ and $\Pi^{-1}(z)$.

The full wave function is the product of the effective wave function (3.16) and the ground state wave function of the fast Hamiltonian. Near

---

[h]The function (3.17) is known from the studies of canonical quantization of pure CS theories.[16,17,30] It also enters the relation (A.7).

the corner $\mathbf{C} = 0$ in the region (3.13), the latter behaves as $\Psi^{\text{fast}} \sim 1/\sqrt{|z|}$ [see Eq. (3.16) of Ref.[2]], which is extended to the behavior

$$\Psi^{\text{fast}} \sim \frac{1}{\sqrt{|\Pi(z)|}} \, . \tag{3.18}$$

in the whole Abelian valley.

Thus, generically, the full wave function thus obtained is singular at the corners,

$$\Psi^{\text{fast}} \chi_m^{\text{eff}}(z, \bar{z}) \sim 1/|\Pi(z)| \, . \tag{3.19}$$

The singularity in $\mathcal{A}_j$ smears out when taking into account the finite core size suggesting that the singularity in the effective wave function smears out too. However, we do not actually have a right to go inside the core in the Abelian BO framework — this approximation breaks down there, as we mentioned.

An accurate corner analysis (it is again a Born-Oppenheimer analysis where we have to treat as slow *all* zero Fourier modes[i] of the fields, both the Abelian and non-Abelian) that involves the matching of the corner wave function to the wave function in the Abelian valley far from the corners was performed in.[2] The result is rather natural. It turns out that the singularity *is* not smeared out when going into the vortex core. In other words, the states whose Abelian BO wave functions exhibit in the massless a singularity at the corners, like in (3.19), stay singular there in the exact analysis with finite mass. Such states are not admissible and should be disregarded.

The admissible wave functions still have the structure (3.16), but theta functions $Q_m^{2k+2}(\bar{z})$ should have zeros at the corners. In other words, they can be presented as $\Pi(\bar{z})$ multiplied by a theta function of level $2k - 2$. This gives

$$\chi_m^{\text{eff}}(z, \bar{z}) \propto Q_m^{2k-2}(\bar{z}) \, \Pi^{3/4}(\bar{z}) \, \Pi^{-1/4}(z) \, , \tag{3.20}$$

The parameter m takes now $2k - 2$ values, which gives $2k - 2$ [rather than $2k + 2$ as would follow naively from (3.16)] "pre-Weyl" vacuum states. After imposing the Weyl-invariance condition, we obtain k states in agreement with (1.7).

The following important remark is of order here. We have obtained $2k - 2$ pre-Weyl states by selecting $2k - 2$ nonsingular states out of $2k + 2$ states in Eq. (3.16). The latter was obtained by taking into account both

---

[i]To be precise, zero Fourier modes are relevant in the corner $C_j = 0$. In other corners in (3.9), the slow modes are characterized by $n_j = (1, 0)$, $n_j = (0, 1)$, and $n_j = (1, 1)$.

gluon-induced and fermion-induced flux lines. However, it is possible to *get rid* of the gluon flux lines altogether.

In the region outside the vortex core where the BO approximation works, one can translate the effective Lagrangian analysis leading to (3.7) and (3.8) to the effective Hamiltonian analysis. The induced vector-potentials are then obtained as Pancharatnam-Berry phases,[31,32]

$$\mathcal{A}_j^{PB} = -i \frac{\int (\Psi^{\text{fast}})^* \frac{\partial}{\partial C_j} \Psi^{\text{fast}} \, dx^{\text{fast}}}{\int (\Psi^{\text{fast}})^* \Psi^{\text{fast}} \, dx^{\text{fast}}}. \tag{3.21}$$

The potentials leading to (3.7) and (3.8) correspond to a particular choice of Ψ^{fast}.

But we can as well modify the definition of the fast wave function by multiplying it by any function of slow variables. In particular, we can multiply it by a factor that is singular at the origin (anyway, the BO approximation is not applicable there) and define

$$\tilde{\Psi}^{\text{fast}} = \Psi^{\text{fast}} \sqrt{\frac{\Pi(\bar{z})}{\Pi(z)}}. \tag{3.22}$$

One cannot decide between Ψ^{fast} and $\tilde{\Psi}^{\text{fast}}$ in the Abelian BO framework. Evaluating (3.21) with $\tilde{\Psi}^{\text{fast}}$ brings about an extra gradient term [cf. (2.4)] giving a negative unit flux line in each corner. It annihilates the fluxes induced by gluon loops. Now, the equation $Q^{\text{eff}} \tilde{\chi}^{\text{eff}} = 0$ reads

$$\left(\frac{\partial}{\partial z} - \frac{1}{4z} \right) \chi^{\text{eff}} = 0 \tag{3.23}$$

with the solution $\chi^{\text{eff}} \sim z^{1/4} F(\bar{z})$. Its extension onto the whole torus is

$$\tilde{\chi}_m^{\text{eff}}(z, \bar{z}) \propto Q_m^{2k-2}(\bar{z}) \sqrt{|\Pi(z)|}. \tag{3.24}$$

When multiplying by $\tilde{\Psi}^{\text{fast}}$, these functions (*all* of them should be taken into account now) give *exactly the same* full wave functions as before. One can thus say that gluon-induced flux lines (more general — any flux line with integer flux) should be disregarded while counting vacua. Such flux lines (kind of Dirac strings) are simply not observable. On the other hand, vortices with fractional fluxes affect vacuum counting. Heuristically, four half-integer flux lines kind of "disturb" this counting making it "more difficult" for the toric vacuum wave functions to stay uniquely defined (a single half-integer flux line would make it just impossible) such that the number of states is decreased.

For all other groups, the index is obtained by substituting in the tree-level result the value of k renormalized by exclusively fermion loops, $k \to k - c_V/2$.[j] We arrive at the result (1.6) for $SU(N)$. For G_2, we obtain

$$I_{\mathcal{N}=1}^{\mathrm{SYMCS}}[G_2] = \begin{cases} \frac{k^2}{4} & \text{for even } k \\ \frac{k^2-1}{4} & \text{for odd } k \end{cases}. \tag{3.25}$$

We also give here for completeness the result for the index for the symplectic groups. For positive k,

$$I_{\mathcal{N}=1}^{\mathrm{SYMCS}}[Sp(2r)] = \begin{pmatrix} k + \frac{r-1}{2} \\ r \end{pmatrix}. \tag{3.26}$$

For negative k, the index is restored via $I(k) = (-1)^r I(-k)$.

4. Theories with matter

In the theories with matter, the index is modified compared to the pure SYMCS theories due to two effects:

- Extra matter-induced renormalization of k.
- Appearance of extra Higgs vacua due to nontrivial Yukawa interactions.

The first effect seems to be rather transparent: extra fermion loops bring about extra renormalization. There are, however, subtleties to be discussed later. As for the extra Higgs vacua, their appearance is not the $3d$ specifics, they appear (and modify the index) also in $4d$ supersymmetric gauge theories. Let us discuss this first.

4.1. 4d theories

Historically, it was argued in Ref.[18] that adding *nonchiral* matter to the theory does not change the estimate for the index. Indeed, nonchiral fermions (and their scalar superpartners) can be given a mass. For large masses, they

---

[j]As we have seen, this renormalization should be understood *cum grano salis* as the renormalized magnetic field is concentrated at the corners invalidating the Abelian BO approximation. That is why it is desirable to extend the rigorous analysis performed for $SU(2)$ in Ref.[2] to other groups. But there is little doubt that the recipe of neglecting the gluon loops and taking into account only the fermion loops would be confirmed during such analysis.

seem to decouple and the index seems to be the same as in the pure SYM theory.[k]

However, it was realized later that, in *some* cases, massive matter *can* affect the index. The latter may change when one adds on top of the mass term also Yukawa terms coupling different matter multiplets. The simplest example[l] is the $\mathcal{N} = 1$ $SU(2)$ theory involving a couple of fundamental matter multiplets Q_f^j ($j = 1, 2$ being the color and $f = 1, 2$ the subflavor index; the indices are raised and lowered with $\epsilon^{jk} = -\epsilon_{jk}$ and $\epsilon^{fg} = -\epsilon_{fg}$) and an adjoint multiplet $\Phi_j^k = \Phi^a(t^a)_j^k$.

Let the tree superpotential be

$$\mathcal{W}^{\text{tree}} = \mu \Phi_k^j \Phi_j^k + \frac{m}{2} Q_f^j Q_j^f + \frac{h}{\sqrt{2}} Q_{jf} \Phi_k^j Q^{kf}, \qquad (4.1)$$

where μ and m are adjoint and fundamental masses, and h is the Yukawa constant.

There is also the instanton-generated superpotential,[37]

$$\mathcal{W}^{\text{inst}} = \frac{\Lambda^5}{V}, \qquad (4.2)$$

where Λ is a constant of dimension of mass and $V = Q_f^j Q_j^f / 2$ is the gauge-invariant moduli. Excluding Φ, we obtain the effective superpotential

$$\mathcal{W}^{\text{eff}} = mV - \frac{h^2 V^2}{4\mu} + \frac{\Lambda^5}{V}. \qquad (4.3)$$

The vacua are given by the solutions to the equation $\partial \mathcal{W}^{\text{eff}} / \partial V = 0$. This equation is cubic, and hence there are *three* roots and *three* vacua.[m]

Note now that, when h is very small, *one* of these vacua is characterized by a very large value, $\langle V \rangle \approx 2\mu m/h^2$ (and the instanton term in the super-potential plays no role here). In the limit $h \to 0$, it runs to infinity and we are left with only *two* vacua, the same number as in the pure SYM $SU(2)$ theory. Another way to see it is to observe that, for $h = 0$, the equation $\partial \mathcal{W}^{\text{eff}} / \partial V = 0$ becomes quadratic having only two solutions.

The same phenomenon shows up in the theory with G_2 gauge group studied in.[39] This theory involves three 7-plets S_f^j. The index of a pure SYM with G_2 group is known to coincide with the adjoint Casimir eigenvalue c_V

---

[k]This does not work for chiral multiplets. The latter are always massless and always affect the index.[33,34]

[l]It was very briefly considered in[35] and analyzed in details in.[36]

[m]These three vacua are intimately related to three *singularities* in the moduli space of the associated $\mathcal{N} = 2$ supersymmetric theory with a single matter hypermultiplet studied in.[38]

of G_2. It is equal to 4. However, if we include in the superpotential the Yukawa term,

$$\mathcal{W}^{\text{Yukawa}} = h\,\epsilon^{fgh} f^{jkl} S_{fj} S_{gk} S_{hl} \tag{4.4}$$

(f^{jkl} being the Fano antisymmetric tensor), two new vacua appear. They run to infinity in the limit $h \to 0$.

The appearance of new vacua when Yukawa terms are added should by no means come as a surprise. This is basically due to the fact that the Yukawa term has a higher dimension than the mass term.

4.2. 3d superspace

We use a variant of the $\mathcal{N} = 1$ $3d$ superspace formalism developed in.[40] The superspace (x^μ, θ^α) involves a real 2-component spinor θ^α. Indices are lowered and raised with antisymmetric $\epsilon_{\alpha\beta}, \epsilon^{\alpha\beta}$, $\theta^2 \equiv \theta^\alpha \theta_\alpha$, $\int d^2\theta\, \theta^2 = -2$. The $3d$ γ-matrices $(\gamma^\mu)^\alpha{}_\beta$ chosen as in (1.3) satisfy the identity

$$\gamma^\mu \gamma^\nu = g^{\mu\nu} + i\epsilon^{\mu\nu\rho} \gamma_\rho. \tag{4.5}$$

Note that $(\gamma^\mu)_{\alpha\beta}$ are all imaginary and symmetric.

Gauge theories are described in terms of the real spinorial superfield Γ_α. For non-Abelian theories, Γ_α represent Hermitian matrices. As in $4d$, one can choose the Wess-Zumino gauge reducing the number of components of Γ_α. In this gauge,

$$\Gamma_\alpha = i(\gamma^\mu)_{\alpha\beta} \theta^\beta A_\mu + i\theta^2 \lambda_\alpha, \tag{4.6}$$

Then the covariant superfield strength is

$$W_\alpha = \frac{1}{2} \mathcal{D}^\beta \mathcal{D}_\alpha \Gamma_\beta - \frac{1}{2}[\Gamma^\beta, \mathcal{D}_\beta \Gamma_\alpha]$$
$$= -i\lambda_\alpha + \frac{1}{2}\epsilon^{\mu\nu\rho} F_{\mu\nu}(\gamma_\rho)_{\alpha\beta}\theta^\beta + \frac{i\theta^2}{2}(\gamma^\mu)^\beta{}_\alpha \nabla_\mu \lambda_\beta. \tag{4.7}$$

In the superfield language, the Lagrangian (1.2) is written as

$$\mathcal{L} = \int d^2\theta \left\langle \frac{1}{2g^2} W_\alpha W^\alpha + \frac{i\kappa}{2}\left(W_\alpha \Gamma^\alpha + \frac{1}{3}\{\Gamma^\alpha, \Gamma^\beta\}\mathcal{D}_\beta \Gamma_\alpha\right)\right\rangle. \tag{4.8}$$

Let us add now matter multiplets. In this talk, we will consider only real adjoint multiplets. (In Ref.,[3] we also treat the theories with complex fundamental multiplets.) Let there be only one such multiplet,

$$\Sigma = \sigma + i\psi_\alpha \theta^\alpha + i\theta^2 D. \tag{4.9}$$

The gauge invariant kinetic term has the form

$$\mathcal{L}^{\text{kin}} = -\frac{1}{2g^2} \int d^2\theta \, \langle \nabla_\alpha \Sigma \, \nabla^\alpha \Sigma \rangle \,, \tag{4.10}$$

where $\nabla_\alpha \Sigma = \mathcal{D}_\alpha \Sigma - [\Gamma_\alpha, \Sigma]$. One can add also the mass term,[n]

$$\mathcal{L}_M = -i\zeta \int d^2\theta \langle \Sigma^2 \rangle \,. \tag{4.11}$$

Adding together (4.8), (4.10), (4.11), expressing the Lagrangian in components, and excluding the auxiliary field D, we obtain

$$\mathcal{L} = \frac{1}{g^2} \left\langle -\frac{1}{2} F_{\mu\nu}^2 + \nabla_\mu \sigma \nabla^\mu \sigma + \lambda \slashed{\nabla} \lambda + \psi \slashed{\nabla} \psi \right\rangle$$
$$+ \kappa \left\langle \epsilon^{\mu\nu\rho} \left(A_\mu \partial_\nu A_\rho - \frac{2i}{3} A_\mu A_\nu A_\rho \right) + i\lambda^2 \right\rangle + i\zeta \langle \psi^2 \rangle - \zeta^2 g^2 \langle \sigma^2 \rangle \,. \tag{4.12}$$

The Lagrangian involves, besides the gauge field, the adjoint fermion λ with the mass $m_\lambda = \kappa g^2$, the adjoint fermion ψ with the mass $m_\psi = \zeta g^2$ and the adjoint scalar σ with the same mass. The point $\zeta = \kappa$ is special. In this case, the Lagrangian (4.12) enjoys $\mathcal{N} = 2$ supersymmetry.

4.3. Index calculations

Consider the theory (4.12). Let first $\zeta > 0$. Then the mass of the matter fermions is positive. To be more precise, it has the same sign as the gluino mass for $k > 0$. The matter loops bring about an extra renormalization of k.

Note that the status of this renormalization is different compared to that due to the gluino loop. We have seen that, for the latter, the induced magnetic field on the dual torus is concentrated in the corners (3.9), which follows from the equality $m_\lambda L \ll 1$. On the other hand, the mass of the matter fields $m_\psi = \zeta g^2$ is an independent parameter. It is convenient to make it *large*, $m_\psi L \gg 1$. For a finite mass, the induced magnetic field has the form like in Eq. (3.7). For small $m_\psi L$, it is concentrated in the corners. But in the opposite limit, the induced flux density becomes constant, as the tree flux density is.

---

[n]This is a so called *real* mass term which can be expressed in terms of $\mathcal{N} = 2$ $3d$ superfields only by adding an explicit θ-dependence in the integrand in $S = \int d^4\theta \cdots$.[41–43] In a theory with two real multiplets (4.9) (which form a chiral $\mathcal{N} = 2$ multiplet), one can also write down Dř complex mass term in the same way as in $4d$ theories. Such terms do not, however, renormalize k and we will not consider them here.

Thus, massive enough matter brings about a true renormalization of k without any qualifications (*sine sale* if you will).

For positive ζ, the renormalization is negative, $k \to k - 1$. The index coincides with the index of the $\mathcal{N} = 1$ SYMCS theory with the renormalized k,

$$I_{\zeta > 0} = k - 1. \tag{4.13}$$

For $k = 1$, the index is zero and supersymmetry is spontaneously broken. For negative ζ, two things happen.

- First, the fermion matter mass has the opposite sign and so does the renormalization of k due to the matter loop. We seem to obtain $I_{\zeta < 0} = k + 1$.

- This is wrong, however, due to another effect. For a positive ζ, the ground state wave function in the matter sector is bosonic. But for a negative ζ, it is fermionic, $\Psi \propto \prod_a \psi^a$, changing the sign of the index.

We obtain

$$I_{\zeta < 0} = -k - 1. \tag{4.14}$$

Supersymmetry is broken here for $k = -1$.

As was mentioned, the Lagrangian (4.12) with $\zeta = \kappa$ enjoys the extended $\mathcal{N} = 2$ supersymmetry. That means, in particular, that ζ changes the sign together with κ and the result is given by

$$I_{\mathcal{N}=2}^{\text{SYMCS}} = |k| - 1 \tag{4.15}$$

in agreement with.[44,45] In contrast to (4.13) and (4.14), this expression is not analytic at $k = 0$, the non-analyticity being due just to the sign flip of the matter fermion mass.

Strictly speaking, the formula (4.15) does not work for $k = 0$. In this case, we should also keep $\zeta = 0$, the matter is massless, massless scalars make the motion infinite, and the index is ill-defined. However, bearing in mind that the regularized theory with $\zeta \neq 0$ gives the result $I_{\mathcal{N}=2 \text{ deformed}}^{\text{SYMCS}}(0) = -1$, irrespectively of the sign of ζ, one can attribute this value for the index also to $I_{\mathcal{N}=2}^{\text{SYMCS}}(0)$.

The three index formulas (4.13), (4.14), and (4.15) are represented together in Fig. 5.

Consider now the theory involving a gauge multiplet (4.6) and *three* $\mathcal{N} = 1$ real adjoint matter multiplets (4.9). If all the masses are equal, the theory enjoys the extended $\mathcal{N} = 2$ supersymmetry provided an extra Yukawa term is added in the Lagrangian.[46] If calling one of the matter

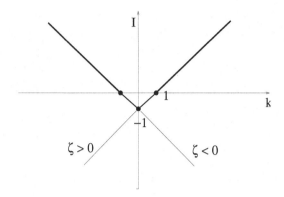

Fig. 5. The indices in the theory (4.12) with $\zeta > 0$, $\zeta < 0$, and $\zeta = \kappa$ (bold lines).

multiplets (the one that forms together with (4.6) a $\mathcal{N} = 2$ superfield) Σ and combine two other real multiplets into the complex $\mathcal{N} = 2$ multiplet Φ, the Yukawa term acquires the form

$$\mathcal{L}_{\text{Yukawa}} = -\frac{2i}{g^2} \int d^2\theta \, \langle \Sigma \Phi \bar{\Phi} \rangle . \tag{4.16}$$

To calculate the index, we will consider, however, the *deformed* $\mathcal{N} = 1$ Lagrangian with different masses M (the mass of the field Σ) and m (the mass of the complex field Φ). We will assume M to be large, but, to make the transition to the $\mathcal{N} = 2$ theory smooth, its sign will coincide with the sign of k.

There are four different cases:[o]

(1) $m > 0$, $k > 0 \Rightarrow M > 0$.

$$k \to k - 1_\Sigma - 2_\Phi = k - 3 . \tag{4.17}$$

This contributes $k - 3$ to the index. Note that, for $k = 1, 2$, this contribution is negative.

(2) $m > 0$, $k, M < 0$.

$$k \to k + 1_\Sigma - 2_\Phi = k - 1 \tag{4.18}$$

---

[o]When comparing with,[24] note the mass sign convention for the matter fermions is *opposite* there compared to our convention. We call the mass positive if it has the same sign as the masses of fermions in the gauge multiplet for positive k (and hence positive ζ). In other words, for positive k, ξ, the shifts of k due to both gluino loop and adjoint matter fermion loop have the negative sign.

Multiplying it by (-1) due to the fermionic nature of the wave function [it involves in this case a fermionic factor associated with the real adjoint matter multiplet Σ — see the discussion before Eq. (4.14)], we obtain $I = -k + 1$.

(3) $m < 0,\ k, M > 0$.

$$k \to k - 1_\Sigma + 2_\Phi = k + 1, \tag{4.19}$$

giving the contribution $I = k + 1$.

(4) $m < 0,\ k, M < 0$.

$$k \to k + 1_\Sigma + 2_\Phi = k + 3. \tag{4.20}$$

The contribution to the index is $-k - 3$.

In contrast to the model with only one real adjoint multiplet, this is not the full answer yet. There are also additional states on the Higgs branch that contribute to the index. Indeed, the superpotential is

$$\mathcal{W} = \frac{-i}{g^2} \left(\frac{M}{2} \Sigma^a \Sigma^a + m \bar{\Phi}^a \Phi^a + i\epsilon^{abc} \Sigma^a \Phi^b \bar{\Phi}^c \right). \tag{4.21}$$

The bosonic potential vanishes provided

$$m\phi^b = i\epsilon^{abc}\sigma^a\phi^c,$$
$$M\sigma^a = i\epsilon^{abc}\bar{\phi}^b\phi^c. \tag{4.22}$$

These equations have nontrivial solutions when both M and m are positive or when both M and m are negative. Let them be positive. Then (4.22) has a unique solution up to a gauge rotation.

$$\sigma^a = m \begin{pmatrix} 0 \\ 0 \\ 1 \end{pmatrix}, \qquad \phi^a = \sqrt{\frac{Mm}{2}} \begin{pmatrix} 1 \\ -i \\ 0 \end{pmatrix}. \tag{4.23}$$

The corresponding contribution to the index is not just equal to 1, however, due to a new important effect that did not take place in $4d$ theory with superpotential (4.1) considered above and would also be absent in a $3d$ theory with fundamental $\mathcal{N} = 2$ matter multiplet.

Indeed, besides the solution (4.23), there are also the solutions obtained from that by gauge transformations. The latter are not necessarily global, they might depend on the spatial coordinates x, y. Note now that, for the theory defined on a torus, one can also apply to (4.23) certain transformations which look like gauge transformations, but are not contractible due to the nontrivial $\pi_1[SO(3)] = Z_2$.[p] An example of such a quasi-gauge transformation is

$$\Omega_1 : O^{ab}(x) = \begin{pmatrix} \cos\left(\frac{2\pi x}{L}\right) & \sin\left(\frac{2\pi x}{L}\right) & 0 \\ -\sin\left(\frac{2\pi x}{L}\right) & \cos\left(\frac{2\pi x}{L}\right) & 0 \\ 0 & 0 & 1 \end{pmatrix}, \qquad (4.24)$$

where L is the length of our box. The transformation (4.24) does not affect $\sigma^a = \sigma \delta^{a3}$ and keeps the fields $\phi^a(\vec{x})$ periodic.[q] There is a similar transformation Ω_2 along the second cycle of the torus.

In $4d$ theories, wave functions are invariant under contractible gauge transformations. In $3d$ SYMCS theories, they are invariant up to a possible phase factor, like in (2.5). But nothing dictates the behaviour of the wave functions under the transformations $\Omega_{1,2}$. The latter are actually *not* gauge symmetries, but rather some global symmetries of the theory living on a torus. We obtain thus four different wave functions, even or odd under the action of $\Omega_{1,2}$.[r] The final result for the index of this theory is

$$I^{\mathcal{N}=2}_{\text{adjoint matter}} = |k| + 1. \qquad (4.25)$$

universally for positive and negative k. Extra Higgs states contribute only for positive k.

The result (4.25) was derived among others in[24] following a different logic. Intriligator and Seiberg did not deform $\mathcal{N} = 2 \rightarrow \mathcal{N} = 1$, but kept the fields in the real adjoint matter multiplet Σ light. Then the light matter fields $\{\sigma, \psi\}$ enter the effective BO Hamiltonian at the same ground as the Abelian components of the gluon and gluino fields. As was mentioned, the fluxes induced by the light fields are not homogeneous being concentrated at the corners. This makes an accurate analysis essentially more difficult. The index (4.25) was obtained in[24] as a sum of *three* rather than just two contributions[s] and it is still not quite clear how it works in the particular case $k = 2$ where k_{eff} as defined in Ref.[24] and including only renormalizations due to complex matter multiplet, $k_{\text{eff}} = k - 2$, vanishes.

Our method is simpler.

---

[p]One should understand $SO(3)$ here not as the orthogonal group itself, but rather as the adjoint representation space.

[q]Note that, for the matter in fundamental representation, the transformation (4.24) is inadmissible: when lifted up to $SU(2)$, it would make a constant solution, the fundamental analog of (4.22), antiperiodic.

[r]The oddness of a wave function under the transformation (4.24) means nonzero *electric flux* in the language of Ref.[47]

[s]On top of the usual vacua with $\phi = \sigma = 0$ and the Higgs vacua with $\phi, \sigma \neq 0$, they had also "topological vacua" with $\phi = 0, \sigma \neq 0$. The latter do not appear in our approach.

We can also add the $\mathcal{N} = 2$ multiplets with higher isospins. Then the counting of Higgs vacuum states becomes more complicated. For example, for $I = 3/2$, there are 10 such states. This number is obtained as a sum of the single state with the isospin projection $1/2$ and $3^2 = 9$ states with the isospin projection $3/2$ (in the latter case, there is a constant solution supplemented by eight $\vec{x}$-dependent quasi-gauge copies). The generic result for the index in the theory involving several $\mathcal{N} = 2$ matter multiplets with different isospins is

$$I = |k| - 1 + \frac{1}{2} \sum_f T_2(I_f) , \qquad (4.26)$$

where

$$T_2(I) = \frac{2I(I + 1)(2I + 1)}{3} \qquad (4.27)$$

is the Dynkin index of the corresponding representation normalized to $T_2(\text{fund}) = 1$.

When deriving (4.26), it was assumed that the matter-induced shift of the index is the sum of the individual shifts due to individual multiplets. This is true provided the Lagrangian does not involve extra cubic $\mathcal{N} = 2$ invariant superpotentials which can bring about extra Higgs vacuum states.

We can observe that the index does not depend on the sign of k, though this universal result is obtained by adding the contributions that look completely different for $k > 0$ and $k < 0$. For an individual multiplet contribution, the Higgs states contribute only for one sign of k (positive or negative depending on the sign of the mass). An interesting explanation for the symmetry with respect to mass sign flip with given k (and hence with respect to the sign flip of k with given m) was suggested in.[24] Basically, they argued that one can add to the mass the size of one of the cycles of the dual torus multiplied by i to obtain a complex holomorphic parameter on which the index of an $\mathcal{N} = 2$ theory should not depend. And hence it should not depend on the real part of this parameter (the mass). To my mind, it is still dangerous to pass the point $m = 0$ where the index is not defined and this argument thus lacks rigour. Anyway, an explicit $SU(2)$ calculation shows that the symmetry with respect to mass sign flip is there, indeed.

The reasoning above can be generalized for the higher unitary groups. Intriligator and Seiberg conjectured the following generalization of (4.26),

$$I^{SU(N)} = \frac{1}{(N-1)!} \prod_{j=-\frac{N}{2}+1}^{\frac{N}{2}-1} \left(|k| + \frac{1}{2} \sum_f T_2(R_f) - \frac{N}{2} - j \right) , \quad (4.28)$$

implying that the overall shift of k is represented as the sum of individual shifts due to indivudual multiplets. For an individual contribution to the shift, this formula can be derived for different signs of k and m when the extra Higgs states do not contribute. It can be extended to k, m of the same sign using the symmetry discussed above.

We checked that the latter works for all $SU(N)$ groups with fundamental matter and for $SU(3)$ with adjoint matter.[t] It would be interesting to construct a rigorous proof of this fact.

Appendix A. Theta functions

We remind here certain mathematical facts concerning the properties of analytical functions on the torus. They are mostly taken from the textbook,[48] but we are using different notations which we find more clear and more appropriate for our purposes.

Theta functions play the same role for the torus as ordinary polynomials for the Riemann sphere. They are analytic, but satisfy certain nontrivial quasiperiodic boundary conditions with respect to shifts along the cycles of the torus. A generic torus is characterized by a complex modular parameter τ, but we will stick to the simplest choice $\tau = i$ so that the torus represents a square $x, y \in [0, 1]$ ($z = x + iy$) glued around.

The simplest θ-function satisfies the boundary conditions

$$\theta(z + 1) = \theta(z),$$
$$\theta(z + i) = e^{\pi(1 - 2iz)}\theta(z).$$
(A.1)

This defines a *unique* (up to a constant complex factor) analytic function. Its explicit form is

$$\theta(z) = \sum_{n=-\infty}^{\infty} \exp\{-\pi n^2 + 2\pi i n z\}.$$
(A.2)

This function (call it theta function of level 1 and introduce an alternative notation $\theta(z) \equiv Q^1(z)$) has only one zero in the square $x, y \in [0, 1]$ — right in its middle, $\theta(\frac{1+i}{2}) = 0$.

For any integer $q > 0$, one can define theta functions of level q satisfying

$$Q^q(z + 1) = Q^q(z),$$
$$Q^q(z + i) = e^{q\pi(1 - 2iz)}Q^q(z).$$
(A.3)

---

[t]We emphasize that this is all an $\mathcal{N} = 2$ specifics. For $\mathcal{N} = 1$ theories, there is no such symmetry, see e.g. Fig. 2.

The boundary conditions (A.3) involve the twist [the exponent in the R.H.S. of Eq. (2.7)] $-2\pi q$ corresponding to the negative *magnetic flux*. [u] The functions $Q^q(z)$ have positive fluxes $2\pi q$. If multiplying $Q^{2k}(\bar{z})$ and $Q^{2k}(z)$ by proper exponentials, one can obtain (not analytic anymore) functions (2.8) satisfying the boundary conditions (2.5).

The functions satisfying (A.3) lie in vector space of dimension q. The basis in this vector space can be chosen as

$$Q_m^q(z) = \sum_{n=-\infty}^{\infty} \exp\left\{-\pi q\left(n + \frac{m}{q}\right)^2 + 2\pi i q z\left(n + \frac{m}{q}\right)\right\},$$
$$m = 0, \ldots, q - 1. \quad \text{(A.4)}$$

$Q_m^q(z)$ can be expressed in Mumford's notation[48] as

$$Q_m^q(z) = \theta_{m/q,0}(qz, iq), \quad \text{(A.5)}$$

where $\theta_{a,b}(z, \tau)$ are theta functions of rational characteristics.

$Q_m^q(z)$ can be called "elliptic polynomials" of order q. Indeed, each $Q_m^q(z)$ has q simple zeros at

$$z_s^{(m)} = \frac{2s + 1}{2q} + i\left(\frac{1}{2} - \frac{m}{q}\right), \qquad s = 0, \ldots, q - 1 \quad \text{(A.6)}$$

(add i to bring it onto the fundamental domain $x, y \in [0, 1]$ when necessary). A product $Q^q(z)Q^{q'}(z)$ of two such "polynomials" of orders q, q' gives a polynomial of order $q + q'$. There are many relations between the theta functions of different level and their products which follow. We can amuse the reader with a relation

$$\frac{Q_5^6(z) - Q_1^6(z)}{(Q_3^4(z) - Q_1^4(z))Q_0^2(z)} = \frac{1}{\eta(i)} = \frac{2\pi^{3/4}}{\Gamma(1/4)}. \quad \text{(A.7)}$$

The ratios of different elliptic functions of the same level give double periodic meromorphic elliptic functions. For example, the ratio of a properly chosen linear combination $\alpha Q_1^2(z) + \beta Q_2^2(z)$ and $[\theta(z)]^2$ is the Weierstrass function.

References

1. A.V. Smilga, *JHEP* **1001** 086 (2010) [arXiv:0910.0803, hep-th].
2. A.V. Smilga, *JHEP* **1205** 103 (2012) [arXiv:1202.6566, hep-th].
3. A.V. Smilga, [arXiv:1308.5951, hep-th].

---

[u]This is a physical interpretation. Mathematicians would call it *monodromy*.

4. J. Bagger and N. Lambert, *Phys. Rev. D* **77** 065008 (2008) [arXiv: 0711.0955, hep-th].
5. O. Aharony, O. Bergman, D.L. Jafferis and J. Maldacena, *JHEP* **0810** 091 (2008) [arXiv:0806.1218, hep-th].
6. J.M. Maldacena, *Adv. Theor. Math. Phys.* **2** 231 (1998).
7. S.S. Gubser, I.R. Klebanov, and A.M. Polyakov, *Phys. Lett. B* **428** 105 (1998).
8. A. Gorsky, *Phys.Usp.* **48** (2005) 1093-1108.
9. G. Romelsberger, *Nucl. Phys. B* **747** 329 (2006) [hep-th/0510060].
10. V.P. Spiridonov and G.S. Vartanov, *Nucl. Phys. B* **824** 192 (2010) [arXiv:0811.1909, hep-th].
11. G.V. Dunne, [hep-th/9902115].
12. E. Witten, in: [Shifman, M.A. (ed.) *The many faces of the superworld*, p.156] [hep-th/9903005].
13. A.J. Niemi and G.W. Semenoff, *Phys. Rev. Lett.* **51** (1983) 2077.
14. N. Redlich, *Phys. Rev. D* **29** (1984) 2366.
15. E. Witten, *Commun. Math. Phys.* **121** 351 (1989).
16. S. Elitzur, G. Moore, A. Schwimmer, and N. Seiberg, *Nucl. Phys. B* **326** 108 (1989).
17. J.M.F. Labastida and A.V. Ramallo, *Phys. Lett. B* **227** 92 (1989).
18. E. Witten, *Nucl. Phys. B* **202** 253 (1982).
19. M. Henningson, *JHEP* **1211** 013 (2012) [arXiv: 1209.1798, hep-th].
20. V.G. Kac and A.V. Smilga, in: [Shifman, M.A. (ed.) *The many faces of the superworld*, p.185] [hep-th/9902029].
21. A. Keurentjes, *JHEP* **9905** (1999) 001 [hep-th/9901154].
22. A. Keurentjes, *JHEP* **9905** (1999) 014 [hep-th/9902186].
23. S. Deser, R. Jackiw, and S. Templeton, *Ann. Phys. (NY)* **140** 372 (1982).
24. K. Intriligator and N. Seiberg, arXiv:1305.1633 [hep-th].
25. B.A. Dubrovin, I.M. Kričever, and S.P. Novikov, *Soviet Math Dokl.* **229** 15 (1976).
26. B.A. Dubrovin and S.P. Novikov, *Soviet Phys. JETP* **79** 1006 (1980).
27. S. Cecotti and L. Girardello, *Phys. Lett. B* **110** 39 (1982).
28. R.D. Pisarski and S. Rao, *Phys. Rev. D* **32** 2081 (1985).
29. H.-S. Kao, K. Lee, and T. Lee, *Phys. Lett. B* **373** 94 (1996) [hep-th/9506170].
30. K. Gawedzki and A. Kupiainen, *Nucl. Phys. B* **320** 625 (1989).
31. S. Pancharatnam, *Proc. Indian Acad. Sci. A* **44** (1956) 247.
32. M. Berry, *Proc. R. Soc. Lond. A* **392** (1984) 45.
33. A.V. Smilga, *Soviet Phys. JETP* **64** 8 (1986).
34. B.Yu. Blok and A.V. Smilga, *Nucl. Phys. B* **287** 589 (1987).
35. K. Intriligator and N. Seiberg, *Nucl. Phys. B* **431** 551 (1994) [hep-th/9408155].
36. A. Gorsky, A. Vainshtein, and A. Yung, *Nucl. Phys. B* **584** 197 (2000) [hep-th/0004087].
37. I. Affleck, M. Dine, and N. Seiberg, *Phys. Lett. B* **137** 187 (1984).
38. N. Seiberg and E. Witten, *Nucl. Phys. B* **431** 484 (1994) [hep-th/9408099].
39. A. Smilga, *Phys. Rev. D* **58** 105014 (1998) [hep-th/9801078].

40. S.J. Gates, M.T. Grisaru, M. Rocek, and W. Siegel, *Front. Phys.* **58** 1-548 (1983) [hep-th/0108200].
41. H. Nishino and S.J. Gates, *Int. J. Mod. Phys. A* **8** 3371 (1993).
42. O. Aharony et al., *Nucl. Phys. B* **499** 67 (1997) [hep-th/9703110].
43. J. de Boer, K. Hori, and Y. Oz, *Nucl. Phys. B* **500** (1997) [hep-th/9703100].
44. K. Ohta, *JHEP* **9910** 006 (1999) [hep-th/9908120].
45. O. Bergman, A. Hanany, A. Karch, and B. Kol *JHEP* **9910** 036 (1999) [hep-th/9908075].
46. E.A. Ivanov, *Phys. Lett. B* **268** 203 (1991).
47. G. 't Hooft, *Nucl. Phys. B* **153** 141 (1979).
48. D. Mumford, *Tata Lectures on Theta* (Birkhäuser Boston, 1983).

CRITICAL NUCLEI
IN A SUPERSTRONG MAGNETIC FIELD

S. I. GODUNOV

Institute for Theoretical and Experimental Physics, Moscow, 117218, Russia,
Novosibirsk State University, Novosibirsk, 630090, Russia,
E-mail: sgodunov@itep.ru

M. I. VYSOTSKY

Institute for Theoretical and Experimental Physics, Moscow, 117218, Russia,
Novosibirsk State University, Novosibirsk, 630090, Russia,
National research nuclear university "MEPhI", Moscow, 115409, Russia,
Moscow Institute of Physics and Technology, Dolgoprudny, Moscow Region, 141700,
Russia,
E-mail: vysotsky@itep.ru

Based on a talk at the memorial seminar devoted to the 100 years birthday of
I. Ya. Pomeranchuk, ITEP, June 5-6, 2013.

1. Introduction

The problem of the critical nuclei was considered for the first time by I. Ya.
Pomeranchuk and Ya. A. Smorodinskii in Ref. 1. They discovered that it
is possible to remove the singularity of the solution of the Dirac equation
for an electron moving in the Coulomb field. This singularity occurs at the
nucleus charge $Z = 137$ when the ground energy level reaches $\varepsilon_0 = 0$. If
the finite size of the nucleus is taken into account, then the solution of the
Dirac equation exists also for larger Z and the ground energy level goes
down until it reaches lower continuum, $\varepsilon_0 = -m_e$, where m_e is the electron
mass. In Ref. 1 the value of a nucleus charge at which it happens was called
critical. According to Ref. 1 $Z_{\mathrm{cr}} = 175 \div 200$ depending on the nucleus
radius.

A physical picture of the phenomenon which takes place at $Z = Z_{\mathrm{cr}} =$
172 was established about 20 years later in Refs. 2. When the charge of
the hydrogen-like ion reaches the critical value two e^+e^- pairs are produced from vacuum. The electrons occupy the ground atomic level while

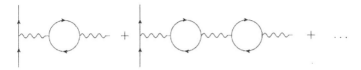

Fig. 1.   Modification of the Coulomb potential due to dressing of the photon propagator.

the positrons are emitted to infinity.

When the results published in Ref. 1 were obtained, Pomeranchuk exclaimed: "It would be great to collide two uranium nuclei" (Ya. A. Smorodinskii[3]).

A natural question arises: if it is possible to obtain criticality at smaller Z which corresponds to nuclei existing in nature. The answer is "yes": in the external magnetic fields[a] $B > B_0 \equiv m_e^2/e$ even the ions with moderate Z are critical.[4]

When B further grows, the Coulomb potential of the nucleus becomes screened[5] due to the radiative corrections. And we are going to study how this screening modifies the dependence of Z_{cr} on the magnetic field.

2. Screening of the Coulomb potential in $d = 1$ and $d = 3$

There is a similarity between the radiative corrections to the Coulomb potential in three space dimensions ($d = 3$) in the strong external magnetic field and in one space dimension ($d = 1$). That is why we are starting from a simpler problem: the Coulomb potential in $d = 1$.[6]

Let us consider $1 + 1$ dimensional QED with massive charged fermions. The electrical potential of the point-like charge with the account of polarization effects (see Fig. 1) equals:

$$\Phi(k) = -\frac{4\pi g}{k^2 + \Pi(k^2)} \ , \tag{1}$$

where $\Pi(k^2)$ is the one-loop expression for the photon polarization operator:

$$\Pi(k^2) = 4g^2 \left[\frac{1}{\sqrt{t(1+t)}} \ln(\sqrt{1+t} + \sqrt{t}) - 1 \right] \equiv -4g^2 P(t) \ , \tag{2}$$

and $t \equiv -k^2/4m^2$, $[g] = $ mass.

---

[a]We use Gauss units, $e^2 = \alpha = 1/137.0359...$

In the coordinate representation for $k = (0, k_\parallel)$ we obtain:

$$\Phi(z) = 4\pi g \int\limits_{-\infty}^{\infty} \frac{e^{ik_\parallel z} dk_\parallel / 2\pi}{k_\parallel^2 + 4g^2 P(k_\parallel^2 / 4m^2)} \quad . \tag{3}$$

With the help of the interpolating formula

$$\overline{P}(t) = \frac{2t}{3 + 2t} \tag{4}$$

the accuracy of which is better than 10% for $0 < t < \infty$ we obtain:

$$\Phi = 4\pi g \int\limits_{-\infty}^{\infty} \frac{e^{ik_\parallel z} dk_\parallel / 2\pi}{k_\parallel^2 + 4g^2 (k_\parallel^2 / 2m^2)/(3 + k_\parallel^2 / 2m^2)} =$$

$$= \frac{4\pi g}{1 + 2g^2/3m^2} \left[-\frac{1}{2}|z| + \frac{g^2/3m^2}{\sqrt{6m^2 + 4g^2}} \exp(-\sqrt{6m^2 + 4g^2}|z|) \right] \quad . \tag{5}$$

In the case of heavy fermions ($m \gg g$) the potential is given by the tree level expression; the corrections are suppressed as g^2/m^2.

In the case of light fermions ($m \ll g$):

$$\Phi(z)|_{m \ll g} = \begin{cases} \pi e^{-2g|z|}, & z \ll \frac{1}{g} \ln\left(\frac{g}{m}\right), \\ -2\pi g\left(\frac{3m^2}{2g^2}\right)|z|, & z \gg \frac{1}{g} \ln\left(\frac{g}{m}\right). \end{cases} \tag{6}$$

The massless case ($m = 0$) corresponds to the Schwinger model; a photon gets a mass due to the photon polarization operator with massless fermions. Light fermions provide the continuous transition from $m > g$ to $m = 0$ case.

To get an expression for the Coulomb potential in $d = 3$ case in the strong external magnetic field we need an expression for the polarization operator. It greatly simplified for $B \gg B_0 \equiv m_e^2/e$. The following result was obtained in Ref. 5:

$$\Phi(k) = \frac{4\pi e}{k_\parallel^2 + k_\perp^2 + \frac{2e^3 B}{\pi} \exp\left(-\frac{k_\perp^2}{2eB}\right) P\left(\frac{k_\parallel^2}{4m^2}\right)} \quad , \tag{7}$$

where P is the same as in $d = 1$ case. A natural question is if the two loop terms are enhanced as $\left(e^3 B\right)^2$. According to Ref. 8, the two loop corrections are very small, and the physical reason for their smallness is nullification of the higher loops in $d = 1$ QED with massless fermions (see e.g. Ref. 9).

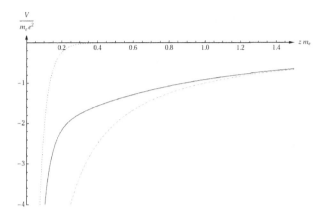

Fig. 2. (Color online) The screened Coulomb potential along the magnetic field ($\rho = 0$) at $B = 5 \cdot 10^4 B_0$. A dashed (green) line corresponds to the Coulomb potential; solid (blue) line corresponds to the screened potential; dotted (red) line corresponds to the asymptotic behaviour of the modified potential at small distances.

In the coordinate representation we obtain:

$$
\Phi(z) = 4\pi e \int \frac{e^{ik_{\parallel}z} dk_{\parallel} d^2 k_{\perp}/(2\pi)^3}{k_{\parallel}^2 + k_{\perp}^2 + \frac{2e^3 B}{\pi}\exp(-k_{\perp}^2/(2eB))(k_{\parallel}^2/2m_e^2)/(3 + k_{\parallel}^2/2m_e^2)}
$$
$$
= \frac{e}{|z|}\left[1 - e^{-\sqrt{6m_e^2}|z|} + e^{-\sqrt{(2/\pi)e^3 B + 6m_e^2}|z|}\right] . \tag{8}
$$

For $B \ll 3\pi m^2/e^3$ the potential is Coulomb up to the small power suppressed terms:

$$
\Phi(z)|_{e^3 B \ll m_e^2} = \frac{e}{|z|}\left[1 + O\left(\frac{e^3 B}{m_e^2}\right)\right] , \tag{9}
$$

in full accordance with the $d = 1$ case, where g^2 plays the role of $e^3 B$.

In the opposite case of the superstrong magnetic fields $B \gg 3\pi m_e^2/e^3$ we get:

$$
\Phi(z) = \begin{cases} \frac{e}{|z|}e^{-|z|\sqrt{2e^3 B/\pi}} , & \frac{1}{\sqrt{2e^3 B/\pi}}\ln\sqrt{\frac{e^3 B}{3\pi m_e^2}} > |z| > \frac{1}{\sqrt{eB}}, \\ \frac{e}{|z|}\left(1 - e^{-|z|\sqrt{6m_e^2}}\right) , & \frac{1}{m_e} > |z| > \frac{1}{\sqrt{2e^3 B/\pi}}\ln\sqrt{\frac{e^3 B}{3\pi m_e^2}}, \\ \frac{e}{|z|} , & |z| > \frac{1}{m_e}. \end{cases} \tag{10}
$$

The potential energy of the electron in the modified potential, $V(z) = -e\Phi(z)$, is shown in Fig. 2.

3. Energy levels of the electron in the modified potential

3.1. Nonrelativistic approach

The following equation determines the energies of the even states of the hydrogen atom in the strong magnetic field with the account of screening of the Coulomb potential:[7]

$$
\ln\left(\frac{B}{m_e^2 e^3 + \dfrac{e^6}{3\pi}B}\right) = \lambda + 2\ln\lambda + 2\psi\left(1 - \frac{1}{\lambda}\right) + \ln 2 + 4\gamma + \psi(1+|m|), \quad (11)
$$

where γ is the Euler constant, $\psi(x)$ is the logarithmic derivative of gamma-function, $m = 0, -1, -2, \ldots$ is the projection of the electron angular momentum on the direction of magnetic field, and the binding energy is defined by λ:

$$
E \equiv -\frac{m_e e^4}{2}\lambda^2. \quad (12)
$$

The analogous equation without screening was derived in Ref. 10. In the limit $B \gg 3\pi m_e^2/e^3$ for the ground level energy we get:

$$
\lambda_{\rm gr} \to 11.2, \quad E_{\rm gr} \to -1.7 keV. \quad (13)
$$

Freezing of the ground level in the limit $B \to \infty$ was discovered by Shabad and Usov.[5]

3.2. Relativistic approach

Without taking screening into account the problem in the framework of the Dirac equation was solved in Ref. 4. Let us follow this paper.

The bispinor of the electron on the lowest Landau level looks like:

$$
\psi_e = \begin{pmatrix} \varphi_e \\ \chi_e \end{pmatrix}, \varphi_e = \begin{pmatrix} 0 \\ g(z)\exp\left(-\rho^2/4a_H^2\right) \end{pmatrix}, \chi_e = \begin{pmatrix} 0 \\ if(z)\exp\left(-\rho^2/4a_H^2\right) \end{pmatrix},
$$
$$(14)$$

and the Dirac equation

$$
\left(\vec{\alpha}\left(\vec{p} - e\vec{A}\right) + V + \beta m_e\right)\psi_e = \varepsilon\psi_e \quad (15)
$$

takes the following form:

$$
\begin{cases} g_z - \left(\varepsilon + m_e - \bar{V}\right)f = 0, \\ f_z + \left(\varepsilon - m_e - \bar{V}\right)g = 0, \end{cases} \quad (16)
$$

where $g_z \equiv dg/dz$, $f_z \equiv df/dz$. The system of equations (16) describes the electron motion in the effective potential $\bar{V}(z)$ (averaged over fast transverse motion):

$$\bar{V}(z) = -\frac{Ze^2}{a_H^2} \times \int_0^\infty \frac{e^{-\rho^2/2a_H^2}}{\sqrt{\rho^2 + z^2}} \rho \, d\rho \ . \tag{17}$$

At the distances $z \gg a_H$ Eq. (17) is greatly simplified: $\bar{V} \approx -Ze^2/|z|$. The solution of the system (16) for $\bar{V} = -Ze^2/|z|$ is well known and it is the linear combination of Whittaker functions (see Ref. 4 for the details).

The solution at small distances was found in Ref. 4 in the limit $\bar{V}(z) \gg 2m_e$, i.e. $|z| \ll Ze^2/(2m_e)$. So in the nonscreened case there is a matching region, $a_H \ll |z| \ll Ze^2/2m_e$, as soon as the condition $B \gg B_0/\left(Ze^2\right)^2$ is satisfied.

Matching the long distance and short distance solutions the equation for the energy levels of the electron in the Coulomb field of the nucleus with the charge Z and the external magnetic field B was obtained in Ref. 4. This equation allows to find the magnetic field B_{cr} at which the ions with the charge Z become critical, i.e. the ground energy level reaches lower continuum, $\varepsilon_0 = -m_e$:

$$\frac{B_{cr}}{B_0} = 2\left(Ze^2\right)^2 \exp\left(-\gamma + \frac{\pi - 2\arg\Gamma\left(1 + 2iZe^2\right)}{Ze^2}\right) . \tag{18}$$

According to this formula uranium becomes critical at $B \approx 10^2 B_0$ and at stronger fields even ions with smaller Z are critical.

To take screening into account, instead of Eq. (17) one should use the following formula for $\bar{V}$:

$$\bar{V}(z) = -\frac{Ze^2}{a_H^2}\left[1 - e^{-\sqrt{6m_e^2}|z|} + e^{-\sqrt{(2/\pi)e^3B + 6m_e^2}|z|}\right] \times \int_0^\infty \frac{e^{-\rho^2/2a_H^2}}{\sqrt{\rho^2 + z^2}} \rho \, d\rho. \tag{19}$$

We see that the screened potential reaches its asymptotic behaviour, $\bar{V} = -Ze^2/|z|$, only at large distances, $|z| \gg 1/m_e$. So it is impossible to match two solutions like it was done in Ref. 4, since the solution at small distances is valid only for $|z| < Ze^2/m_e$. That is why the analytic equation for the ground level in the screened potential has not been derived yet.

That is why we have solved the problem numerically. Following V.S. Popov,[2] we reduce the Dirac equation to the effective Schrödinger equation:

$$\frac{d^2\chi}{dz^2} + 2m_e(E - U)\chi = 0, \tag{20}$$

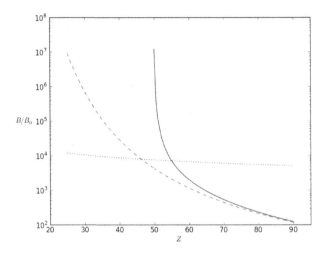

Fig. 3. (Color online) The critical magnetic field in units of B_0. The dashed (green) line corresponds to formula (18) originally obtained in Ref. 4; the solid (blue) line corresponds to numerical results with the account of screening. The dotted (black) line corresponds to the magnetic field at which Landau radius a_H becomes smaller than the size of the nucleus.

$$E = \frac{\varepsilon^2 - m_e^2}{2m_e},$$

$$U = \frac{\varepsilon}{m_e}\bar{V} - \frac{1}{2m_e}\bar{V}^2 + \frac{\bar{V}''}{4m_e(\varepsilon + m_e - V)} + \frac{3/8(\bar{V}')^2}{m_e(\varepsilon + m_e - V)^2}.$$

For $B \ll B_0$ relativistic corrections are small and the binding energy $E \approx \varepsilon - m_e$ is defined by the nonrelativistic equation. However, for $B \gg B_0$ relativistic corrections grow as powers of (B/B_0) and correction terms have different signs. It considerably complicates the numerical calculations.

For a hydrogen atom we have found that the relativistic corrections are small even for very strong fields and the value of the freezing energy practically does not change. We have also considered the ions with larger Z and obtained the effect of freezing in the relativistic domain. For example, the ground energy level for the ions with $Z = 40$ freezes at $\varepsilon_0 \approx -m_e/2$.

The freezing of the ground energy level is crucial for the phenomenon of the critical nucleus charge. We have found that the ions with $Z < 50$ never become critical and calculated the values of the critical magnetic field B_{cr} for the ions with larger Z. These results are shown in Fig. 3. The ions with $Z \lesssim 55$ reach criticality in such a strong magnetic field that a_H becomes

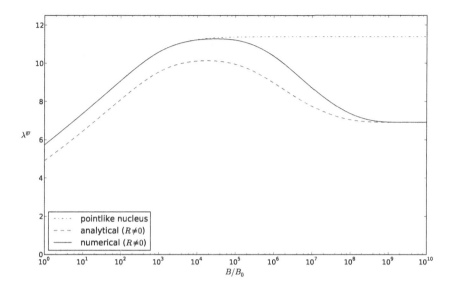

Fig. 4. (Color online) The dependence of λ^{gr} on the magnetic field. The dot-dashed (red) line corresponds to the pointlike nucleus; the dashed (green) line — to the analytical formula, Eq. (22), for $h(|z|) = 1$; the solid (blue) line — to the numerical solution for $h(|z|) = 1$.

smaller than the size of the nucleus. That is why the finiteness of the nucleus radius should be taken into account.

Without screening the magnetic field B appears in Eq. (17) only through Landau radius $a_H \equiv 1/\sqrt{eB}$. When a_H becomes smaller than the nucleus size R, one should substitute a_H by R. It means that the spectrum we are looking for coincides with the spectrum in the magnetic field $B = 1/(eR^2)$ which corresponds to $a_H = R$.

However, the magnetic field B appears directly in the expression for the screened potential Eq. (19). That is why in the case of screening special consideration is needed.

4. Finite nucleus size and the ground energy level

The three-dimensional formula for the screened potential has not been derived yet and the distribution of the electric charge inside a nucleus in such a strong magnetic fields is not known. Thus, one cannot find the analytic formula for the potential of the nucleus which has the finite size. However, we have found the approximate expression for the potential along the magnetic field (see Ref. 12 for the details). In case of proton $(Z = 1)$ it looks

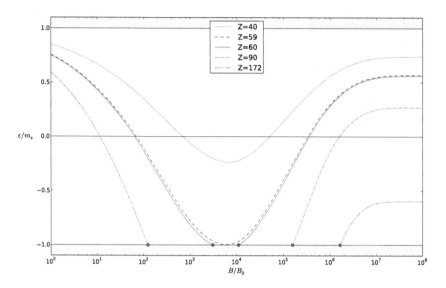

Fig. 5. (Color online) The dependence of the ground energy level on magnetic field for $Z = 40, 59, 60, 90, 172$. The correspondence between charge Z and linestyle (color) is shown in the legend.

like:

$$\Phi(z) = \begin{cases} \dfrac{e}{|z|}\left(1 - e^{-|z|\sqrt{6m_e^2}} + h(R)e^{-\mu|z|}\right), & |z| \geq R, \\[2ex] \dfrac{e}{R}\left(1 - e^{-R\sqrt{6m_e^2}} + h(|z|)e^{-\mu R}\right), & |z| < R, \end{cases} \qquad (21)$$

where $h(|z|)$ is determined by the charge distribution inside the proton, $\mu \equiv \sqrt{6m_e^2 + (2e^3 B/\pi)}$, and $R = 0.877$fm is the proton charge radius.

Eq. (21) allows us to derive a nonrelativistic analytical formula for the hydrogen energy levels analogous to Eq. (11):

$$\ln \frac{a_B}{\sqrt{R^2 + a_H^2}} - E_1\left(\sqrt{R^2 + a_H^2}\sqrt{6m_e^2}\right) + h(R)E_1\left(\mu\sqrt{R^2 + a_H^2}\right) \qquad (22)$$

$$= \frac{\lambda}{2} + \ln \lambda + \psi\left(1 - \frac{1}{\lambda}\right) + 2\gamma + \ln 2,$$

where $a_B \equiv 1/(m_e\alpha)$ is the Bohr radius, λ defines the binding energy, $E \equiv -\left(m_e e^4/2\right)\lambda^2$, and

$$E_1(x) \equiv \int\limits_x^\infty \frac{e^{-t}}{t}\,dt. \qquad (23)$$

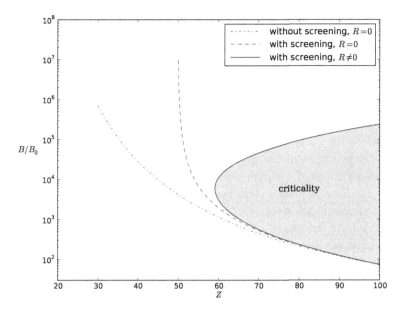

Fig. 6. (Color online) The values of the critical magnetic field. The dot-dashed (red) line corresponds to Eq. (18); the dashed (green) line corresponds to the numerical results for the screened potential of the pointlike charge; the solid (blue) line corresponds to the numerical results with the account of both screening and the finite nucleus size.

According to Eq. (22) the value of λ in the limit $B \to \infty$ equals 6.9 instead of $\lambda = 11.2$ which was obtained for the point-like proton. The dependence of λ^{gr} (corresponding to the ground energy level) on the magnetic field for $h(|z|) = 1$ is shown[b] in Fig. 4.

We see that the ground energy level goes up (or the binding energy diminishes) until it reaches the final freezing energy. This effect is even more pronounced for the heavier ions (see Fig. 5). Due to the rising of the ground energy level the ions with $Z = 60 \div 210$ stop being critical in the strong enough magnetic field. Even the ion with $Z = 172$ becomes noncritical at $B/B_0 \gtrsim 2 \cdot 10^6$, while it is critical in the absence of the magnetic field. At $Z \approx 210$ the final freezing energy reaches lower continuum and the nuclei with $Z > 210$ are critical regardless of the magnetic field strength. In Fig. 6 the dependence of the critical nucleus charge on the magnetic field is shown.

---

[b]The function $h(|z|) = 1$ was chosen for the simplicity, and it was checked that other distributions (like that for the homogeneously charged sphere) lead to rather close results (see Ref. 12 for details).

5. Conclusions

The influence of screening and the finite nucleus size on the energy levels of the hydrogen-like ions has been studied. Both screening and the finite nucleus size push the ground energy level up. The screening starts at $B \sim m_e^2/e^3 \approx 6 \cdot 10^{15}$G and leads to the freezing of the ground energy level. The finite nucleus radius R comes into play at $B \sim 1/(eR^2) = 10^{17} \div 10^{18}$ G and the ground energy level goes up until in the magnetic fields $B \sim 1/(e^3 R^2) = 10^{19} \div 10^{20}$ G it reaches the final freezing energy.

The dependence of the ground energy level on the magnetic field was calculated analytically using a nonrelativistic approach and numerically by solving the Dirac equation. Our main result is the calculation of the critical nucleus charge in the magnetic field shown in Fig. 6.

The effects discussed appear only in the superstrong magnetic fields which have not been found in Nature yet. However, considering of such an asymptotic behaviour is in the spirit of I. Ya. Pomeranchuk. According to Ya. A. Smorodinskii,[3] looking for the asymptotic behaviour of different quantities was Pomeranchuk's approach to physics (extremely low temperatures, extremely high energies).

The authors were partially supported by the RFBR under the Grants No. 11-02-00441, 12-02-00193 and by the Russian Federation Government under Grants No. 11.G34.31.0047, NSh-3172.2012.2.

References

1. I. Ya. Pomeranchuk, Ya. A. Smorodinsky, *J. Phys. USSR* **9**, 97 (1945).
2. S. S. Gershtein, Ya. B. Zeldovich, *Zh. Eksp. Teor. Fiz.* **57**, 654 (1969), *Nuovo Cimento Lett.* **1**, 835 (1969);
   V.S. Popov, *Pis'ma Zh. Eksp. Teor. Fiz.* **11**, 254 (1970);
   V.S. Popov, *Zh. Eksp. Teor. Fiz.* **59**, 965 (1970);
   Ya.B. Zeldovich, V.S. Popov, *UFN* **105**, 403 (1971);
   V.S. Popov, *Yad. Fiz.* **14**, 458 (1971).
3. *Vospominaniya o I.Ya. Pomeranchuke*, Moskva:Nauka, 1988 (ISBN 5-02-000656-4).
4. V.N. Oraevskii, A.I. Rez, V.B. Semikoz, *Zh. Eksp. Teor. Fiz.* **72**, 820 (1977) [*Sov. Phys. JETP* **45**, 428 (1977)].
5. A.E. Shabad, V.V. Usov, *Phys. Rev. Lett.* **98**, 180403 (2007);
   A.E. Shabad, V.V. Usov, *Phys. Rev. D* **77**, 025001 (2008).
6. M. I. Vysotsky, *Pis'ma Zh. Eksp. Teor. Fiz.* **92**, 22 (2010) [*JETP Lett.* **92**, 15 (2010)].
7. B. Machet, M. Vysotsky *Phys. Rev. D* **83**, 025022 (2011).
8. S. I. Godunov, *Phys. Atom. Nucl.* **76**, 901–918 (2013).
9. V. B. Berestetskii, *Proceedings of LIYaF Winter School* **9**, part 3, 95 (1974).

10. B.M. Karnakov, V.S. Popov, *Zh. Eksp. Teor. Fiz.* **124**, 996 (2003) [*J. Exp. Theor. Phys.* **97**, 890 (2003)];
    B.M. Karnakov, V.S. Popov, *Zh. Eksp. Teor. Fiz.* **141**, 5 (2012) [*J. Exp. Theor. Phys.* **114**, 1 (2012)].
11. S. I. Godunov, B. Machet, M. I. Vysotsy *Phys. Rev. D* **85**, 044058 (2012).
12. S. I. Godunov, M. I. Vysotsy *Phys. Rev. D* **87**, 124035 (2013).

ANOMALY AND LONG-RANGE FORCES

V.P. KIRILIN

ITEP, B. Cheremushkinskaya 25, Moscow, 117218, Russia
Department of Physics, Princeton University, Princeton, NJ 08544, USA
E-mail: vkirilin@princeton.edu

A.V. SADOFYEV

ITEP, B. Cheremushkinskaya 25, Moscow, 117218, Russia
Center for Theoretical Physics, Massachusetts Institute of Technology, Cambridge,
MA, 02139, USA
E-mail: sadofyev@mit.edu

V.I. ZAKHAROV

ITEP, B. Cheremushkinskaya 25, Moscow, 117218, Russia
Max-Planck Institut für Physik, München, 80805, Germany
Moscow Inst Phys & Technol, Dolgoprudny, Moscow Region, 141700, Russia
E-mail: vzakharov@itep.ru

We consider infrared dependences of chiral effects, like chiral magnetic effect, in chiral media. The main observation is that there exist competing infrared-sensitive parameters, sometimes not apparent. The value of the chiral effects depends in fact on the actual hierarchy of the parameters. Some examples have been already given in the literature. We argue that magnetostatics of chiral media with a non-vanishing chiral chemical potential $\mu_5 \neq 0$ is also infrared sensitive. In particular, the system turns to be unstable if the volume is large enough. The instability is with respect to the decay of the system into domains of non-vanishing magnetic field with non-trivial helicity.

Keywords: Chiral Anomaly, Chiral Magnetic Effect, Infrared Effects.

1. Introduction

Relation of the chiral anomaly to long-range interactions is actually an old topic, with rich literature existing. The point is that the chiral anomaly can be derived both in terms of ultraviolet and infrared sensitive regulators. Namely, divergence of the axial current j_μ^5 of massless fermions of charge e

is given by a polynomial in the photonic field A_μ:[1]

$$\partial_\mu j_\mu^5 = \frac{e^2}{8\pi^2} F_{\mu\nu} \tilde{F}_{\mu\nu} , \tag{1}$$

where $F_{\mu\nu} = \partial_\mu A_\nu - \partial_\nu A_\mu$, $\tilde{F}_{\mu\nu} = 1/2\epsilon_{\mu\nu\alpha\beta}F_{\alpha\beta}$. Being a polynomial, the divergence of the current is ultraviolet sensitive.[1]

On the other hand, if one turns to the matrix element of the axial currrent itself, $\langle\gamma\gamma|j_\mu^5|0\rangle$ it exhibits a pole:[2]

$$\langle\gamma\gamma|j_\mu^5|0\rangle = \frac{iq_\mu}{q^2} \frac{e^2}{2\pi^2} \epsilon_{\rho\nu\alpha\beta} k_\rho^{(1)} k_\nu^{(2)} e_\alpha^{(1)} e_\beta^{(2)} , \tag{2}$$

where q_μ is the 4-momentum brought in by the axial current, $k_\alpha^{(1)}$, $k_\beta^{(2)}$ and $e_\mu^{(1)}, e_\nu^{(2)}$ are the 4-momenta and wave functions of the photons, respectively. Note that Eq. (2) is only valid for photons on mass shell, $(k^{(1)})^2 = (k^{(2)})^2 = 0$. The emergence of the pole in q^2 in Eq. (2) is a pure perturbative phenomenon and a reflection of the vanishing fermionic mass.

However, the pole is to be exhibited also in the massless limit of strongly interacting quarks when the perturbation theory does not apply. In a confining theory this requirement results in the 't Hooft matching condition:[3]

$$\frac{e^2}{8\pi^2} N_c (Q_u^2 - Q_d^2) = f_\pi \cdot f_{\pi\to\gamma\gamma} , \tag{3}$$

where N_c is the number of colors, $Q_u = 2/3, Q_d = -1/3$ are the quark charges and $f_\pi, f_{\pi\to\gamma\gamma}$ are the constants related to the pion decays into a lepton pair and two photons, respectively.

Thus, within this picture the divergence of the axial current is driven by the ultraviolet physics alone while the matrix elements of the current are infrared sensitive. However, this is correct only in the limit of exact chiral symmetry and in this sense is an oversimplification. Indeed, if one introduces small but finite quark masses the matrix element of the divergence of the current becomes also infrared sensitive:

$$\langle\gamma\gamma|\partial_\mu j_\mu^5|0\rangle \approx f_\pi f_{\pi\to\gamma\gamma} \cdot \left(1 - \frac{m_\pi^2}{(q^2 + m_\pi^2)}\right) F_{\mu\nu} \tilde{F}_{\mu\nu} , \tag{4}$$

where m_π^2 is the pion mass squared. We see that the result now depends strongly on the ratio of the two infrared sensitive parameters involved, q^2 and m_π^2. The exclusively ultraviolet-sensitive origin of the non-trivial divergence of the current, see Eq. (1) appears to be property of a particular hierarchy of the infrared-sensitive parameters, $q^2 \gg m_\pi^2$. In the opposite limit of $m_\pi^2 \gg q^2$ the matrix element (4) is vanishing.

The actual focus of our attention is the infrared sensitivity of chiral effects in chiral media, that is media whose constituents are massless fermions. The chiral media attracted a lot of attention recently, for a review and further references see, e.g., Ref. 4. One of main reasons is that one expects that in such media the chiral anomaly (1) which is a pure loop, or quantum effect has macroscopic manifestations. The best known example of such manifestations is the chiral magnetic effect:[5]

$$j_\mu^{el} = \sigma_M \mu_5 B_\mu , \tag{5}$$

where j_μ^{el} are components of the electromagnetic current, μ_5 is the chiral chemical potential, σ_M can be called magnetic conductivity, while B_μ is defined in terms of external electromagnetic field $F_{\alpha\beta}$ and four-velocity u_ν of an element of the chiral liquid, $B_\mu \equiv (1/2)\epsilon_{\mu\nu\alpha\beta}u_\nu F_{\alpha\beta}$. Moreover, σ_M is fixed[6] in the hydrodynamic approximation by the chiral anomaly (1) and equals to:

$$\sigma_M = \frac{e^2}{2\pi^2} , \tag{6}$$

where e is the electric charge of the constituents.

We will discuss also the parity-reflected cousin of Eq (5):

$$j_\mu^5 = \sigma_5 \mu \cdot B_\mu , \tag{7}$$

where μ is the chemical potential, conjugated to the electric charge. According to Ref. 7 the coefficient σ_5 is again uniquely determined in terms of the anomaly:

$$\sigma_5 = \frac{e}{2\pi^2}. \tag{8}$$

Another effect which we have in mind is the so called chiral vortical effect:

$$j_\mu^5 = \frac{1}{2}\sigma_\omega \epsilon_{\mu\nu\alpha\beta} u_\nu \partial_\alpha u_\beta , \tag{9}$$

where σ_ω, to the lowest order in the chemical potentials μ, μ_5 can be expressed, again, in terms of the anomaly:[6,8]

$$\sigma_\omega = \frac{\mu^2 + \mu_5^2}{2\pi^2} + O(\mu_{(5)}^3). \tag{10}$$

As for the corrections of order $O(\mu_{(5)}^3)$ they also arise in the hydrodynamic approximation, generally speaking, see, e.g., Ref. 6. However, these corrections are sensitive to details of infrared regularization, see, e.g. Ref. 9, and can be removed by proper choice of the coordinate frame.

All the chiral effects we mentioned refer to equilibrium, as first emphasized in Ref. 10. As a result, the linear-response relation for the transport
coefficient σ_M looks so as if we were considering a static effect:

$$\sigma_M = \lim_{k_n \to 0} \epsilon_{ijn} \frac{i}{2k_n} \langle j_i, j_j \rangle, \tag{11}$$

where $\langle j_i,, j_j \rangle \equiv \Pi_{ij}$ is the current-current correlator in the momentum
space. Similarly, for the chiral separation effect one gets:

$$\sigma_5 = \lim_{k_n \to 0} \epsilon_{ijn} \frac{i}{2k_n} \langle j_i^{(5)}, j_j \rangle . \tag{12}$$

For derivation of Eqs (11), (12) see, e.g., Refs. 5, 8, 11 and references therein.
Note that in case of the standard Kubo formulae the limiting procedure is
different. Namely, in that case the spatial momentum is vanishing identically, $k_i \equiv 0$ while the frequency tends to zero, $\omega \to 0$.

Eq. (11) is a convenient starting point to explain, what kind of problems
we will address here. Eq. (11) relates the chiral conductivity to a polynomial
in the current-current correlator, $\Pi_{ij} \sim \epsilon_{ijn} k_n$. The polynomial, in turn,
is determined by ultraviolet-sensitive regulators. One can argue, therefore,
that it is crucial to regularize the product of the two electromagnetic currents, entering Π_{ij} at coinciding points. A detailed derivation of σ_M along
these lines can be found in Ref. 11 and does reproduce the standard result
(6). Note that the ultraviolet behavior of the correlator of the currents is
determined by the subtraction constant and fixed uniquely by the theory.

We feel, however, that ascribing the chiral magnetic effect entirely to
the ultraviolet physics is an oversimplification. Namely, in analogy with the
discussion above we expect that there exist alternative derivations of the
chiral effects which are sensitive to infrared physics. Moreover, we expect,
that the final result depends on hierarchy of infrared-sensitive parameters,
see for a discussion above. To uncover such a hierarchy it is useful to introduce again (compare Eq. (4)) a finite fermion mass, violating the chiral
symmetry. Then one can argue[12] that the standard value of the chiral conductivity (6) corresponds to the following hierarchy of the infrared-sensitive
parameters:

$$e|k| \gg e\omega \gg m_f . \tag{13}$$

Further examples of infrared dependences of the chiral magnetic effect can
be found in Ref. 13.

2. From a non-local function to a polynomial

At first sight, the non-local expression (2) for the matrix element of an anomalous current is very different from polynomials which, according to (11), (12) determine the chiral effects. As a preliminary remark, we argue in this section that in some, physically motivated limits the non-local functions do reduce to polynomials.

As first example, consider quantum chromodynamics at finite temperature in the Euclidean formulation of the theory. Moreover, consider the limit of exact chiral symmetry. The chiral anomaly at finite temperature has been considered, of course, in many papers, see for example.[14] Here we consider only the resolution of an apparent contradiction between existence of a massless pion and absence of long range-forces at finite temperature.

In more detail, as far as the temperature is small compared to the temperature of the deconfining phase transition T_c the chiral symmetry remains spontaneously broken and there is a massless pion. Thus, we expect the pole (2) to be present in the matrix elements of the axial current. On the other hand, the 4d theory reduces now to a sequence of 3d theories with finite fermion masses:

$$m_f^{(3d)} = 2\pi T(n + 1/2) , \qquad (14)$$

This implies that perturbatively the triangle graph does not correspond any longer to an infinite-range interaction, in an apparent contradiction with existence of the pole (2). This is the paradox we would like to address.

Let us consider the kinematics in more detail. The extension of space in the time direction is now limited to

$$\tau \leq 1/T ,$$

where τ is the Euclidean time coordinate. Thus, one can probe propagation to large distances only in spatial directions. Moreover, the pion is massless only in case of the 3d theory corresponding to the Matsubara frequency $\omega_M = 0$ and we concentrate on this case. We can choose $q_i \sim (0, 0, q_3)$ and rewrite the non-trivial component of the matrix element (2) as:

$$\langle j_3^{(5)} \rangle_{external\ fields} = \frac{q_3 q_3}{q_3^2} \frac{e^2}{2\pi^2} \epsilon_{3ij} A_0 i k_i A_j = \frac{e\mu}{2\pi^2} B_3 \qquad (15)$$

where we replaced the external potential eA_0 by the chemical potential μ and B_3 is the external magnetic field. Clearly, we succeeded to rewrite Eq. (4) in the form which coincides with the equation (7) for the chiral separation effect.

The central point of this simple exercise is that the pole exhibited in Eq (4) disappears in case of the specific kinematics relevant to the chiral effects. The reason is that we do not observe Lorentz covariance any longer since the rest frame of the liquid is singled out on physical grounds. The cancellation of the pole we observed is universal for any matrix element associated with the pion exchange. Indeed, it is a general property of interactions of Goldstone particles that at small momenta all the vertices are proportional to the momentum of the massless particle. And if the momentum of the pion has only a single non-vanishing component the cancellation is obvious.

Note also that we can readily read off Eq. (12) from Eq. (15). In this sense, the both approaches are equivalent to each other in case of exact chiral symmetry. However, if we start from the non-local equation (4) we are better equipped to study dependence on extra infrared-sensitive parameters. In particular, turning on a finite pion mass turns off the chiral separation effect at distances $d \gg m_\pi^{-1}$:

$$\lim_{q_3/m_\pi \to 0} \langle j_3^{(5)} \rangle = 0 , \tag{16}$$

in analogy with Eq. (4).

Another example of this type is provided by evaluation of matrix element of the axial charge over a photon state:

$$Q_\gamma^{axial} = \langle \gamma | \widehat{Q}^{axial} | \gamma \rangle, \quad \widehat{Q}^{axial} = \int d^3x \, \bar{\Psi} \gamma_0 \gamma_5 \Psi . \tag{17}$$

Definition of the charge assumes that

$$\vec{q} \equiv 0, \quad q_0 \to 0 ,$$

where q_μ is the momentum carried in by the axial current. The 4-momentum of the photon, k_ν can be arbitrary, on the other hand. We can use now the result (2) for the matrix element (17):

$$Q_\gamma^{axial} = -i \frac{e^2}{8\pi^2} \frac{q_0}{q_0^2} F_{\mu\nu} \tilde{F}_{\mu\nu} = \frac{e^2}{4\pi^2} \epsilon_{0ijk} A_i \partial_j A_k , \tag{18}$$

where A_i is the vector potential. We see again that, once we do not impose Lorentz covariance, the non-locality vanishes in the chiral limit and we come to a polynomial.

Equation (18) implies that non-vanishing axial charge is to be ascribed to certain configurations of external magnetic fields. More specifically, let us introduce helicity of electromagnetic field as

$$\mathcal{H} = \int \vec{A} \cdot \vec{B} d^3x . \tag{19}$$

Then the chiral anomaly implies, by passing from Fourier to coordinate space, that we have to ascribe axial charge

$$Q_\gamma^{axial} = \frac{e^2}{4\pi^2}\mathcal{H} \tag{20}$$

to classical magnetic field configuration.

3. Evaluation of the chiral effects

In the presence of a chemical potential $\mu_5 \neq 0$ the Hamiltonian of the system, H_0 is redefined as

$$H_0 \rightarrow H_0 - \mu_5 Q^{axial} \ .$$

Moreover, in the Lagrangian language $\delta L = -\delta H$. Eq (18) then implies a modification of the effective action:

$$\delta S_{eff} = \int dt d^3x \ \mu_5 \frac{e^2}{4\pi^2}\epsilon_{ijk}A_i\partial_j A_k \ . \tag{21}$$

Electromagnetic current can be evaluated by varying the effective action with respect to the potential A_i. As a result we get, as is expected:

$$\mathbf{j}_{el} = \mu_5 \frac{e^2}{2\pi^2}\mathbf{B} \ , \tag{22}$$

where $\mathbf{B}$ is the external magnetic field.

The simplest way[9] to derive the chiral vortical effect is to utilize the analogy between gauge potential A_μ of field theory and local 4-velocity of an element of liquid u_μ:

$$e \cdot A_\mu \rightarrow \mu \cdot u_\mu \ . \tag{23}$$

Indeed in case of liquid the effective interaction is given by:

$$L_{int}^{eff} = \mu u_\mu \bar{\Psi}\gamma_\mu\Psi + eA_\mu\bar{\Psi}\gamma_\mu\Psi \ . \tag{24}$$

Discussion of the validity of this analogy and references can be found in Ref. 9. It should be noted that to take into account also the μ_5 contribution one has to introduce effective axial "gauge" field.

Applying substitution (23) to (20) , one concludes that helical macroscopic motion of the liquid is associated with a non-vanishing axial charge:

$$Q_\omega^{axial} = \frac{\mu^2}{4\pi^2}\int d^3x \ \epsilon_{ijk}u_i(x)\partial_j u_k(x) \ . \tag{25}$$

An alternative derivation of the chiral vortical effect can be given in terms of the gravitational, or mixed chiral anomaly, see, e.g., Ref. 11.

To summarize, both the chiral magnetic effect and chiral vortical effect have clear connection with the chiral anomaly. Namely, the axial charge of the (massless) constituents is not conserved because of the anomaly. However, one can introduce a conserved axial charge by ascribing it to electromagnetic field configurations with non-vanishing helicity and to a helical motion of chiral liquids. It is worth emphasizing that so far we assumed the chiral symmetry to be exact. If we introduce finite fermion mass but keep μ_5 time-independent, then there is no manifestations of the chiral anomaly. Also, the reduction of the chiral magnetic effect to the current commutator (11) is lost generally speaking.

Turn now to consideration of the chiral separation effect (7). Apparently, the effect is associated with the triangle graph, generated by the interactions μQ^{el}, $eA_\mu j_\mu^{el}$, $g_A j_\mu^{axial}$ with the coupling $g_A = 1$. In more detail, consider first the kinematics similar to the one considered above, with substitution of the vertex $\mu_5 Q^{axial}$ by μQ^{el}. Namely, let the 3-momentum carried in by the axial current be equal to the 3-momentum carried by the electromagnetic potential, and the vertex proportional to $\mu \cdot u_\mu$ be associated with 3-momentum equal to zero, $|\mathbf{q}| \equiv 0$ while the q_0 component tends to zero, $q_0 \to 0$. Since the electric current is not anomalous,

$$Q^{el}_{\gamma_A - \gamma} \equiv \langle \gamma_A | \widehat{Q}^{el} | \gamma \rangle = 0 \;, \tag{26}$$

where γ_A is a fictitious axial photon coupled to the axial current. From (26) we would conclude that

$$\sigma_5 = 0 \;. \tag{27}$$

which is in apparent contradiction with evaluations of σ_5 in Ref. 7.

One of the reasons is that our treatment of the anomaly is asymmetric with respect to the vector and axial currents. The vector current in our case is distinguished by its coupling to a physical massless vector particle, photon. If one concentrates on, say, vector current associated with the baryonic quantum number, as in Ref. 7, then the anomaly can be treated in a different way. For further discussion see, e.g., Ref. 9.

We could consider also another kinematics which is exactly the same as above, with small momentum $q_0, q_0 \to 0$ carried in by the axial current. Moreover we consider now the average value of spatial components of the axial current. Then the Lorentz-covariant completion of (18) would bring the result

$$\langle j_i^{axial} \rangle = \frac{\mu e}{2\pi^2} \epsilon_{ijk} \partial_j A_k \;, \tag{28}$$

which is equivalent to (7).

For non-interacting fermions one can also evaluate σ_5 directly, in terms of the Landau levels.[7] The calculation goes through for massive fermions as well. The only change in σ_5 to be made is:

$$\mu \rightarrow \sqrt{\mu^2 - m_f^2} \,, \qquad (29)$$

where m_f is the fermion mass. Clearly, any relation of the matrix element $< j_i^{axial} >$ to the chiral anomaly is lost unless μ not much larger than m_f.

It follows from these remarks that, in any case, the status of the chiral separation effect (7) is different from the status of the chiral magnetic effect (5).[a] In the latter case we consider axial charge and there apply various non-renormalization theorems. In the former case we consider spatial component of the axial current and there are no reasons to expect that any non-renormalization theorems exist. Indeed, there are no theorems on non-renormalizability of magnetic moments of fermions. Also, the hierarchy of infrared-sensitive parameters for the two types of defects is different. Namely, the constraint (13) on the frequency ω is related to the fact that the chiral magnetic effect is associated with production of chiral fermions. This is not true in case of the chiral separation effect. As is noted in Ref. 7 for non-relativistic fermions the chiral separation effect reduces to evaluation of the average spin value:

$$\langle j_i^5 \rangle \rightarrow \langle \sigma_i \rangle \,.$$

Thus, there is no actual flow of chirality along the magnetic field.

4. Infrared instabilities

4.1. Infrared divergences due to massless charged particles

Chiral magnetic effect arises as a result of interplay between quantum field theory and phenomenological, hydrodynamic approach. As far as one continues with evaluation of quantum corrections in field theory it is quite obvious that further infrared singularities are encountered. Indeed, we are considering now field theory of massless charged particles which is actually not well in defined on the mass shell and results of measurements in such a theory would depend on the resolution, or experimental set up, for discussion and references see, e.g., Ref. 15.

On the other hand, assuming hydrodynamics to apply, one postulates that results are not dependent, say, on the volume of the system. In other

---

[a]This point was elaborated in collaboration with Sergey Vavilov.

words, one assumes in fact that the infrared regularization is somehow provided by the constituents interaction, without destroying symmetries and the non-renormalization theorems proven within the field theory approach, see, e.g., Ref. 6. Since there is no explicit mechanism of such an infrared cut off known, there is no guarantee that matching of the field theory and hydrodynamics results in fact in a self-consistent picture. There are some hints in the literature on possible inconsistencies and let us mention a few examples.

As is expected, evaluation of the radiative correction to the chiral separation effect in the approximation of non-interacting fermions results in an infrared unstable expression:[16]

$$\delta \langle j^{axial} \rangle \;=\; -\,\frac{\alpha_{el} e B \mu}{2\pi^3}\left(\ln \frac{2\mu}{m_f} + \ln \frac{m_\gamma^2}{m_f^2} + \frac{4}{3} \right), \tag{30}$$

where fermion mass is taken to be $m_f \ll \mu$ and m_γ is the fictitious photon mass.

The problem with the infrared divergence in the photon mass m_γ in expression (30) could be settled by considering higher orders in the magnetic field. Then the fermions occupy actually the Landau levels and are off mass shell. This would bring also $m_\gamma^2 \neq 0$. It is less clear how to make sense of expression (30) in the limit $m_f \to 0$. As is mentioned above, it is a highly non-trivial problem how to define a massless charged particle on the mass shell.

Turning to the effective action (21), it is calculable in field theory. Phenomenologically, one could introduce terms which contain higher orders in derivatives. In the hydrodynamic approximation, they are assumed to be small. The question is, what is the characteristic mass scale in the hydrodynamic expansion in derivatives. For example, if one considers the 2d Hall liquid, then the underlying field theory provides the scale for the hydrodynamic approximation, which is the energy gap to the next Landau level. If one considers a 3d liquid, then, to the best of our knowledge, there is no field-theoretic mechanism to provide a gap. Nevertheless, one postulates validity of the expansions both in magnetic field and in the derivatives. Under such assumptions the coefficient σ_ω, for example, receives corrections of order μ^3:[6]

$$\sigma_\omega \;=\; \frac{\mu^2}{2\pi^2}\left(1 + \frac{2}{3}\frac{\mu \cdot n}{\epsilon + P} \right), \tag{31}$$

where n is the density of particles, ϵ and P are the energy density and pressure, respectively. Note that the enthalpy, $\epsilon + P$ plays the role of a "hydrodynamic mass" in some other cases as well.

Equation (31) refers to the temperature independent part of σ_ω. There is also contribution proportional to T^2, σ_ω^T. Moreover, using the underlying field theory, one can evaluate the first radiative correction to σ_ω^T[17,18] in terms of the interaction coupling g^2:

$$\sigma_\omega = \frac{T^2}{2\pi^2}\left(1 + \text{const}\frac{g^2}{48\pi^2}\right), \tag{32}$$

where the (const) depends on the fermionic representation. The central point is that if one considers Yang-Mills theory, the coupling g^2 is to be replaced at high temperature by the running coupling $g^2(T)$. However, any logarithmic dependence, brought by the running of the coupling, would not fit the thermodynamic expansion which does not reserve for any non-analyticity in temperature.

Another reservation suggested by field theory, is that the current (5) would, generally speaking, radiate. On the other hand, in classical approximation the chiral magnetic effect is dissipation free since the current flows along the magnetic field, and the magnetic field produces no work.[19]

4.2. Back-reaction of chiral medium

In this subsection we will discuss another inconsistency between field theoretic and hydrodynamic approaches. In hydrodynamics, one simply postulates that medium with $\mu_5 \neq 0$, placed in external magnetic field can be in equilibrium state, and demonstrates then that there exists current (5). Field theory tells us that actually such a medium is unstable. One of the explanations is that the current (5) produces extra magnetic field and this effect should be taken into account. Since this back-reaction of the medium is of next order in electromagnetic interaction, large volumes are needed for the effect to become important. This instability has been discussed in Refs. 20, 21. Moreover, similar instabilities were discussed earlier, see, e.g., Refs 22–25 in some other sets up. We will follow the paper in Ref. 21. Moreover, even in the absence of external magnetic field the medium with $\mu_5 \neq 0$ seems to be unstable against spontaneous generation of the magnetic field, or, probably, domains of it.

It is convenient to start with the effective action (21) and concentrate on the static case with no dependence on time. Then, the 3d (Euclidean) looks as:

$$S_{eff}^{3d} = \int d^3x\left(\frac{1}{2}\sigma_M\epsilon_{ijl}A_i\partial_j A_k + eA_i j_i^{el} - \frac{1}{4}(F_{ik})^2\right). \tag{33}$$

Furthermore, introduce propagator of the photon which in general takes the form:

$$D_{ij}(\vec{k}) = D_S(k)(\delta_{ij} - \widehat{k}_i\widehat{k}_j) + D_A(k)i\epsilon_{ijl}\widehat{k}^l + a\frac{\widehat{k}_i\widehat{k}_j}{k^2} , \qquad (34)$$

where $\widehat{k}$ is the unit vector along the 3d momentum and the last term is the gauge fixing.

Explicit summation of the bubble-type graphs associated with the action (33) provides the following result:

$$D_{ij}(k_i) = \frac{1}{k^2 - \sigma_M^2}\left(\delta_{ij} - \frac{k_ik_j}{k^2}\right) + \frac{i\epsilon_{ijl}\sigma_M k_l}{k^2(k^2 - \sigma_M^2)} + a\frac{k_ik_j}{k^4} . \qquad (35)$$

A conspicuous feature of the expression (35) is the pole in the physical region at $k^2 = \sigma_M^2$. This means instability of the perturbative vacuum with $\langle A_i \rangle_{class} = 0$. The basic assumption is the validity of the effective action (33). This action, on the hand, is commonly derived in hydrodynamic approximation, see, e.g., Ref. 8.

Equation (35) indicates that the true vacuum has non-vanishing magnetic field with fixed $k^2 = \sigma_M^2$. Moreover, there is also a pole in front of the structure $\epsilon_{ijl}k^l$ which means that one expects that the magnetic field in the true vacuum possesses non-trivial helicity (19).

These features look exotic. However, they can be readily appreciated if we look at the action (33) in somewhat different way. Namely, instead of evaluating propagator for the gauge field on the trivial background we can start with the classical equation corresponding to the action (33):

$$\mathbf{curl\ B} = \sigma_M\mathbf{B} . \qquad (36)$$

Equation (36) is nothing else but the well known Beltrami equation, which has been studied since long in various physical frameworks, see, in particular, Ref. 26. And, indeed, the simplest solution of the Beltrami equation represents a standing wave with $k^2 = \sigma_M^2$ and non-trivial helicity (19).

This concludes our brief review of the instability of then chiral medium with $\mu_5 \neq 0$ with respect to spontaneous generation of helical classical magnetic fields. As far as we know, there are no estimates of the life-time of the unstable vacuum, if we start with vanishing magnetic field.

Another result seems worth mentioning. As can readily be seen, in the limit of vanishing momentum, $k_i \to 0$, the propagator (35) tends to:

$$D_{ij}(\vec{k}) \to -\frac{i\epsilon_{ijl}k^l}{\sigma_M k^2}, \quad \vec{k} \to 0 . \qquad (37)$$

One can prove[21] that this result survives with account of all the electromagnetic corrections. On the other hand the term (37) results in a topological type interaction of static current loops ($k \ll \sigma_M$), proportional to their linking number. Thus, this topological interaction of external Wilson loops is not renormalized.

In this brief review we have illustrated infrared instabilities which are inherent to macroscopic manifestations of the chiral anomaly. From phenomenological point of view of view, the most crucial question is whether the anomaly is relevant at all. In particular, the electromagnetic fields should be intense and satisfy conditions like $\sqrt{eB} \gg m_f$ where m_f is the scale of explicit chiral symmetry breaking. This condition is difficult to satisfy in realistic sets up.

4.3. *Decoherence of fermionic wave function*

From theoretical point of view, it is most intriguing that one predicts[6] the effect of the quantum anomaly to survive in the classical, hydrodynamic limit. Moreover, the corresponding current (5) is predicted to be dissipation free since it exists in the equilibrium.[10,19]

There is explicit derivation of the chiral magnetic effect in terms of the Berry phase in the collision-less approximation to chiral plasma.[27] However, it seems far from being obvious that decoherence of the wave function, due to interaction with the medium, does not damp down the effect of the topological phase. It is worth mentioning in this connection[b] that there is an analog of the chiral magnetic effect in (1+1) dimensional quantum wires, see Ref. 10 and references therein:

$$\langle j_{el} \rangle = e(\mu_L - \mu_R) \,, \tag{38}$$

where μ_L, μ_R are the chemical potentials for left- and right-movers, respectively. Equation (38) expresses the so called universal conductance and it works for quantum wires as far as the wire is not too long, to avoid scattering, $l_{wire} \ll l_{m.f.p.}$. One could imagine that scattering in (3+1) dimensions plays a similar role. Then one would need, say, superfluid to observe the chiral magnetic effect. For discussion of a particular model of this type see Ref. 28.

5. Conclusions

Thus, we have demonstrated that infrared sensitivities are inherent to chiral media. Eventually all the infrared effects go back to the fact that chiral

---

[b]The remark is due to D.E. Kharzeev.

285

media introduce massless charged particles. The actual mechanism of infrared regularization is not known and may vary from case to case. In this
sense, the physics of the chiral media could be even richer than one would
think.

Mostly, these notes are a kind of a mini-review since many points have
already been made in literature. In conclusion, let us emphasize original
points which might be in variance with conclusions of other papers:

• The effective action for photon in chiral media is commonly derived
  from the chiral anomaly and in static limit takes the form of the 3d
  topological photon mass. In Ref. 21 it was demonstrated however that
  this mass turns to be imaginary and signals instability of magnetostatics
  of the chiral media.
• Moreover, field theory does not provide any apparent mass scale for
  the hydrodynamic expansion for chiral media and, in principle, higher
  orders could be the same important as the terms fixed by the anomaly.
• We have demonstrated that standard derivations of the chiral magnetic
  effect and chiral separation effect refer in fact to different infrared limits.
• Chiral effects have been derived in collisionless approximation and re-
  flect topological contributions to phases of fermionic wave functions.
  We emphasized that decoherence arising due to interactions among the
  constituents could wipe the effects out. One of the ways out is to concentrate on superfluids.

Acknowledgments

The authors are grateful to A.S. Gorsky, Z. Khaidukov, D.E. Kharzeev, M.
Stephanov for useful discussions. We are thankful to the organizers of the
seminar devoted to the memory of Academician I.Ya. Pomeranchuk for the
invitation to the seminar. The work of VPK and AVS has been supported
by grant IK-RU-002 of Helmholtz Association.

References

1. S. Adler, Phys. Rev. 177, 2426 (1969);
   J.S. Bell and R. Jackiw, Nuovo Cim. A 60, 47 (1969)
2. A.D. Dolgov, V.I. Zakharov, Nucl. Phys. B 27, 525 (1971)
3. G. 't Hooft, in "Recent Developments in Gauge Theories", ed. G. 't Hooft
   et al. (Plenum, N.Y., 1980)
4. D. E. Kharzeev, K. Landsteiner, A. Schmitt, and H.-U. Yee, Lect. Notes
   Phys. 871, 1 (2013)
5. D. E. Kharzeev, L. D. McLerran, and H. J. Warringa, Nucl. Phys. A 803,
   227 (2008);

K. Fukushima, D. E. Kharzeev and H. J. Warringa, Phys. Rev. D 78, 074033 (2008)

D. Kharzeev, Phys. Lett. B 633, 260 (2006)

6. D. T. Son and P. Surowka, Phys. Rev. Lett. 103, 191601 (2009)
7. M. A. Metlitski, A. R. Zhitnitsky, Phys. Rev. D 72 (2005) 045011
8. K. Jensen, M. Kaminski, P. Kovtun, R. Meyer, A. Ritz, A. Yarom, Phys. Rev. Lett. 109 (2012) 101601

   K. Jensen, Phys. Rev. D 85, 125017 (2012)

   N. Banerjee, J. Bhattacharya, S. Bhattacharyya, S. Jain, S. Minwalla, and T. Sharma, JHEP 1209, 046 (2012)
9. A. V. Sadofyev, V. I. Shevchenko, and V. I. Zakharov, Phys. Rev. D 83, 105025 (2011)
10. A. Yu. Alekseev, V. V. Cheianov, J. Froelich, Phys. Rev. Lett. 81, 3503 (1998)
11. K. Landsteiner, E. Megias, and F. Pena-Benitez, Lect. Notes Phys. 871, 433 (2013)
12. V. I. Zakharov, Lect. Notes Phys. 871 (2013)
13. Defu Hou, Hui Liu, Hai-cang Ren, JHEP 1105, 046 (2011)
14. R. D. Pisarski, T. L. Trueman, M. H. G. Tytgat, Phys. Rev. D 56 (1997) 7077-7088
15. T. D. Lee and M. Nauenberg, Phys. Rev. 133, B1549 (1964);

    H. Georgi, Phys. Rev. Lett. 98, 221601 (2007)
16. E. V. Gorbar, V. A. Miransky, I. A. Shovkovy, X. Wang, Phys. Rev. D 88 (2013) 025025
17. S. Golkar and D. T. Son, arXiv:1207.5806 [hep-th]
18. D.-F. Hou, H. Liu, H.-C. Ren, Phys. Rev. D 86, 1A (2012)
19. D. E. Kharzeev and H.-U. Yee, Phys. Rev. D 84, 045025 (2011)
20. Y. Akamatsu and N. Yamamoto, Phys. Rev. Lett. 111 052002 (2013)
21. Z. V. Khaidukov, V. P. Kirilin, A. V. Sadofyev, V. I. Zakharov, e-Print: arXiv:1307.0138 [hep-th]
22. D. V. Deryagin, D. Yu. Grigoriev, V. A. Rubakov, Int. J. Mod. Phys. A 7, 659 (1992)
23. A. Ballon-Bayona, K. Peeters, and M. Zamaklar, JHEP 1211, 164 (2012)
24. M. Krusius, T. Vachaspati, G. E. Volovik, e-Print: cond-mat/9802005
25. A. Boyarsky, O. Ruchayskiy, M. Shaposhnikov. Phys. Rev. Lett. 109, 111602 (2012)
26. V. I. Arnold, C. R. Acad Sci. Paris 261, 17 (1965);

    S. Childress, J. Math. Phys. 11, 3063 (1970);

    N. Kleeorin, I. Rogachevskii, D. Sokoloff, and D. Tomin, Phys. Rev. E 79, 046302 (2009)
27. D. T. Son, N. Yamamoto, Phys. Rev. Lett. 109, 181602 (2012)

    D. T. Son, N. Yamamoto, Phys. Rev. D 87, 085016 (2013)

    J. -W. Chen, S. Pu, Q. Wang, and X. -N. Wang, Phys. Rev. Lett. 110, 262301 (2013)
28. V. P. Kirilin, A. V. Sadofyev, V. I. Zakharov, Phys. Rev. D 86 025021 (2012)

LOCALIZATION AT LARGE N

J. G. RUSSO

Institució Catalana de Recerca i Estudis Avançats (ICREA),
Pg. Lluis Companys, 23, 08010 Barcelona, Spain
Department ECM, Institut de Ciències del Cosmos,
Universitat de Barcelona,
Martí Franquès, 1, 08028 Barcelona, Spain

K. ZAREMBO

Nordita, KTH Royal Institute of Technology and Stockholm University,
Roslagstullsbacken 23, SE-106 91 Stockholm, Sweden
Department of Physics and Astronomy, Uppsala University
SE-751 08 Uppsala, Sweden
Institute of Theoretical and Experimental Physics, B. Cheremushkinskaya 25
117218 Moscow, Russia

We review how localization is used to probe holographic duality and, more generally, non-perturbative dynamics of four-dimensional $\mathcal{N} = 2$ supersymmetric gauge theories in the planar large-N limit.

1. Introduction

String theory on $AdS_5 \times S^5$ gives a holographic description of the superconformal $\mathcal{N} = 4$ Yang-Mills (SYM) through the AdS/CFT correspondence.[1–3] This description is exact, it maps correlation functions in SYM, at any coupling, to string amplitudes in $AdS_5 \times S^5$. Gauge-string duality for less supersymmetric and non-conformal theories is at present less systematic, and is mostly restricted to the classical gravity approximation, which in the dual field theory corresponds to the extreme strong-coupling regime. For this reason, any direct comparison of holography with the underlying field theory requires non-perturbative input on the field-theory side.

There are no general methods, of course, but in the basic AdS/CFT context a variety of tools have been devised to gain insight into the strong-coupling behavior of $\mathcal{N} = 4$ SYM, notably by exploiting integrability of this theory in the planar limit.[4] Another approach is based on supersymmetric localization.[5] Applied to the $\mathcal{N} = 4$ theory, localization provides direct

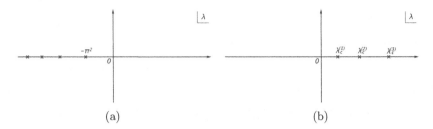

Fig. 1.   Possible analytic structure in the coupling constant plane: (a) No phase transitions (all singularities lie on the negative part of the real axis). This case is realized in $\mathcal{N} = 4$ SYM. (b) The singularities lie on the positive real axis, leading to phase transitions at $\lambda = \lambda_c^{(n)}$.

dynamical tests of the AdS/CFT correspondence, but localization does not require as high supersymmetry as $\mathcal{N} = 4$ and more importantly does not rely on conformal invariance thus allowing one to explore a larger set of models including massive theories. Some of localizable theories have known gravity duals, opening an avenue for direct comparison of holography with the first-principle field-theory calculations. In this contribution we concentrate on $\mathcal{N} = 2$ theories in four dimensions. A review of localization in $D = 3$ and its applications to the AdS_4/CFT_3 duality can be found in;[6] early results for $\mathcal{N} = 4$ SYM[7,8] are reviewed in.[9]

Although our main goal is holography and hence strong coupling, localization gives access to more general aspects of non-perturbative dynamics. An example of a non-perturbative phenomenon captured by localization is all-order OPE.[10] Suppose that we integrate out a heavy field of mass M in an asymptotically free theory with a dynamically generated scale $\Lambda_{\text{eff}} \ll M$. We then expect that any observable will have an expansion

$$\mathcal{A} = \Lambda_{\text{eff}}^{\Delta} \sum_{n=0}^{\infty} C_n \left(\frac{\Lambda_{\text{eff}}}{M} \right)^{2n} , \tag{1}$$

where Δ is the scaling dimension of $\mathcal{A}$. Expansions of this type make prominent appearance in the ITEP sum rules.[11,12] The mass M in the denominator arises from expanding the effective action in local operators and powers of Λ_{eff} in the numerator come from the condensates, the vacuum expectation values of local operators generated by the OPE. The coefficients in this expansion carry non-perturbative information, and are usually difficult to calculate, but for observables amenable to localization it is possible to compute these coefficients to all orders.

The OPE of the form (1) is of course expected on general grounds in the regime $\Lambda_{\text{eff}} \ll M$. What is less expected, but appears to be generic, is the emergence of large-N phase transitions when $\Lambda_{\text{eff}} \gtrsim M$.[10,13] Large-$N$ phase transitions are very familiar from matrix models[14,15] as singularities reflecting the finite radius of convergence of planar perturbation theory.[16] For any UV finite theory, including $\mathcal{N} = 4$ SYM, planar perturbation theory also has a finite radius of convergence. But in the $\mathcal{N} = 4$ case all singularities lie on the negative real axis of the 't Hooft coupling $g_{\text{YM}}^2 N \equiv \lambda$ (the leading singularity appears at $\lambda = -\pi^2$). Interpolation from weak to strong coupling is thus continuous as illustrated in Fig. 1(a). There are no distinct "perturbative" and "holographic" phases. A priori this does not follow from any fundamental principle. A possibility that $\mathcal{N} = 4$ SYM undergoes a strong-weak phase transition was in fact contemplated in the early days of AdS/CFT,[17] but subsequent developments showed that such a transition does not occur. Is it still possible that theories different from $\mathcal{N} = 4$ SYM have a structure of singularities shown in Fig. 1(b)? And, if yes, what are the implications for the holographic duality? Localization gives partial answers to these questions: phase transitions do occur in massive theories, but it is not clear at the moment how to describe them holographically.

2. Localization in $\mathcal{N} = 2^*$ SYM and large-N limit

Our prime example will be the $\mathcal{N} = 2^*$ theory, a massive deformation of $\mathcal{N} = 4$ SYM which preserves half of the supersymmetry:

$$\mathcal{L}_{\mathcal{N}=2^*} = \mathcal{L}_{\mathcal{N}=4} + M\mathcal{O}_3 + M^2\mathcal{O}_2. \tag{2}$$

The dimension two and dimension three operators $\mathcal{O}_{2,3}$ give masses to four out of six adjoint scalars and to half of the fermions, and also contain certain tri-linear couplings. Two scalars that remain massless belong to the vector multiplet of $\mathcal{N} = 2$ supersymmetry: $(A_\mu, \psi_\alpha^1, \psi_\alpha^2, \Phi, \Phi')$, while massive fields combine to the complex hypermultiplet $(\phi, \chi_\alpha, \tilde{\chi}_\alpha, \tilde{\phi})$, also in the adjoint representation of the $SU(N)$ gauge group.

This theory inherits finiteness of the $\mathcal{N} = 4$ SYM. The holographic description of $\mathcal{N} = 2^*$ SYM at strong coupling[18,19] is based on the solution of type IIB supergravity in ten dimensions that was obtained by Pilch and Warner[18] by perturbing $AdS_5 \times S^5$ with constant sources dual to relevant operators in the Lagrangian (2).

The moduli space of vacua of $\mathcal{N} = 2^*$ SYM is parameterized by the diagonal expectation value of the adjoint scalar in the vector multiplet:

$$\langle \Phi \rangle = \text{diag}\,(a_1, \ldots, a_N)\,. \tag{3}$$

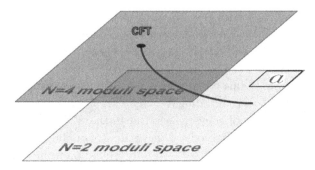

Fig. 2. $\mathcal{N} = 2^*$ flow trajectory in the moduli space of $\mathcal{N} = 2$ theories.

The same vacuum degeneracy exists in $\mathcal{N} = 4$ SYM as well, where any of the six scalars can take on a vev. However, when talking about $\mathcal{N} = 4$ SYM, we will always assume that the theory is at the conformal point with zero vevs. We also assume that the $\mathcal{N} = 2^*$ flow starts with this conformal theory at the origin of the moduli space. In other words, by $\mathcal{N} = 2^*$ SYM we really mean a line of theories that at $M = 0$ degenerate to the conformal state of $\mathcal{N} = 4$ SYM with $\langle \Phi \rangle = 0$. The flow so defined traces a trajectory in the moduli space of vacua as schematically illustrated in Fig. 2. At the IR end of the flow, hypermultiplets become very heavy and can be integrated out leaving behind pure $\mathcal{N} = 2$ SYM with the dynamically generated scale $\Lambda_{\text{eff}} = M \exp(-4\pi^2/\lambda)$. It is important to keep in mind that the $(\mathcal{N} = 4) \to (\mathcal{N} = 2)$ flow picture is only correct at sufficiently weak coupling when the scales M and Λ_{eff} are largely separated. For finite coupling, Λ_{eff} is of the form $\Lambda_{\text{eff}} = g(\lambda)M$, with calculable $g(\lambda)$, which starts off exponentially small, but becomes much greater than one at $\lambda \to \infty$, where Λ_{eff} exceeds M.

Normally the scalar vev (3) can be chosen at will, but in the flow that starts at the origin of the moduli space, the vev is fixed by dynamics. Since in practice we will use a definition of the $\mathcal{N} = 2^*$ SYM that is different from conformal perturbation theory, it is worthwhile to discuss the vacuum selection mechanism in some more detail. Consider, for the sake of illustration, a Heisenberg ferromagnet whose degenerate set of vacua is a two-sphere. A commonly used definition of the theory consists in slightly lifting the vacuum degeneracy, for instance by switching on an external magnetic field, and then adiabatically relaxing the field to zero. The system will end up with magnetization aligned with the original direction of the field, or, depending on the temperature and details of the spin interactions, in the disordered state with no magnetization. These two regimes are separated by a phase transition.

The vacuum selection in the $\mathcal{N} = 2^*$ theory can also be defined by switching on infinitesimal external perturbation that lifts the vacuum degeneracy. A natural choice would be temperature, which generates a potential on the moduli space. The path integral on $\mathbb{R}^4$ would then be defined by taking the zero-temperature limit. Temperature here plays the same rôle as the magnetic field for the ferromagnet. While selecting thermalizable vacuum is physically appealing, supersymmetry breaking makes this procedure difficult to implement in practice. An alternative procedure, the one that we are going to use, consists in compactifying the theory on S^4. The curvature couplings generate a scalar potential which completely lifts the vacuum degeneracy, while the path integral on S^4 includes integration over zero modes of all fields and is thus well defined without specifying any boundary conditions. Sending subsequently the radius of the sphere to infinity defines a unique vacuum in the decompactification limit. We conjecture that the S^4 vacuum actually coincides with the thermalizable state, and that both definitions are equivalent to conformal perturbation theory in $\mathcal{N} = 4$ SYM with zero scalar vev. The advantage of the S^4 compactification is that the path integral on the sphere can be computed exactly using localization.[5] From this point of view the radius of the sphere is just an IR regulator, playing the same rôle as the magnetic field in the ferromagnet, but it is also of some interest to keep the radius finite introducing an extra parameter into the theory.

At strong coupling all three ways to define the $\mathcal{N} = 2^*$ flow are manifestly equivalent. Indeed, the Pilch-Warner solution is obtained by perturbing $AdS_5 \times S^5$ with the hypermultiplet mass and thus by construction is dual to the flow trajectory that ends in the conformal point on the $\mathcal{N} = 4$ moduli space (Fig. 2). On the other hand, the Pilch-Warner background can be thermalized[20] as well as compactified on S^4.[21] The latter case can be directly compared to field theory through localization[21,22] thus providing further evidence that the S^4 vacuum is dual to the Pilch-Warner solution at strong coupling.

In the planar limit the vacuum is characterized by the master field, a distribution of the eigenvalues of the adjoint scalar from the vector multiplet:

$$\rho(x) = \left\langle \frac{1}{N} \sum_{i=1}^{N} \delta(x - a_i) \right\rangle. \tag{4}$$

Using the localization results of,[5] we can explicitly calculate the large-N master field on a sphere of any radius. The S^4 vacuum is obtained by sending the radius to infinity. However, keeping the radius finite is interesting in its own right, and we will also discuss dependence of the master field on the radius of compactification.

In theories with a fermionic symmetry the exact functional integral may localize in a subset of field configurations plus a one-loop contribution. The $\mathcal{N} = 2^*$ theory belongs to this class, as its partition function on S^4 reduces to a finite-dimensional integral over the eigenvalues of the adjoint scalar in the vector multiplet (3). The result can be expressed as follows:[5]

$$
Z_{\mathcal{N}=2^*} = \int \prod_i da_i \, \delta \left(\sum_i a_i \right) \prod_{i<j} \frac{(a_i - a_j)^2 H^2(a_i - a_j)}{H(a_i - a_j - M)H(a_i - a_j + M)}
$$
$$
\times \, e^{-\frac{8\pi^2 N}{\lambda} \sum_i a_i^2} \, |Z_{\text{inst}}|^2 \,, \tag{5}
$$

where $H(x)$ encodes the product over the spherical harmonics of all field fluctuations:

$$
H(x) \equiv \prod_{n=1}^{\infty} \left(1 + \frac{x^2}{n^2} \right)^n e^{-\frac{x^2}{n}}. \tag{6}
$$

The one-loop contribution of the vector multiplet combines into H^2 in the numerator, while that of the hypermultiplet gives the two H's in the denominator. The instanton factor Z_{inst} is the Nekrasov partition function[23-25] with the equivariant parameters given by $\epsilon_1 = \epsilon_2 = 1$. In the large-$N$ limit the instantons are suppressed by a factor[a] $\exp(-8\pi^2 N|k|/\lambda)$, where k is the instanton number. Since we are interested in the planar limit, we will just set $Z_{\text{inst}} = 1$.

For notational convenience the radius of the sphere has been set to one. The dependence on R is reinstated by rescaling $a_i \to a_i R$ and $M \to MR$. We mostly use dimensionless units throughout the paper, but will recover the dependence on R when discussing the decompactification limit $R \to \infty$. In the dimensionless units, decompactification is equivalent to the infinite-mass limit $M \to \infty$.

---

[a]The volume of the instanton moduli space could potentially compensate the exponential suppression,[26] but explicit evaluation of the one-instanton weight at large N indicates that this never happens in the $\mathcal{N} = 2^*$ theory, the instantons always remaining suppressed at $N \to \infty$.[10] And indeed the results of localization with instantons neglected are in perfect agreement with the supergravity predictions at strong coupling.[21,22]

The eigenvalue integral (5) is (literally!) infinitely simpler than the path integral of quantum field theory, making localization a very powerful computational tool, especially at large N when instantons can be neglected and the eigenvalue integral can be analyzed by methods familiar from random matrix theory.[16] This simplicity comes at a price of dealing with a limited number of observables. One quantity that can be calculated with the help of localization is the free energy

$$F = -\frac{1}{N^2} \ln Z. \tag{7}$$

Another one is expectation value of the circular Wilson loop. This couples to the gauge field and the scalar from the vector multiplet as follows:

$$W(C) \equiv \left\langle \frac{1}{N} \operatorname{tr} \mathrm{P} \exp \left[\oint_C d\tau \, (i\dot{x}^\mu A_\mu + |\dot{x}|\Phi) \right] \right\rangle. \tag{8}$$

If the contour C goes around the equator of the four-sphere, the fields can be replaced by their classical values, $A_\mu = 0$ and Φ given by (3). Therefore

$$W(C) = \left\langle \frac{1}{N} \sum_i e^{2\pi a_i} \right\rangle. \tag{9}$$

For the circular Wilson loop the expectation value maps to the average of the exponential operator in the matrix model (5).

In the large N limit, the eigenvalue integral (5) is of the saddle-point type. The saddle-point equations are the force balance conditions for N particles on a line with pairwise interactions which are subject to a common external potential:

$$\sum_{j \neq i} \left(\frac{1}{a_i - a_j} - \mathcal{K}(a_i - a_j) + \frac{1}{2}\mathcal{K}(a_i - a_j + M) + \frac{1}{2}\mathcal{K}(a_i - a_j - M) \right)$$
$$= \frac{8\pi^2 N}{\lambda} a_i. \tag{10}$$

where

$$\mathcal{K}(x) \equiv -\frac{H'(x)}{H(x)} = 2x \sum_{n=1}^{\infty} \left(\frac{1}{n} - \frac{n}{n^2 + x^2} \right). \tag{11}$$

These equations are equivalent to a singular integral equation for the eigenvalue density (4):

$$\fint_{-\mu}^{\mu} dy \, \rho(y) \left(\frac{1}{x - y} - \mathcal{K}(x - y) + \frac{1}{2}\mathcal{K}(x - y + M) + \frac{1}{2}\mathcal{K}(x - y - M) \right)$$
$$= \frac{8\pi^2}{\lambda} x, \qquad x \in [-\mu, \mu]. \tag{12}$$

Here we have assumed that eigenvalues are distributed in one cut along the interval $[-\mu, \mu]$, where the density $\rho(x)$ is unit-normalized.

Once the integral equation is solved, the Wilson loop can be computed from the Laplace transform of the density:

$$W(C) = \int_{-\mu}^{\mu} dx\, \rho(x)\, \mathrm{e}^{2\pi x} \equiv \left\langle \mathrm{e}^{2\pi x} \right\rangle. \tag{13}$$

The free energy is given by a double integral. It is actually easier to compute its derivatives, for instance,

$$\frac{\partial F}{\partial \lambda} = -\frac{8\pi^2}{\lambda^2} \left\langle x^2 \right\rangle. \tag{14}$$

3. Strong coupling and holography

Having the exact coupling dependence for the planar theory, one can study the important limit $\lambda \gg 1$, which explores the deep quantum regime of the theory. An extra motivation for studying the $\lambda \gg 1$ limit is that this is precisely the limit where super Yang-Mills theories are expected to have a holographic description in terms of a weakly-curved supergravity dual. In this section we shall discuss two examples: the superconformal $\mathcal{N} = 4$ SYM and its mass deformation preserving $\mathcal{N} = 2$ supersymmetry.

3.1. $\mathcal{N} = 4$ *SYM*

Since $\mathcal{N} = 4$ SYM is a conformal theory and the sphere is conformally equivalent to $\mathbb{R}^4$, one can use localization to compute a circular Wilson loop in flat space. The answer does not depend on the radius of the circle by conformal invariance, and is given by the exponential average in the Gaussian matrix model:[5,7,8]

$$W(C_{\mathrm{ircle}}) = \left\langle \frac{1}{N} \operatorname{tr} \mathrm{e}^{2\pi\Phi} \right\rangle, \qquad Z = \int d\Phi\, \mathrm{e}^{-\frac{8\pi^2 N}{\lambda} \operatorname{tr} \Phi^2}. \tag{15}$$

This is equivalent to (5) with $M = 0$, upon gauge fixing the matrix measure to eigenvalues.

In the localization approach, the random matrix Φ is the zero mode of the adjoint scalar on S^4. By construction it has a constant propagator. If we start directly from field theory on $\mathbb{R}^4$, the result (15) can be understood as resummation of rainbow diagrams – all possible diagrams without internal vertices (Fig. 3).[7,8] One can argue that other diagrams do not contribute to the circular loop average.[8] The constant propagator then arises from partial cancellation between the scalar and gluon exchanges. The numerator in the

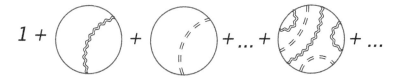

Fig. 3.   Rainbow diagrams for the circular Wilson loop.

one-loop correction (the first two diagrams in Fig. 3) contains the gauge boson $\dot{x}_1 \cdot \dot{x}_2$ and scalar $|\dot{x}_1||\dot{x}_2|$ contributions. For the circular loop, they combine into $|\dot{x}_1||\dot{x}_2| - \dot{x}_1 \cdot \dot{x}_2 = 1 - x_1 \cdot x_2$ and cancel the denominator $(x_1 - x_2)^2 = 2 - 2x_1 \cdot x_2$, leaving behind a constant propagator $\lambda/8\pi^2$. This argument extends to rainbow graphs of any order in perturbation theory, and the problem effectively reduces to combinatorics, taken into account by the matrix integral. There is no Feynman-diagram derivation for more complicated localization matrix models, but in the case of $\mathcal{N} = 2$ superconformal QCD one can check that the first vertex correction that appears at three loops[27] can be consistently reproduced from the skeleton graphs of the zero-dimensional matrix integral.[28]

For later reference we do a simple exercise of solving the Gaussian model (15) at large N. The saddle-point equation (12) at $M = 0$ becomes

$$\fint_{-\mu}^{\mu} dy \, \frac{\rho(y)}{x - y} = \frac{8\pi^2}{\lambda} \, x. \tag{16}$$

The eigenvalue density is then given by the Wigner's semi-circle law:

$$\rho(x) = \frac{8\pi}{\lambda} \sqrt{\frac{\lambda}{4\pi^2} - x^2}, \tag{17}$$

and from (9) we get the vacuum expectation value of the circular Wilson loop:[7]

$$W(C_{\text{circle}}) = \int dx \, \rho(x) \, e^{2\pi x} = \frac{2}{\sqrt{\lambda}} I_1 \left(\sqrt{\lambda} \right). \tag{18}$$

The free energy can be inferred from (14), or calculated directly by doing the Gaussian integral in (15):

$$F = -\frac{1}{2} \ln \lambda. \tag{19}$$

In the strong coupling $\lambda \gg 1$ regime, the Wilson loop has the behavior

$$W \simeq \sqrt{\frac{2}{\pi}} \lambda^{-\frac{3}{4}} e^{\sqrt{\lambda}} \qquad (\lambda \gg 1). \tag{20}$$

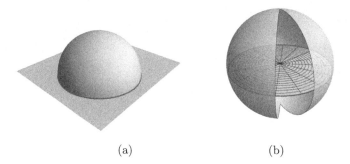

(a)   (b)

Fig. 4. Minimal surface in AdS_5 that describes the circular Wilson loop: (a) in Poincaré coordinates; (b) in the S^4 slicing.

These results, obtained by directly computing the path integral in $\mathcal{N} = 4$ SYM (or summing up Feynman diagrams) can be compared to the holographic predictions of string theory in $AdS_5 \times S^5$.

According to the AdS/CFT dictionary, Wilson loop expectation values obey the area law at strong coupling:[29,30]

$$W(C) \simeq e^{-\frac{\sqrt{\lambda}}{2\pi} A_{\text{reg}}(C)}, \tag{21}$$

where $A_{\text{reg}}(C)$ is the regularized area of the minimal surface in AdS_5 that ends on the contour C, placed at the boundary, as illustrated in Fig. 4(a). The minimal surface ending on a circular contour was constructed in[31,32] in the Poincaré coordinates. For computing its area it is easier though to deal with the S^4 slicing of AdS_5:

$$ds^2_{AdS_5} = \frac{dr^2}{1 + r^2} + r^2 d\Omega^2_{S^4}. \tag{22}$$

If the Wilson loop is chosen to run along the big circle of S^4, as appropriate for comparing to localization (but the answer will be the same for any circular loop, by conformal invariance), the minimal surface will coincide with the equatorial plane, Fig. 4(b). The area can be readily computed from (22):

$$A(C_{\text{circle}}) = 2\pi \int_0^{\Lambda_0} \frac{dr\, r}{\sqrt{1 + r^2}} = 2\pi\Lambda_0 - 2\pi, \tag{23}$$

where $\Lambda_0 \to \infty$ is a UV regulator. The first, linearly divergent term should be removed by regularization. We thus get for the Wilson loop expectation value:

$$W(C_{\text{circle}}) \simeq e^{\sqrt{\lambda}}, \tag{24}$$

in agreement with the direct field-theory calculation.[7,8]

The area law arises in the leading, semiclassical order of the strong-coupling expansion and gets corrections once the string fluctuations are taken into account. Since the disc partition function contains a factor of $\alpha'^{3/2}$ from gauge fixing the residual $SL(2,\mathbb{R})$ conformal symmetry, the prefactor of the Wilson loop expectation value should be proportional to $\lambda^{-3/4} \sim \alpha'^{3/2}$,[8] which is indeed the case for (20). The numerical coefficient in (20) is the one-loop contribution of string fluctuations, also potentially calculable from string theory.[33,34]

The holographic free energy of $\mathcal{N} = 4$ SYM is given by the on-shell gravitational action,

$$F = -\frac{1}{8\pi^2} \int_{AdS_5} d^5x \sqrt{g} \left(\mathcal{R} + 12\right), \qquad (25)$$

evaluated on the metric (22). We have taken into account here that the five-dimensional Newton's constant in the dimensionless units that we use is given by $G_N = \pi/2N^2$. The factor of N^2 is already included in the definition of the free energy in (7). The substitution of (22) into (25) results in a badly divergent integral:

$$F = \frac{8}{3} \int_0^{\Lambda_0} \frac{dr\, r^4}{\sqrt{1 + r^2}} = \frac{2}{3} \Lambda_0^4 - \frac{2}{3} \Lambda_0^2 + \frac{1}{2} \ln \Lambda_0 - \frac{7}{12}. \qquad (26)$$

The gravitational action therefore has to be regularized by adding boundary counterterms,[35] much like in the calculation of the Wilson loop. There is one crucial difference though. The gravitational action, in contradistinction to the minimal area, requires a log-divergent counterterm. The log has to be treated with care, as the answer will depend on the precise definition of the UV cutoff. The free energy, computed directly in field theory on S^4, is log-divergent too,[36] and to match the two expressions it is necessary to use the same subtraction scheme in both cases. The radial coordinate of AdS_5 differs from the energy scale of $\mathcal{N} = 4$ SYM by a factor of $\sqrt{\lambda}$, and so the UV cutoffs do.[37,38] To see this, one can compare the divergent part of the string action, given by the first term in (23) multiplied by $\sqrt{\lambda}/2\pi$, with the action of a heavy probe in the $\mathcal{N} = 4$ theory, equal to $2\pi\Lambda$, where we identified the mass of the probe with the energy cutoff. Equating the two expressions we find that $\Lambda_0 = 2\pi\Lambda/\sqrt{\lambda}$. In the field-theory regularization scheme, one subtracts $\ln \Lambda$ from the free energy (this subtraction is implicit in the matrix model). We thus conclude that the AdS/CFT prediction for the free energy, regularized in the field-theory scheme, is[39]

$$F = -\frac{1}{2} \ln \lambda + \text{const}, \qquad (27)$$

which coincides with the matrix-model prediction (19) up to an unimportant constant shift. For the Wilson loop, the string calculation gives the leading order of the strong-coupling expansion. The α'-corrections to the leading order result organize into an (asymptotic) power series in $1/\sqrt{\lambda}$. Interestingly, the gravity prediction for the free energy is exact, there are no α' corrections.

3.2. $\mathcal{N} = 2^*$ theory

The integral equation (12) has a complicated kernel and, at present, cannot be solved analytically in the closed form. Following the work on related $\mathcal{N} = 2$ theories,[28,39–42] approximate solutions have been constructed in various regimes in.[10,13,22,43] The asymptotic large-λ solution[22] is particularly simple. In writing down (12) we have assumed that eigenvalues are distributed in one cut along the interval $[-\mu, \mu]$, where the density $\rho(x)$ is unit-normalized. The one-cut solution can be justified by starting at very weak coupling and gradually increasing λ. The linear force $8\pi^2 x/\lambda$ in the saddle-point equation (12) is attractive and pushes the eigenvalues towards the origin. When λ is small, this force is very strong and eigenvalues are distributed in a small interval. As λ is gradually increased, the linear force $8\pi^2 x/\lambda$ becomes weaker, and the eigenvalue distribution expands to larger intervals. For a sufficiently large λ, the width of the eigenvalue distribution becomes much larger than the bare mass of the $\mathcal{N} = 2^*$ theory:[b] $\mu \gg M$. In this limit, most of the eigenvalues will satisfy $|x - y| \gg M$, $|x - y| \gg 1$, which justifies the following approximation:

$$\frac{1}{2} \mathcal{K}(x-y+M) + \frac{1}{2} \mathcal{K}(x-y-M) - \mathcal{K}(x-y) \approx \frac{1}{2} \mathcal{K}''(x-y)M^2 \approx \frac{M^2}{x-y}, \qquad (28)$$

where we have used the asymptotic formula for $\mathcal{K}(x)$,

$$\mathcal{K}(x) = x \ln x^2 + 2\gamma x + O(x^{-1}). \qquad (29)$$

As a result, the net effect of the complicated terms in the saddle-point equation reduces to multiplicative renormalization of the Hilbert kernel:

$$\frac{1}{x-y} \longrightarrow \frac{1+M^2}{x-y}. \qquad (30)$$

---

[b]Notice that μ plays the rôle of the effective IR scale of the theory.

The solution to the saddle-point equation is again the Wigner' semicircle

$$\rho(x) = \frac{2}{\pi\mu^2} \sqrt{\mu^2 - x^2}, \tag{31}$$

but now with

$$\mu = \frac{\sqrt{\lambda(1 + M^2)}}{2\pi}. \tag{32}$$

For the Wilson loop this result implies the same behavior as in $\mathcal{N} = 4$ SYM, with λ rescaled by $1 + M^2$:

$$\ln W(C) \simeq \sqrt{\lambda(1 + M^2)}. \tag{33}$$

As far as the free energy is concerned, the dependence on M is more complicated, and cannot be inferred just from (14). A more accurate calculation that keeps track of the λ-independent constant gives:[22]

$$F = -\frac{1 + M^2}{2} \ln \frac{\lambda(1 + M^2) e^{2\gamma + \frac{1}{2}}}{16\pi^2} + \gamma + \frac{1}{4} - \ln 4\pi. \tag{34}$$

The free energy can be compared with the gravitational action of the solution dual to $\mathcal{N} = 2^*$ theory on S^4. Such solution was recently constructed,[21] and its action perfectly matches the scheme-independent part of the matrix model prediction[c] after implementing holographic renormalization to cancel the UV divergences, similar to those that appear in (26). To compare the Wilson loop, the solution of the five-dimensional supergravity obtained in[21] has to be uplifted in ten dimensions, which has not been done so far.

However, we can compare the matrix model prediction (33) with generic expectations on the field theory in flat space, for which the supergravity dual is known in the full ten-dimensional form.[18] The flat-space limit can be reached by restoring the dependence on the radius of the four-sphere: $M \to MR$, $\mu \to \mu R$, and subsequently taking $R \to \infty$. The circular Wilson loop behaves as $\sqrt{\lambda}MR$ in this limit, and although we cannot compute Wilson loops for any other contour using localization, it is quite natural

---

[c]The constant term is chosen here to match the $\mathcal{N} = 4$ result (19) at $M = 0$, but otherwise it is obviously scheme-dependent. The term proportional to M^2 also depends on regularization, since the free energy is log-divergent. We follow the regularization scheme used by Pestun in deriving the matrix model from the path integral.[5] It is unclear to us how to implement precisely this scheme in the supergravity calculation. A pragmatic point of view, taken in,[21] consists in comparing the third derivatives of the free energy to remove scheme-dependent ambiguities altogether.

to assert that any sufficiently big Wilson loop in $\mathcal{N} = 2^*$ SYM obeys the perimeter law:

$$W(C) \simeq e^{\frac{\sqrt{\lambda}ML}{2\pi}}, \tag{35}$$

where L is the length of the contour C. The coefficient $\sqrt{\lambda}M/2\pi$ is chosen to match localization prediction for the circle, and can be interpreted as finite mass renormalization of a heavy external probe. This result is expected to hold for any loop on $\mathbb{R}^4$ and can be compared to the minimal area law in the Pilch-Warner geometry.[22]

The Pilch-Warner solution[18] asymptotes to $AdS_5 \times S^5$ near the boundary. The mass scale M is set geometrically by a domain wall placed at the distance M along the radial direction. Beyond the domain wall the metric deviates substantially from that of AdS_5. The slice of the Pilch-Warner geometry, that is necessary for computation of the Wilson loop (8), has the following metric in the string frame:

$$ds = \frac{BM^2 dx_\mu^2}{c^2 - 1} + \frac{dc^2}{B(c^2 - 1)^2}, \tag{36}$$

where

$$B = c + \frac{c^2 - 1}{2} \ln \frac{c - 1}{c + 1}. \tag{37}$$

This metric has the same asymptotics at $c \to 1$ as (22) has at $r \to \infty$, upon the coordinate transformation[d]

$$c = 1 + \frac{M^2}{2r^2}. \tag{38}$$

The domain wall is located at $c \sim 1$, and thus $r \sim M$. Beyond the domain wall the metric behaves as

$$ds^2 \simeq \frac{1}{c^3} \left(\frac{2}{3} M^2 dx_\mu^2 + \frac{3}{2} dc^2 \right) \qquad (c \gg 1). \tag{39}$$

We would like to compute the minimal area for the surface that ends on a space-like contour $x^\mu(s)$ at $c = 1$. The surface will start off as a vertical wall, and then will shrink gradually some distance away from the boundary, Fig. 5(a). In other words, near the boundary the solution for the minimal surface behaves as

$$X^\mu(s, \tau) \simeq x^\mu(s), \qquad C(s, \tau) \simeq \tau. \tag{40}$$

---

[d]The Pilch-Warner solution should be more appropriately compared to the slicing of AdS_5 in which the boundary is flat $\mathbb{R}^4$. Since we only look at $R = \infty$ limit, the difference is immaterial here.

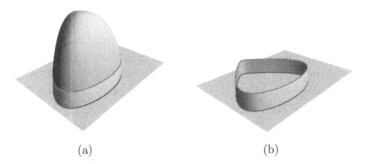

(a)   (b)

Fig. 5. (Color online) (a) A minimal surface in the Pilch-Warner geometry. The portion of the surface beyond the domain wall is shown in yellow. (b) The vertical part of the minimal surface described by the near-boundary asymptotics (40).

The bigger the contour is, the farther in the bulk will the surface extend. For a very big contour the surface extends far beyond the domain wall, and the asymptotic solution (40) is a good approximation up to $c \sim c_0 \gg 1$. Beyond the domain wall the metric takes a simple scaling form^e (39). For a sufficiently big contour, the largest part of the surface will lie in this region (shown in yellow in Fig. 5(a)), reaching up to $c \sim L$, where L is the size of the contour. But the area element for the metric (39) scales as $1/c$, and the contribution of the beyond-the-domain region to the area actually goes to zero as $1/L$. This somewhat counterintuitive conclusion can be substantiated by an explicit calculation for the circle of a large radius.[22] The largest contribution to the area still comes from $c \sim 1$ even for very large contours. But for large loops the near-boundary solution (40) is a good approximation for $1 \ll c \ll L$. For the sake of computing the area we can thus replace the full surface by its cylindric truncation, Fig. 5(b), with the induced metric

$$ds^2_{\text{ind}} = \frac{BM^2 x'^2 ds^2}{\tau^2 - 1} + \frac{d\tau^2}{B\left(\tau^2 - 1\right)^2}, \tag{41}$$

for which we find

$$A[x] = ML \int_{1+\frac{M^2}{2\Lambda_0^2}}^{\infty} \frac{d\tau}{\left(\tau^2 - 1\right)^{\frac{3}{2}}} = \Lambda_0 L - ML. \tag{42}$$

---

^{e}A curious consequence of this geometric picture, observed in,[44] is that the theory becomes effectively five-dimensional in deep IR.

Strictly speaking, the upper limit of integration should have been $c_0 \gg 1$, but since the integral converges well, we have extended integration to infinity. The lower cutoff is chosen according to the relationship (38) between the radial coordinates in AdS and in the Pilch-Warner geometry. The subtraction of the linear divergence then is the same as in $AdS_5 \times S^5$, which insures the continuity of the $M \to 0$ limit for small loops. After that we get $A_{\text{reg}} = -ML$ and then perimeter law (35) follows from the area law (21).[22]

We conclude that the free energy and the perimeter law for Wilson loops, computed holographically, reproduce the results obtained by direct path-integral calculation in field theory. Interestingly, the eigenvalue distribution itself can be also compared to supergravity, where it is determined by the D-brane probe analysis of the Pilch-Warner background. The distribution obtained this way appears to satisfy the Wigner law (31) with $\mu = \sqrt{\lambda}M/2\pi$.[19] This is reproduced by the matrix-model result (32) in the decompactification limit, when both M and μ are rescaled by R and R is subsequently sent to infinity.

4. Large-N phase transitions: Super-QCD in Veneziano limit

The decompactification limit of $R \to \infty$ can be regarded as a way to define the theory in flat space, as we discussed in Sec. 2. From this point of view, R is an IR regulator that should be sent to infinity at the end of the calculation. The localization matrix model of $\mathcal{N} = 2^*$ SYM simplifies dramatically in this limit and can be solved exactly for a finite interval of λ.[13] The solution terminates at the point of a fourth-order phase transition which happens at $\lambda_c \approx 35.4$ and is associated with new light states that appear in the spectrum.[10,13] Indeed, the mass of the hypermultiplet is not just M, but gets a contribution from the vacuum condensate (3), such that the mass squared of the component Φ_{ij}^{hyper} equals to $(a_i - a_j \pm M)^2$, and ranges from $(M - 2\mu)^2$ to $(M + 2\mu)^2$ as the eigenvalues a_i and a_j scan the interval $[-\mu, \mu]$. The correction due to the condensate is small compared to the bare mass only if $\mu \ll M$, which is true at weak coupling.[13] But as μ grows with λ and eventually reaches $M/2$, a massless hypermultiplet contributes to the saddle-point, triggering the transition to the strong-coupling phase. As shown in,[13] the theory undergoes secondary transitions each time the largest eigenvalue satisfies the resonance condition $2\mu(\lambda_c^{(n)}) = nM$. Since at strong coupling μ grows as $\sqrt{\lambda}M/2\pi$, there are infinitely many critical points asymptotically approaching $\lambda_c^{(n)} \simeq \pi^2 n^2$. The resulting phase diagram looks like the one in Fig. 1(b). We are not going to discuss the phase structure of the $\mathcal{N} = 2^*$ SYM in more detail, because of the technical complications (the details can be found in[10,13]), and will instead consider a simpler model, where the same phenomena are under better analytical control.

The model that we are going to consider is $\mathcal{N} = 2$ super-QCD, by which we mean supersymmetric gauge theory with $2N_f$ massive hypermultiplets of equal mass M. We shall assume that $N_f \leq N$, in which case the theory is asymptotically free. The model interpolates between pure $\mathcal{N} = 2$ SYM at $N_f = 0$ and the mass deformation of superconformal YM at $N_f = N$. We will study $\mathcal{N} = 2$ SQCD in the Veneziano limit $N \to \infty$, $N_f \to \infty$ with N_f/N fixed,[45] starting with the partition function on S^4 and subsequently taking the radius of the sphere to infinity.

A neat way to define the partition function of $\mathcal{N} = 2$ SQCD is to complement the theory with $2N - 2N_f$ additional hypermultiplets of mass $M_0 \gg M$. The theory with $2N$ hypermultiplets is a massive deformation of the $N_f = N$ superconformal theory, therefore it is finite. The partition function of this regularized version of $SQCD_{N_f}$ is

$$
Z_{N_f}^{\text{SQCD}} = \int d^{N-1}a \, \frac{\prod\limits_{i<j} (a_i - a_j)^2 \, H^2(a_i - a_j)}{\prod\limits_i H(a_i + M)^{N_f} H(a_i - M)^{N_f}}
$$
$$
\times \, e^{-\frac{8\pi^2 N}{\lambda_0} \sum\limits_i a_i^2} \prod_i \frac{1}{H(a_i + M_0)^{N-N_f} H(a_i - M_0)^{N-N_f}} \,, \quad (43)
$$

where the last factor is the contribution of the heavy fields that regularizes the UV divergences. The heavy mass M_0 acts as a UV cutoff and should be sent to infinity. Taking $M_0 \to \infty$, we expand

$$
\ln H(a + M_0) + \ln H(a - M_0) \simeq 2 \ln H(M_0) - 2 \left(\ln M_0 + \gamma + 1 \right) a^2
$$

where we used (11) and (29). The large logarithm combines with the bare coupling into the dynamically generated scale

$$
\Lambda = M_0 \, e^{-\frac{4\pi^2}{\lambda_0(1-\zeta)}} \,, \quad (44)
$$

where ζ is the Veneziano parameter

$$
\zeta = \frac{N_f}{N} \,. \quad (45)
$$

After the cutoff is removed, the partition function can be written as

$$Z_{N_f}^{\text{SQCD}} = \int d^{N-1}a \, \frac{\prod_{i<j}(a_i - a_j)^2 \, H^2(a_i - a_j)}{\prod_i H^{N_f}(a_i + M) H^{N_f}(a_i - M)}$$
$$\times \, e^{2(N-N_f)(\ln \Lambda + \gamma + 1) \sum_i a_i^2}, \tag{46}$$

omitting an unimportant normalization constant.

As before, in the large-N limit (with ζ fixed) the dynamics is described by the saddle-point equation:

$$2\fint_{-\mu}^{\mu} dy \rho(y) \left(\frac{1}{x-y} - \mathcal{K}(x-y) \right) = -4\,(1-\zeta)\,(\ln \Lambda + \gamma + 1)\, x$$
$$-\zeta \mathcal{K}(x+M) - \zeta \mathcal{K}(x-M). \tag{47}$$

The model depends on two parameters Λ and M, and greatly simplifies in the decompactification regime obtained by multiplying Λ, M, μ, x and y with R, which restores their canonical mass dimensions, and then sending R to infinity. In this limit the arguments of the function $\mathcal{K}$ are large and we can use the asymptotic formula (29). Differentiating the resulting equation in x we get:

$$2 \int_{-\mu}^{\mu} dy \, \rho(y) \ln \frac{(x-y)^2}{\Lambda^2} = \zeta \ln \frac{(x^2 - M^2)^2}{\Lambda^4}. \tag{48}$$

Differentiating once more we obtain a singular integral equation that can be easily solved:[f]

$$2\fint_{-\mu}^{\mu} dy \, \frac{\rho(y)}{x-y} = \frac{\zeta}{x+M} + \frac{\zeta}{x-M}. \tag{49}$$

The driving term in this equation has poles at $x = \pm M$ which may or may not lie within the eigenvalue distribution. The poles inside the distribution are due to resonances on massless hypermultiplets that appear when $a_i \pm M = 0$. Depending on whether massless hypermultiplets appear or not, the model has two phases: (1) the weak-coupling phase with $\mu < M$, in which all hypermultiplets are heavy, and (2) the strong-coupling phase at $\mu > M$, where light hypermultiplets appear in the spectrum. The solution to the saddle-point equations changes discontinuously when the pole at $x = M$ crosses the endpoint at $x = \mu$, and we need to consider the weak and strong coupling regimes separately.

---

[f]The same equation appears in the matrix model for zero-dimensional open strings,[46] albeit with different boundary conditions.

Strong-coupling phase (M < μ)

When $\mu > M$, the normalized eigenvalue density that solves the integral equation (49) is given by

$$\rho(x) = \frac{1 - \zeta}{\pi\sqrt{\mu^2 - x^2}} + \frac{\zeta}{2}\delta(x + M) + \frac{\zeta}{2}\delta(x - M). \tag{50}$$

We still need to fix μ in terms of M and Λ. This is done by substituting the solution into (48). The latter equation is satisfied if

$$\mu = 2\Lambda. \tag{51}$$

The endpoint of the eigenvalue distribution turns out to be independent of the hypermultiplet mass.

The solution (50) is valid as long as $M < \mu$. When M exceeds μ, the delta-functions jump out of the interval $[-\mu, \mu]$ rendering the solution inconsistent. As a result, the solution changes at the critical point and the system undergoes a transition to the weak-coupling regime. The phase transition thus happens at

$$M_c = 2\Lambda. \tag{52}$$

Weak-coupling phase (M > μ)

Assuming $\mu < M$ one finds the solution

$$\rho(x) = \frac{1}{\pi\sqrt{\mu^2 - x^2}}\left(1 - \zeta + \frac{\zeta M\sqrt{M^2 - \mu^2}}{M^2 - x^2}\right). \tag{53}$$

The integrated form of the saddle-point equation (48) then leads to a transcendental equation for μ:

$$2\left(1 - \zeta\right)\ln\frac{\mu}{2M} + 2\zeta\ln\frac{\mu}{M + \sqrt{M^2 - \mu^2}} = (1 - \zeta)\ln\frac{\Lambda^2}{M^2}. \tag{54}$$

The solution can be conveniently expressed in a parametric form:

$$\mu = M\sqrt{1 - u^2}, \tag{55}$$

$$\left(\frac{2\Lambda}{M}\right)^{2 - 2\zeta} = (1 + u)^{1 - 2\zeta}(1 - u). \tag{56}$$

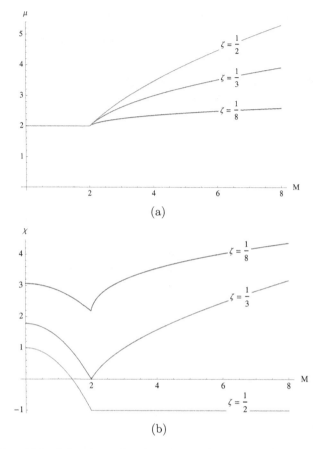

Fig. 6. The width of the eigenvalue distribution (upper panel) and susceptibility $\chi = -\partial^2 F/\partial \Lambda^2$ (lower panel) as functions of the quark mass at fixed Λ for different values of ζ. The units on the M and μ axes are normalized to Λ. The susceptibility is dimensionless.

Critical behavior

From the solutions in the two phases we can calculate the Wilson loop and the free energy. The Wilson loop obeys the perimeter law, with the coefficient of proportionality equal to the largest eigenvalue of the matrix model:

$$W(C) \simeq \mathrm{e}^{\mu L(C)}. \tag{57}$$

As for the free energy, it is easier to compute its first derivative:

$$\Lambda \frac{\partial F}{\partial \Lambda} = -2\left(1 - \zeta\right)\left\langle x^2 \right\rangle. \tag{58}$$

We find:

$$-\Lambda \frac{\partial F}{\partial \Lambda} = (1 - \zeta) \times \begin{cases} 4\left(1 - \zeta\right)\Lambda^2 + 2\zeta M^2, & M < 2\Lambda \\ (1 - u)\left[1 + \zeta + (1 - \zeta)u\right]M^2, & M > 2\Lambda \end{cases} \tag{59}$$

where u is the variable defined in (56). Comparing the two expressions at $M = 2\Lambda$, one can show that F is continuous across the transition together with its first and second derivatives, while the third derivative experiences a finite jump.[10] The transition thus is of the third order, as it usually happens in matrix models.[14,15] The perimeter-law coefficient has discontinuous first derivative, as clear from Fig. 6.

Operator product expansion

When the mass scale and the dimensional transmutation scale are widely separated, $M \gg \Lambda$, hypermultiplets can be integrated out. What remains is pure gauge $\mathcal{N} = 2$ SYM without matter. Taking into account the difference in the beta functions of pure $\mathcal{N} = 2$ SYM and $\mathcal{N} = \in$ SQCD, the dynamical scale of the low-energy effective field theory must be

$$\Lambda_{\text{eff}} = \Lambda^{1-\zeta}M^\zeta. \tag{60}$$

We may expect that the free energy in this regime has an OPE-type expansion of the form (1). On the other hand, the free energy can be calculated from the solution of the matrix model.

Integrating (59), the free energy in the weak-coupling phase can be written explicitly in terms of the variable u defined in (56):

$$F = -\frac{M^2}{2}\left\{(1 - \zeta)(1 - u)\left[1 + 5\zeta + (1 - \zeta)u\right] + 4\zeta(1 - 2\zeta)\ln\frac{1 + u}{2}\right\}. \tag{61}$$

We have chosen the integration constant such that the free energy vanishes at $u = 1$, which corresponds to the limit we are interested in, $\Lambda \ll M$. This expression can be now expanded in $1 - u$. To facilitate this expansion it is convenient to introduce a new variable $v = (1 - u)/2$. The equations (55), (56), (61) then become

$$\mu = 2M\sqrt{v\left(1 - v\right)} \tag{62}$$

$$F = -2M^2\left\{(1 - \zeta)\left[1 + 2\zeta - (1 - \zeta)v\right]v + \zeta\left(1 - 2\zeta\right)\ln\left(1 - v\right)\right\} \tag{63}$$

$$v\left(1 - v\right)^{1-2\zeta} = \frac{\Lambda_{\text{eff}}^2}{M^2}. \tag{64}$$

It is obvious from these expressions that the free energy has a power series expansion in $\Lambda_{\text{eff}}^2/M^2$:

$$F = \Lambda_{\text{eff}}^2 \left[-2 + \zeta \, \frac{\Lambda_{\text{eff}}^2}{M^2} + \frac{2}{3} \zeta \, (1 - 2\zeta) \, \frac{\Lambda_{\text{eff}}^4}{M^4} + \frac{1}{6} \zeta \, (1 - 2\zeta) \, (5 - 8\zeta) \, \frac{\Lambda_{\text{eff}}^6}{M^6} + \ldots \right]$$

$$(65)$$

Likewise,

$$\mu^2 = 4\Lambda_{\text{eff}}^2 \left[1 - 2\zeta \, \frac{\Lambda_{\text{eff}}^2}{M^2} - 3\zeta \, (1 - 2\zeta) \, \frac{\Lambda_{\text{eff}}^4}{M^4} + \frac{4}{3} \zeta \, (1 - 2\zeta) \, (5 - 8\zeta) \, \frac{\Lambda_{\text{eff}}^6}{M^6} + \ldots \right]$$

$$(66)$$

As we discussed in the introduction this expansion can be identified as arising from the OPE of the effective action induced be integrating out heavy hypermultiplets. The equations (62)–(64), which are exact, thus resum OPE to all orders.

Interestingly, in the particular case $\zeta = 1/2$, the OPE truncates at the first order:

$$F_{\zeta=1/2} = -2\Lambda_{\text{eff}}^2 + \frac{1}{2} \frac{\Lambda_{\text{eff}}^2}{M^2} \tag{67}$$

$$\mu_{\zeta=1/2}^2 = 4\Lambda_{\text{eff}}^2 - 4 \frac{\Lambda_{\text{eff}}^4}{M^2} \, . \tag{68}$$

This suggests that superselection rules must exist in $\mathcal{N} = 2$ SQCD with $N_f = N/2$ which set to zero the vevs of higher dimensional operators.

5. Conclusions

We have shown above how to solve for the large-N master field of $\mathcal{N} = 2^*$ SYM, and $\mathcal{N} = 2$ SQCD with $2N_f$ flavors using supersymmetric localization. One of the important lessons that one can draw from these calculations is the existence of quantum weak/strong coupling phase transitions, which seem to be generic features of massive $\mathcal{N} = 2$ theories. There is a single third-order phase transition in SQCD, while $\mathcal{N} = 2^*$ theory exhibits an infinite number of large-N phase transitions occurring as λ is increased and accumulating towards $\lambda = \infty$.[10,13] At large N, the functional integral is dominated by a saddle point. Our calculation shows that, when the coupling overcomes a certain critical value (or several critical values, as in the case of $\mathcal{N} = 2^*$), this saddle-point includes field configurations with extra massless hypermultiplets, thus producing discontinuities in vacuum expectation values of gauge invariant observables.

The free energy and the expectation values of large Wilson loops have only non-perturbative terms in their weak-coupling expansion. We have shown how to compute the expansion coefficients for SQCD to any order (the results for $\mathcal{N} = 2^*$ theory can be found in[13]). Non-perturbative series of this type can be understood as OPE in the underlying field theory, arising due to large separation of scales.

The results of localization at strong coupling can be compared to prediction of the holographic duality. The results of explicit field-theory calculations perfectly agree with predictions of holography for the eigenvalue distribution, the vev of large Wilson loops[22] and the free energy on S^4.[21] We demonstrated this for the $\mathcal{N} = 4$ and $\mathcal{N} = 2^*$ SYM theories, holographic duals of which are explicitly known.

We conclude by mentioning a number of open problems. One important problem concerns additional checks of holographic duality. In particular, the recent construction of the five-dimensional supergravity solution dual to $\mathcal{N} = 2^*$ compactified on S^4[21] illustrates the way to construct *euclidean* gravity solutions representing supersymmetric gauge theories on spaces of positive curvature. This new type of solutions would permit one to perform a number of new tests and thereby achieve a deeper understanding of gauge/gravity duality in non-conformal settings.

At strong coupling, the phase transitions occur at $\sqrt{\lambda} \sim n\pi$, with positive integer $n \gg 1$. It would be extremely interesting to find a string-theory interpretation of these special values of λ. It is conceivable that some signs of the non-analyticity at $\sqrt{\lambda} \sim n\pi$ could be manifested for semiclassical strings in the Pilch-Warner geometry.[47]

In the case of pure $\mathcal{N} = 2$ SYM, it was shown in[39] that in the decompactification limit localization reproduces the same eigenvalue distribution that arises from the Seiberg-Witten solution.[48,49] This distribution arises in the limit of maximally degenerate curves. It would be interesting to reproduce the results of $\mathcal{N} = 2^*$ $SU(N)$ localization from the corresponding Seiberg-Witten solution studied in.[50,51] It seems plausible that, like in pure $\mathcal{N} = 2$, there is a suitable limit that reproduces the same eigenvalue density found by localization and hence the same pattern of quantum phase transitions discussed here.

Acknowledgments

The work of K.Z. was supported in part by People Programme (Marie Curie Actions) of the European Union's FP7 Programme under REA Grant Agreement No 317089. J.R. acknowledges support by MCYT Research Grant No. FPA 2010-20807.

References

1. J. M. Maldacena, *Adv. Theor. Math. Phys.* **2**, 231 (1998).
2. S. S. Gubser, I. R. Klebanov and A. M. Polyakov, *Phys. Lett. B* **428**, 105 (1998).
3. E. Witten, *Adv. Theor. Math. Phys.* **2**, 253 (1998).
4. N. Beisert, C. Ahn, L. F. Alday, Z. Bajnok, J. M. Drummond *et al.*, *Lett. Math. Phys.* **99**, 3 (2012).
5. V. Pestun, *Commun. Math. Phys.* **313**, 71 (2012).
6. M. Marino, *J. Phys. A* **44**, p. 463001 (2011).
7. J. K. Erickson, G. W. Semenoff and K. Zarembo, *Nucl. Phys. B* **582**, 155 (2000).
8. N. Drukker and D. J. Gross, *J. Math. Phys.* **42**, 2896 (2001).
9. G. W. Semenoff and K. Zarembo, *Nucl. Phys. Proc. Suppl.* **108**, 106 (2002).
10. J. Russo and K. Zarembo, *JHEP* **1311**, p. 130 (2013).
11. M. A. Shifman, A. Vainshtein and V. I. Zakharov, *Nucl. Phys. B* **147**, 385 (1979).
12. M. A. Shifman, A. Vainshtein and V. I. Zakharov, *Nucl. Phys. B* **147**, 448 (1979).
13. J. G. Russo and K. Zarembo, *JHEP* **1304**, p. 065 (2013).
14. D. Gross and E. Witten, *Phys. Rev. D* **21**, 446 (1980).
15. S. R. Wadia (2012).
16. E. Brezin, C. Itzykson, G. Parisi and J. B. Zuber, *Commun. Math. Phys.* **59**, 35 (1978).
17. M. Li, *JHEP* **9903**, 004 (1999).
18. K. Pilch and N. P. Warner, *Nucl. Phys. B* **594**, 209 (2001).
19. A. Buchel, A. W. Peet and J. Polchinski, *Phys. Rev. D* **63**, 044009 (2001).
20. A. Buchel and J. T. Liu, *JHEP* **0311**, 031 (2003).
21. N. Bobev, H. Elvang, D. Z. Freedman and S. S. Pufu (2013).
22. A. Buchel, J. G. Russo and K. Zarembo, *JHEP* **1303**, 062 (2013).
23. N. A. Nekrasov, *Adv. Theor. Math. Phys.* **7**, 831 (2004).
24. N. Nekrasov and A. Okounkov (2003).
25. T. Okuda and V. Pestun, *JHEP* **1203**, 017 (2012).
26. D. J. Gross and A. Matytsin, *Nucl. Phys. B* **429**, 50 (1994).
27. R. Andree and D. Young, *JHEP* **09**, 095 (2010).
28. F. Passerini and K. Zarembo, *JHEP* **1109**, 102 (2011).
29. J. M. Maldacena, *Phys. Rev. Lett.* **80**, 4859 (1998).
30. S.-J. Rey and J.-T. Yee, *Eur. Phys. J. C* **22**, 379 (2001).
31. N. Drukker, D. J. Gross and H. Ooguri, *Phys. Rev. D* **60**, 125006 (1999).
32. D. E. Berenstein, R. Corrado, W. Fischler and J. M. Maldacena, *Phys. Rev. D* **59**, 105023 (1999).
33. M. Kruczenski and A. Tirziu, *JHEP* **0805**, 064 (2008).
34. C. Kristjansen and Y. Makeenko, *JHEP* **1209**, 053 (2012).
35. K. Skenderis, *Class. Quant. Grav.* **19**, 5849 (2002).
36. C. Burgess, N. Constable and R. C. Myers, *JHEP* **9908**, 017 (1999).
37. A. W. Peet and J. Polchinski, *Phys.Rev.* **D59**, 065011 (1999).
38. M. Bianchi, D. Z. Freedman and K. Skenderis, *JHEP* **0108**, 041 (2001).

39. J. Russo and K. Zarembo, *JHEP* **1210**, 082 (2012).
40. S.-J. Rey and T. Suyama, *JHEP* **01**, 136 (2011).
41. J.-E. Bourgine, *J. Phys. A* **45**, 125403 (2012).
42. B. Fraser and S. P. Kumar, *JHEP* **1203**, 077 (2012).
43. J. G. Russo, *JHEP* **1206**, 038 (2012).
44. C. Hoyos, *Phys. Lett. B* **696**, 145 (2011).
45. G. Veneziano, *Nucl. Phys. B* **117**, 519 (1976).
46. V. Kazakov, *Phys. Lett.B* **237**, 212 (1990).
47. H. Dimov, V. G. Filev, R. Rashkov and K. Viswanathan, *Phys. Rev. D* **68**, 066010 (2003).
48. M. R. Douglas and S. H. Shenker, *Nucl. Phys. B* **447**, 271 (1995).
49. F. Ferrari, *Nucl. Phys. B* **612**, 151 (2001).
50. R. Donagi and E. Witten, *Nucl. Phys. B* **460**, 299 (1996).
51. E. D'Hoker and D. Phong, *Nucl. Phys. B* **513**, 405 (1998).

Printed in the United States
By Bookmasters